BLOOD
IN THE
MACHINE

ALSO BY BRIAN MERCHANT

The One Device: The Secret History of the iPhone

Terraform (as co-editor with Claire L. Evans)

BLOOD
IN THE
MACHINE

THE ORIGINS OF THE REBELLION
AGAINST BIG TECH

BRIAN MERCHANT

Little, Brown and Company

New York • Boston • London

Little, Brown and Company
Hachette Book Group
1290 Avenue of the Americas, New York, NY 10104
littlebrown.com

First Edition: September 2023

Little, Brown and Company is a division of Hachette Book Group, Inc. The Little, Brown name and logo are trademarks of Hachette Book Group, Inc.

The publisher is not responsible for websites (or their content) that are not owned by the publisher.

The Hachette Speakers Bureau provides a wide range of authors for speaking events. To find out more, go to hachettespeakersbureau.com or email hachettespeakers@hbgusa.com.

Little, Brown and Company books may be purchased in bulk for business, educational, or promotional use. For information, please contact your local bookseller or the Hachette Book Group Special Markets Department at special.markets@hbgusa.com.

ISBN 9780316487740

Printing 1, 2023

LCCN is available at the Library of Congress

LSC-C

Printed in the United States of America

For Russell, Aldus, and Corrina — now and
future machine breakers.

Chant no more your old rhymes about bold Robin Hood,
His feats I but little admire
I will sing the Achievements of General Ludd
Now the Hero of Nottinghamshire.

— *"General Ludd's Triumph," sung to the tune of "Poor Jack"*

Killer Mike: Are jobs still necessary? Should we still be pushing that agenda of capitalism that forces people to work at the lowest possible wage to enrich the top?

El-P: I mean we'll find out soon when the robots really rise. You think you've got problems with your jobs now, wait until Plucky the Robot takes over your gig.

A soul is of more value than work or gold.

— *George Mellor*

PERSONS OF INTEREST

Primary

B: A spy embedded in the Manchester Workers Committee.

Charles Ball: Enslaved cotton worker and soldier in the American South.

George Mellor: Cloth dresser, or cropper, in Huddersfield, England.

Gravener Henson: Writer, agitator, reformer, and framework knitter in Nottingham.

Lord Byron: George Gordon, sixth Baron Byron; poet, lord of Newstead Abbey.

Mary Godwin: Writer, student, teenager in London, daughter of Mary Wollstonecraft and William Godwin.

Ned Ludd: The fictional figurehead of the machine-breaker movement, based on the myth of an apprentice who smashed his master's device.

Prince George: The son of King George III; Prince Regent and Ruler of England after George III is incapacitated in 1811; and later King George IV (1820–1830).

Robert Blincoe: Indentured factory worker and orphan.

William Cartwright: Entrepreneur, cloth manufacturer, and owner of Rawfolds Mill in the West Riding of Yorkshire.

William Horsfall: Entrepreneur, businessman, factory owner in the West Riding. Operated Ottiwells Mill, which manufactured woollen cloth, using automated machinery.

Secondary

Andrew Ure: Business theorist and early advocate of automation.

Anna Lætitia Barbauld: Children's learning advocate; poet, author of *Eighteen Hundred and Eleven* (1812).

Ben Bamforth: Son of a weaver; George Mellor's cousin.

Ben Walker: Cropper in Huddersfield.

Caroline of Brunswick: Princess of Wales, estranged wife of the Prince Regent, reformer, friend of Byron.

Colonel Ralph Fletcher: Magistrate of Bolton, Lancashire; employed the spy known as B.

Ed Baines: Prodemocracy reformer and organizer of the republican club in Halifax.

Edmund Cartwright: Clergyman, poet; inventor of the power loom in 1785.

Edward Maitland: General, leader of the occupying army.

Francis Raynes: Captain in the Crown's army, sent to occupy the Midlands.

Isabella Baxter: Mary Godwin's best friend.

John Booth: Apprentice worker, Owenite, and political radical from the West Riding.

Joseph Radcliffe: Magistrate, landowner, and businessman in Huddersfield.

Lady Caroline Lamb: poet, novelist, Byron's love interest.

Lady Ludd: Name given to women who organize direct actions to seize food and distribute it illegally, reducing the price so the poor can afford it.

Lord Holland: Whig leader in the House of Lords, mentor to Lord Byron.

Lord Hume: Whig leader in the House of Lords, laissez-faire advocate.

Mary Wollstonecraft: Writer, philosopher, founding figure of feminism; Mary Godwin's mother.

Percy Bysshe Shelley: Poet, radical writer, heir.

Princess Charlotte: Daughter of Lady Caroline and the Prince Regent.

Richard Arkwright: Businessman, entrepreneur, "father of the factory." Patented the water frame and carding mill and built the first factory operations in the Midlands.

PERSONS OF INTEREST

Richard Ryder: Secretary of the Home Office in the Perceval administration.

Robert Owen: Radical factory owner and influential utopianist.

Spencer Perceval: Prime minister of England (1809–1812); leader of the Tories.

Thomas Smith: Cropper in Huddersfield.

Viscount Sidmouth: Henry Addington; succeeded Richard Ryder as Home Office secretary.

Will Thorpe: Cropper in Huddersfield.

William Felkin: Apprentice weaver turned historian in Nottingham.

William Godwin: Writer, philosopher, founding proponent of anarchism; Mary Godwin's father.

THE MACHINERY

Stocking frame: Machine used by framework knitters to produce stockings, lace, and knit goods. Invented in Nottingham in 1589.

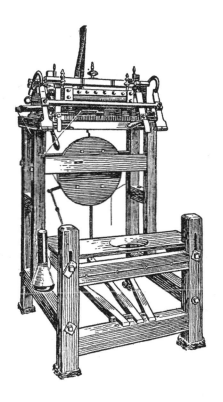

A STOCKING FRAME, *c.* 1784

Handloom: Machine used to weave woollen and cotton cloth. The earliest looms date to the fifth century BCE.

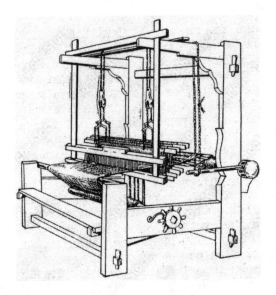

Spinning jenny: Invented by James Hargreaves in 1764, the device reduced the labor required to spin wool into yarn.

Water frame: Patented by Richard Arkwright in 1769, the machine applied waterpower to the design of the spinning jenny and further automated the process of producing yarn.

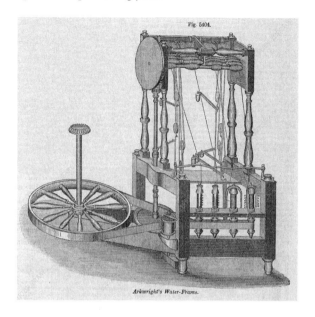

Arkwright's Water-Frame.

Gig mill: Also called the gigging machine, this device automatically raised the nap of woven fabric to draw out the ends of the fibers, making it easier to finish, or smooth, the cloth.

Power loom: Invented in 1786 and gradually improved in following decades, this machine automated the task of weaving.

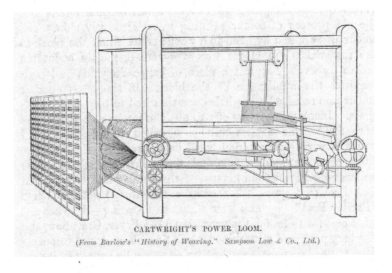

CARTWRIGHT'S POWER LOOM.

(From Barlow's "History of Weaving." Sampson Low & Co., Ltd.)

A BRIEF NOTE ON SOURCES

This book draws both from original reportage and a rich body of research on a notoriously difficult subject. As we'll see, the participants in the Luddite uprising had good reason to keep a low profile, and to keep information and communications about their inner workings off the record and out of public view. The sections that examine the rebellion of 1811 are indebted to the work of historians like F. O. Darvall, Barbara Hammond and J. L. Hammond, E. P. Thompson, Malcolm Thomis, Maxine Berg, Adrian Randall, Kevin Binfield, Katrina Navickas, Kirkpatrick Sale, Leslie Kipling, and Alan Brooke, as well as Richard Holland's archival curation.

This book also draws from firsthand sources — newspaper articles, handbills, memoirs, and letters, many of which are housed in Britain's Public Record Office, Home Office Papers, and in regional archives around the nation — as well as two works of oral history that warrant a quick mention here. The 1880 oral history *The Risings of the Luddites,* by Frank Peel, is based on interviews with those who participated in and had firsthand knowledge of the rebellion. These interviews were recorded years after the fact, and so dialogue in particular must be taken with a grain of salt. Similarly, *Ben o' Bill's, The Luddite: A Yorkshire Tale* (1898), by the local West Riding historians D. F. E. Sykes and George Henry Walker, is based on participant interviews and historical research. However, Sykes and Walker ultimately chose to present this book as a novel, a decision that has puzzled some historians. The major events it portrays and the conditions it describes are largely accurate, as corroborated by other accounts, news articles, and research. The dialogue is rich and faithful to the style of the times, but it also contains folklore and embellishments; it must be read with this foreknowledge. Whenever either of these books is cited, it will be noted in the text.

The historians Adrian Randall and Alan Brooke both read this manuscript with an eye to its accuracy, and I am immensely grateful for their help. Any errors that remain, of course, are my own. A full bibliography is included in the back.

CONTENTS

CONTENTS

PART II
Metropolis of Discontent

PART III
Breaking Frames, Breaking Bones

CONTENTS

PART IV
More Value than Work or Gold

CONTENTS

PART V
The Modern Prometheus

PART VI
The Owners of the New Machine Age

CONTENTS

INTRODUCTION

Imagine millions of ordinary people plagued by a fear that technology is accelerating out of control. They worry that machines are coming to take away their jobs, erode their status, threaten their futures, and upend the order of their lives. Inequality is rampant, and power is wielded by those commanding wealth and new technologies. Every sign points to immense social and economic upheaval on the horizon.

This could be today. It could also be two hundred years ago, in the early days of the Industrial Revolution, when the story of the rebellion against the use of those machines — and against the first tech titans — began.

WINTER 1812

The West Riding of England

A few dozen men began to assemble after dark. They had gathered like this many times before, and they would gather again. Some nights the meeting place was behind a pub they all knew, or a landmark outside town. Tonight it was a muddy hillside just off the main road.

The contingent grew as the night deepened. It was filled with workingmen — artisans, weavers, and machinists among them. They formed ranks on the damp sod. It was frigid, cold enough that the men could see their breath in the air.

Their faces were blackened with coal dust or covered by masks, and they moved into military formation with an ease born of hours spent running drills. They armed themselves with axes and pistols. Some were anxious, some hungry, and some — most, maybe — brimmed with righteous indignation. All were gathered under common cause: to destroy the machinery that entrepreneurs had built to replace them.

These machines were now doing the work, cheaply and less artfully, that had for generations formed the foundation of their lives and their communities. Because of those machines — or, rather, the entrepreneurs and factory owners who deployed them — many on those dark fields could not earn enough money to feed their families. Some of those families were now on the brink of starvation.

Others there were not starving but saw too clearly a future approaching in which such devices and their owners would grind away the security, status, and dignity they held in their lives too. Others had loftier aspirations. They saw the glint of opportunity in an uprising against machines — a chance to change not just how men worked, but how they lived, organized, and governed themselves.

When the last of the men had arrived, their leader, their *general,* gave the word. Roll was called. All accounted for, the troop moved out. One among them carried the great hammer they called Enoch. The machines, idle at this late hour, and their masters, sleeping lightly or keeping watch, awaited.

Before the machine breakers were through, thousands of devices would lie shattered. Factories would burn. And the blood of men, women, and children, of rich and of poor — though mostly of poor — would stain the ground upon which the gears of industry turned.

⚙o

This book is about why so much blood was spilled over machines in the early nineteenth century — the first time that technology was used to replace human jobs en masse — and about the thunderous uprising that followed. It's also a story about the twenty-first century, when calls of "The robots are coming for our jobs" and "Big Tech is becoming too powerful" are everywhere, and economic conditions look all too familiar to historians of the Industrial Revolution.

In the eighteenth century, merchants, business owners, and industrialists started investing in automated machinery and creating the first factories. In the process, they began to displace the workers whose livelihoods had existed for centuries, and, by the nineteenth century, had concentrated wealth, power, and technological advantage in a relatively small number of hands.

In the twenty-first century, corporate executives, start-up founders, and tech giants like Amazon and Uber are using a new version of the same concept to entrench new automation-heavy and digitally organized models of work. In the process, they've disrupted Main Street businesses and traditional livelihoods — and concentrated wealth, power, and technological advantage in an ever smaller number of hands.

In 1812, skilled weavers feared that a factory filled with power looms would automate the job of weaving cloth and put them out of work.

In 2012, veteran taxi drivers feared that Uber's algorithm would flood the market with independent contractors beholden to an app and put them out of work.

In 1812, those fears drove Britain, then the richest and most technologically advanced nation in the world, to the brink of civil war.

There, at the dawn of the Industrial Revolution, a small but growing number of industrialists took the work that artisans had done in their own homes or in small shops — on their terms, with their families, for generations — and automated it with machinery and moved it into factories. These workers faced a future in which they would have to toil at a manager's machine, putting in long hours in toxic, suffocating environments, for the owners' profit and not their own. If, that is, they could find any work at all.

Now, two hundred years later in the United States, we're edging up to the brink again.

Working people are staring down entrepreneurs, tech monopolies, and venture capital firms that are hunting for new forms of labor-saving tech — be it AI, robotics, or software automation — to replace them. They are again faced with the prospect of losing their jobs to the machine.

Today, the entrepreneur who takes a risk to disrupt an industry is widely celebrated; venture capital firms look for unicorn ideas that shrink a market while consolidating their control. And automation is often presented as an unfortunate but unavoidable outcome of "progress," a reality that workers in any technologically advancing society must simply learn to adapt to.

But in the 1800s, automation was not seen as inevitable, or even morally ambiguous. Working people felt it was wrong to use machines to "take another man's bread," and so thousands rose up in a forceful, decentralized resistance to smash them. The public cheered these rebels, and for a time they were bigger than Robin Hood, and more powerful.

They were known as the Luddites. And they launched what was "perhaps the purest of English working-class movements, truly popular of the people," as one historian put it. Their targets were the entrepreneurs, factory owners, and other facilitators and profiteers of automation.

Even though that uprising took place two centuries ago, it contained the seeds of a conflict that continues to shape our relationship to work and technology today. In many ways, our future still depends on the outcome of that conflict.

I set out to write this book with the hope that, by telling the stories of those who experienced the first furious rise of "job-killing" technologies and the incredible and violent reaction it inspired, we might better grasp the ways that modern machines and their owners are shaping the working present.

The present, after all, is rife with predictions that AI, robots, and software automation will degrade or eliminate large swaths of jobs all over again. One notorious study carried out by researchers at the University of Oxford concluded that nearly half of all American jobs are ripe for technological replacement. Critics have taken issue with that forecast, but our largest tech companies would love to see it come true. Robots, the old corporate adage goes, never call in sick.

Tech companies like Amazon, Uber, Facebook, OpenAI, and Microsoft have accumulated vast power and influence. In ways large and small, they are already remaking our working lives. We now face a future where work — even for the so-called middle class, even for white-collar workers — is increasingly informal, precarious, and organized by inscrutable and unaccountable technologies. Big Tech's algorithms help dictate whether we get hired for a job, how much we'll earn, and whether we'll keep it. These companies invest huge sums in automation initiatives and worker-surveillance software. Their preference for app-controlled gig work is driving a shift away from salaried jobs with benefits to a precarious, technologically arbitrated part-time model. Their platforms determine the rules, or lack thereof, that govern our speech, entertainment, and communities.

But if you raise concerns about any of this, you're apt to be labeled anti-technology, or anti-progress. You're apt to be called a Luddite. Today, the term is used as an epithet for someone who hates or doesn't understand new technology — but that couldn't be any further from the historical truth. The Luddites understood technology all too well; they didn't hate it, but rather the way it was used against them. And as we'll soon see, the technologies that people are ridiculed for protesting tend to be the ones designed to profit at the protester's expense.

So, as Amazon expanded its vast automated factories while its workers peed in bottles instead of being able to take bathroom breaks; as Uber's user-friendly algorithms crowded out veteran cab drivers; as chatbots and voice automation took on the work of customer service reps, and cheap AI art flooded the market, I looked back at the first artisans to resist the robots of their day. Were they right to do so? What could be learned from their example? How does their legacy live on?

In pursuit of answers to those questions, I spent three years researching the harsh conditions and violent riots that attended the early days of the Industrial Revolution. I traveled to the birthplaces of modern industry in England, where the technological systems that shape our lives even now first sprang up — and where they incited sustained and bloody conflict. I dove into the archives, interviewed historians, spent time with Luddite scholars — I even lodged with one, in an old weaver's cottage — and tried to picture this world in the throes of change. To bring the past into dialogue with the present, I interviewed historians of automation and artificial intelligence, attended worker protests against Uber and Lyft, and stopped by then presidential candidate Andrew Yang's campaign headquarters to hear him raise the specter of the robot jobs apocalypse.

I found a ripping tale of unlikely heroes and antiheroes, of fearsome rebellion, of oath-bound secret societies and the spies sent to infiltrate them, of the fickle allegiances of celebrities like Lord Byron, and of the violent royal rage that would culminate in the largest domestic military occupation in Britain's history.

This uprising not only sent shock waves through the most powerful nation in the world, it laid the foundation for trade unions, the modern welfare state, and science fiction. By understanding this story, what led to that first rebellion, and the cataclysm that followed, perhaps we will be able to avert the need for another.

This book follows the artisans and machine workers as they rose up in that rebellion, the inventors devising the machines, the entrepreneurs and industrialists building the factories to put them in, the poets and writers sounding the alarm about the growing crisis, the Prince Regent and an authoritarian-leaning government ignoring that alarm, and the reformers fighting for a democratic resolution. In the early 1800s, Britain was the world leader in industrial power and technological development; that's where the narrative will unfold. In the twenty-first century, the United States, home to Silicon Valley and the globe's largest economy, occupies that role — so that will be the focus when I interrupt the narrative to add context and commentary examining the parallels between this history and the gathering storm today.

As the scales tipped against working men and women, as machines reaped profits for their owners at the expense of their neighbors, some of those neighbors decided to fight back. They showed us whose blood runs in the machines that move the world.

PRELUDE

Power Looms

1786

Edmund Cartwright

It was a simple conviction that set everything in motion, to hear the inventor tell the story. Edmund Cartwright had come to believe that any work a human did with a tool, a machine could do too, with the right upgrades and a dependable power source. Even, say, the act of weaving cloth on a loom — the work carried out by hundreds of thousands of England's skilled artisans. Yet when Cartwright championed the idea of a power loom one night at a dinner gathering, his companions scoffed at the notion. They insisted that weaving was just too complex.

Cartwright had met his friends, a group of businessmen and clergymen, in Matlock, in the Midlands of England. There, they walked along the emerald-green banks of the River Derwent to visit the immense new mill built by Richard Arkwright, a barbershop owner turned entrepreneur who had mechanized the process of spinning cotton into yarn with a device called a water frame.

As Cartwright and his friends dined, they discussed being impressed, even alarmed, by the innovation on display there, by the might of Arkwright's water-powered factory and the sheer volume of yarn it produced.

But *spinning* yarn was one thing, Cartwright's friends insisted, and *weaving* was another entirely. Weaving was a skilled art, and building a machine to replicate its complexity was impossible, "on account of the variety of movements required," they said. Cartwright, who had never actually examined a weaver's work before that point, argued that it certainly *was* possible to mechanize. It *had* to be.

"There's no real impossibility in applying power to any part of the most complicated machine," Cartwright declared, "and whatever variety of movements the art of weaving might require, the skillful application of mechanism might produce them." He believed, in other words, in the potential of automation: the technique of making an apparatus or a system work automatically, or with as little human input as possible. The term *automation* would not be coined until the 1940s, when a vice president of engineering at Ford used it to describe the roboticization of the assembly line at car factories, but the concept was largely the same in the 1780s. And Cartwright believed its potential was nearly limitless.

After all, Cartwright had watched an automaton, a chess-playing mechanical Turk, as it beat human players. Weaving certainly couldn't be more complicated than that. (Probably unbeknownst to Cartwright, there were real men hidden inside those early chess-playing androids, executing each move.)

Cartwright was no industrialist or engineer, and certainly no textile expert; he was a cleric and a poet. But he'd always been searching and sharp. Born in 1743 to a family of wealthy landowners outside Nottingham, Cartwright was forbidden from following his brothers into the navy. Instead, his father sent him into the clergy. By fourteen, he was studying at Oxford. By forty, he was the prebendary of Lincoln Cathedral, in the East Midlands. He'd published some well-received poetry, including "The Prince of Peace," which voiced pity for the American revolutionaries, who, at the time of its writing, seemed hopelessly outgunned.

There had been a buzz around Richard Arkwright's massive factory complex in Matlock for years, and Cartwright had been keen to pay it a visit. The first buildings of the factory were erected in 1771, and by the 1780s the complex towered over the riverside. Inside its walls churned machines that tirelessly spun cotton into yarn. Historically, spinning had been done by teams of women, and it took more hours to prepare the yarn than to weave it into cloth. But Arkwright used hydropower instead, directing water from the river into a giant wheel attached to the four-story building to run his patented water frame. He no longer needed those teams of people; instead, he could spin cotton into yarn nearly around the clock.

But that innovation, Cartwright's dinner companions pointed out, could unintentionally hurt England's economy.

"If this new mode of spinning by machinery should be generally adopted, so much more yarn would be manufactured than our own weavers could work up," one of his friends said. So much yarn could be made that it would have to be exported abroad, "where it might be woven into cloth so cheaply, as to injure the trade in England."

"The only remedy for such an evil," Cartwright replied, "would be to apply the power of machinery to the art of weaving as well as to that of spinning, by contriving looms to work up the yarn as fast as it was produced by the spindle" — in other words, automate the other half of the process, the part that required skilled labor, too (or at least the part historically done by men, and so described as being more complex and valuable).

At that, Cartwright threw himself into building an automated loom. There was, of course, another tantalizing motivator: "the splendid fortunes that some ingenious mechanics…were supposed to be realizing," as his biographer Mary Strickland put it. Arkwright, born a middle-class man, had begun to accumulate wealth to rival the richest nobles in England, thanks to his water wheel. But before Cartwright could aspire to similar rewards, he would have to educate himself on the region's textile industry.

It all started with wool.

For centuries, weavers had made woollen garments in the damp regions north of London. In the Middle Ages, "when the center of Western civilization shifted northward with the collapse of the Roman empire," the historian W. G. Crump explained, "wool became its main clothing fiber." Demand soared and wool became the most important industry in Britain. By the end of the eighteenth century, the wool trade employed a million people, a tenth of the population of Britain. "Weavers were, and probably had been for hundreds of years, the largest single group of industrial workers in England," as the historian E. P. Thompson put it.

An industry of such size was ripe for innovation. Experimentation on cloth-making machinery, and the organization of the labor force that used it, unfolded slowly, then all at once. In 1589, William Lee of Calverton developed one of history's most quietly revolutionary technologies. The legend goes that Lee was upset that his wife spent more time knitting than with him, so he devised the stocking frame to speed up the process. Lee's machine, about the size of a large desk, allowed its operator to use pedals and bars to automatically mimic the movements of a hand knitter, making it much easier, and faster, to produce stockings, socks, tights, and other knit garments. (At the time, men wore tights, not pants.)

The machine worked so well that he tried to commercialize it. But Queen Elizabeth refused to grant Lee a patent, and left him with a foreboding rebuttal: "You aim high, Master Lee," she said, before expressing concern for the hand knitters his device would affect. "Consider…what the invention could do to my poor subjects. It would assuredly bring to them ruin by depriving them of employment, thus making them beggars." Lee died broke, oblivious that he'd sowed some of the earliest seeds of the Industrial Revolution. His brother James Lee pressed on with his invention, however, and it later became a key tool in England's booming textile industry.

For two hundred years, the stocking frame reigned. With its counterpart for weaving, the handloom, it helped England become known around the world for finely produced textiles. Along the way, the queen's prediction did come true — like the cotton spinners Arkwright's mill was replacing in the 1780s, hand knitters were largely replaced by the stocking frame operator. Customs, standards, and traditions developed around the machines, which were worked in the weavers' or knitters' homes, or in small shops. The massive demand for English cloth continued to inspire innovations in both the machinery that made a worker's life easier, and more disruptive technologies that promised a dramatic increase in the rate of production.

In 1733, an apprentice craftsman named John Kay patented the flying shuttle, a device that let a single weaver, instead of two, work a broadloom.

In the 1760s, a weaver and cotton spinner named James Hargreaves invented the spinning jenny, which let a single worker spin multiple threads into yarn with the crank of a handle — previously it took six. Just a couple years later, Arkwright built his water frame, applying hydropower to the spinning device and assuring the production of vast volumes of yarn. To cap off the remarkable decade, James Watt patented his steam engine, in 1769, promising a more efficient and affordable power source for the turning spindles and clattering shafts all across the country.

Coal mines were dug to feed the engines, canals were blasted to connect them, and new brick edifices began to dot the pastoral skyline. The next generation of cloth workers would watch smoke plumes begin to rise over the sheep pastures, heralding the arrival of the modern factory and the industrial age.

That Edmund Cartwright felt so confident that he could mechanize the weaving process is not necessarily surprising, given the surge of interest in automation and technological innovation surrounding him. The 1760s and '70s had seen a boom in disruptive commercial tech rivaled by any in history. But his first effort was a disaster. Cartwright built his prototype without ever observing a weaver in action, and wound up with a machine that required twice the manpower to produce a single garment.

Humbled, he arranged a trip to Manchester, where weavers had flocked to tend to the surplus of yarn. The town was becoming a bustling city, a gateway for industrialization. Wool was now giving way to cotton, which was cheaper and more machine-friendly, and boatfuls of the stuff were streaming in from India and the United States. Manchester would soon boast the nickname Cottonopolis — a city built on the nation's fastest-growing, and soon to be most important, export.

Cartwright's trip had two objectives. First, he wanted to raise interest in his machine among local businessmen. Second, he wanted to watch the weavers in detail. He sent his prototype ahead, asking the workingmen to examine his "power loom" and offer ideas for improvement.

The workers flatly refused. They immediately recognized the implications of the machine: if it worked, it could put them out of a job.

"Indeed, the workmen who had undertaken it despaired of ever making it answer the purpose it was intended for," Cartwright wrote to a friend. He would have to automate the device on his own.

There in a cotton mill, with his soft, rounded features and faintly aristocratic air, Cartwright cut a strange figure. He was finely dressed in the fashions of the upper class. As he studied the weavers, he might have marveled at the simplicity and staying power of the tools that made such garments possible.

He would have seen the intensity and intricacy of weaving work up close: the weaver's darting hands, flicking wrists, and endlessly rising and falling knees all syncing up with the apparatus before him. The weaver depressing and releasing those pedal-like levers called treadles, one under each foot; throwing the flying shuttle; and angling the lathe, each step rapid-fire, creating a fine woven garment inch by inch. Movements so assured, they could perhaps have been said to be machinelike.

Amidst the clanking din, Edmund Cartwright watched and made notes. He refined his prototype, not fully comprehending that the schemes bouncing around his skull would hit the gas on the Industrial Revolution, or what that might mean for men and women like the weaver he was watching at work.

1797

Mary Godwin

The Polygon was in a state of chaos. Mary Wollstonecraft and William Godwin's home, No. 29 in a sixteen-sided building complex in London, was abuzz with doctors trying desperately to save the life of the ailing radical, writer, and new mother. Over the course of nine excruciating hours, she had given birth to her first child, a daughter. But now something was direly wrong and the physicians didn't know what.

William, a muckraking journalist and radical philosopher, first met Mary, the celebrated author of *A Vindication of the Rights of Men,* at a dinner held for the American revolutionary Thomas Paine in 1791. The two did not think much of each other at first. They spent the entire night arguing philosophy and politics, subjects about which there was plenty to argue. The French Revolution had set the world reeling, and much seemed possible for the freethinkers and reformers who chafed under a retrograde monarchy. If France, which lay just twenty-two miles away, across the English Channel, could dispense with its king, why couldn't England? An industrial transformation was already underway; a political one could follow.

Mary had furiously written *Rights of Men* in 1790 to defend the French Revolution from the conservative gripes of philosopher Edmund Burke, who described working people as a "swinish multitude" unworthy of the vote. *Men* was an instant bestseller. Paine's 1791 *Rights of Man,* which was similarly sympathetic to the Revolution, was an even bigger hit; it sold as many as a million copies. It inspired, among many other things, the pioneering Romantic poet Robert Burns to write "A Man's a Man for A' That," about a classless future that's "coming yet," one where "Man to Man the warld o'er / Shall brothers be." Romanticism was a burgeoning cultural movement; its poets and artists tended to celebrate a heroic individualism while rejecting the dull rationalism of the Enlightenment. They cheered the French Revolution. Burns's song became world famous.

Democratic reform movements and radical clubs began to spring up across England. The London Corresponding Society, one of the first working-class organizations, was founded in 1792 by the shoemaker Thomas Hardy and eight compatriots to rally for parliamentary reform and demand true democracy. Their rolls exploded into the thousands, and chapters opened across England. Its founding principle was "that the number of our members be unlimited" — anyone at all could join — a radical and inclusive principle for the day.

Mary and William went their separate ways after that first contentious dinner, and both produced the works that would cement their fame. Mary followed up *Rights of Men* with *A Vindication of the Rights of Woman* in 1792; it was an influential work of early feminism, calling for drastic upgrades in gender equality and asserting the importance of education for girls. William published *An Enquiry Concerning Political Justice* the year after that, in 1793. *Justice* was an early blueprint for anarchism, arguing for a fair, egalitarian, and marriage-free society where wealth is distributed to those who need it most.

It was a hopeful time; change seemed imminent.

Then the whipsaw of history snapped backward. The French Revolution grew bloody and devoured its children. England declared war on France. The Jacobins, for a time the largest and most influential faction in Revolutionary France, espoused a radical working-class politics, and it was a Jacobin, Maximilien Robespierre, who led the Reign of Terror and delivered King Louis XVI to the guillotine. All this struck fear in the heart of George III, England's "mad king," whose Tory prime minister, William Pitt the Younger, grew paranoid about a revolution at home. They cracked down on the press, free speech, and political organizing. They passed the Combination Acts making unionizing illegal, and banned the London Corresponding Society clubs and the Tom Paine pamphlets.

Hardy and his peers were tried for treason. They only barely escaped the gallows, thanks in part to Godwin, who'd sprung to their defense, publishing a furious and widely read diatribe in the *Morning Chronicle* in 1794 assailing the state's overreach. Still, the reformers and radicals were beaten back, and the personal fortunes of Mary Wollstonecraft and William Godwin were drawn back with them.

When the two met again, in 1796, Mary had a fraught trip to revolutionary France behind her, and William had a ledger of crushing debts. But this time, their mutual unorthodoxy and agile minds clicked. They were married in March the next year.

Now, in September 1797, in the Polygon, their medical crisis worsened. Mary's placenta had not fallen during birth, and their physician had removed it with his hands. Her condition deteriorated by the day. She grew feverish and weak, probably owing to an infection introduced by the birthing doctor. The doctors came in and out and in and out. One broke the news that Mary should stop breastfeeding; she now risked poisoning the baby.

Less than a month after giving birth to the child they had named Mary after her — the "little animal," as William had affectionately called the baby in utero — Mary Wollstonecraft died.

But there was little Mary, bearing the name of an icon, animated from the start by the sweep and tumult of history, born to the keening over loss amid progress.

These currents would continue to move young Mary, as she grew up among radical minds during some of England's most turbulent days, and they would help grant her the capacity to transform the charge of popular culture, forever.

1799

Robert Blincoe

When the towering building came into view from the road, the seven-year-old orphan Robert Blincoe thought it must be a church. It was enormous, imposing, and surmounted with a cupola. Some of his excitement had waned since he'd set out for Nottingham by wagon four days earlier, with eighty of his peers from the St Pancras orphanage in London, but the scene was striking — windmills dotted the skyline over lush green fields and meadows — and he was optimistic. He'd been the first in line to go, dressed in his Sunday best, when the wagons arrived at the grounds of St Pancras. He leapt into his carriage and cheered loudly as the convoy set off.

The boys and girls played in the hay that lined the wagon for a few hours. But the road was bad, the ride bumpy, and the rough travel started to sicken the kids, who soon realized they were locked inside the carriages.

Just weeks before, Robert had stood tall, proudly allowing himself to be examined alongside his peers, hoping to be among those chosen to leave the orphanage and apprentice as stocking weavers at a mechanized cotton factory. Robert had never known his mother or father. He was brought to St Pancras when he was four years old, and he couldn't recall the circumstances. He hated the place, and wanted badly to get out of there. So when word spread that a far-off cotton mill in the Midlands would choose some children to apprentice in the knitting trade, he was "in a manner intoxicated with joy."

The parish officers fanned rumors that in Nottinghamshire, the children "would be fed on roast beef and plum-pudding — be allowed to ride their masters' horses, and have silver watches, and plenty of cash in their pockets." They would be given fine clothes, taught stocking and lace work, and treated like gentlemen and ladies. They just had to agree to indenture themselves for apprenticeship at Lowdham mill until they were twenty-one, at which time they could take up the trade professionally. For young Robert Blincoe, that would amount to fourteen years of unpaid work in a cotton factory. Scores of the children signed up. Within days, they were arguing over who would get to ride the master's horse first.

By the time they reached Nottingham, most of the kids were pale, sick, and unsettled. While they were being ushered out of the wagons, Robert overheard someone refer to them as "livestock." As the children were led to their quarters a half mile from the factory, the local townspeople came out to watch.

"God help the poor wretches," Robert heard a woman say. "What a fine collection of children," said another; "little do they know to what a life of slavery they are doomed."

"The Lord have mercy on them," said a third.

That may have resonated with Robert, who had just commented on that "large and lofty" structure that looked like a church down the way. The men driving the wagon jeered. They said he would soon know what sort of service was performed there.

There were not yet many large factories standing in England at the turn of the nineteenth century, but there were a few famous ones. Arkwright's factories in Cromford were best known, for their mechanical ingenuity and their rotten conditions, and who was made to work there. Of the nearly two thousand workers, two thirds were children, many as young as six years old.

The very concept of the factory—of laborers organized to do repetitive, unskilled work for wages, for long days, indoors, under the command of an overseer—was despised, for essentially the same reason the townspeople in Nottingham feared for the future of those orphans. The factory forecasted an oppressive life of monotony, danger, and illness. It represented a new concentration of power, produced not by bloodlines or by force of arms but from devices like Cartwright's power loom. The locals in Nottinghamshire saw firsthand what factory life did to the children whom mill owners had shipped in from workhouses and orphanages across England, and what they saw repulsed them.

The first meal at Lowdham was served in a large dining hall, and Robert finally laid eyes on his new peers and fellow apprentices. The last scraps of the dream sold to him, of horseback rides, freedom, and nice clean clothes, were gone. These kids were clad in rags, many without shoes or jackets, and tufts of cotton stuck in their dirty, tangled hair. The mood was sour, and the air was thick with the smell of burnt grease. After a signal was given, each child went to receive their dinner, pulling out their shirttails, into which the attendants dropped a single potato. They looked gaunt, sallow, even injured. At Lowdham, child apprentices worked fourteen hours a day, six days a week, with a lone day off on Sunday.

Bad food, Robert would find, was among the least pressing of his concerns. He was about to begin his life on the factory floor, among the ceaselessly spinning cotton machinery. Those machines, and the whip-toting managers who tended them, would subject him to daily torments and an endless regime of work that would swallow his youth.

Most working people of England had not yet smelled the toxic fumes inside the mills, but they shared a fear with the orphans arriving at Lowdham: that the factory would swallow their futures, too.

1803

Ned Ludd

At the dawn of the Industrial Revolution, before the dominance of the machines was certain, a legend grew about a young man who broke them.

In 1779, the story goes, a boy named Edward Ludd lived in a small town in the Midlands of England. Young Ned was an apprentice in the cloth trade, learning to operate a knitting frame. He did not enjoy the work. His boss was short-tempered and the work was repetitive, intricate, and difficult. And he was just a boy. So, Ned did what boys tend to do when they are forced to do boring tasks: he slacked off.

"Ned being rather averse to the confinement of the frame," went one early recorded version of the story, he "did not exert himself to the satisfaction of his master." For the crime of being insufficiently productive, the local magistrate recommended the boy be whipped. After his lashings, young Ned, furious at the abuse, "took the first opportunity of getting a great hammer" and set about "entirely demolishing the machine."

Ludd smashed the vessel of his misfortune to pieces and took to Sherwood Forest, where it was said that he hid out like the woods' last heroic inhabitant, Robin Hood. In due time, he would inspire other oppressed men to smash the instruments of their masters' profit. He was not the first, and he would certainly not be the last.

The spirit of the legend, its relatability, gave it advantage. Ludd likely never existed at all. But his avatar would become a mascot, an organizing tool, and a winking joke—a powerful nineteenth-century meme. Like Robin Hood, Ned Ludd lent his name to a broader struggle that was very real, whether or not he was. Every good legend needs an origin story.

To those being ground under the gears of the Industrial Revolution, Ludd would become a captain, a general, and even a king.

A gig factory in Littleton burned to the ground in spectacular fashion in the summer of 1802. The factory's manager, a man named Clyde Heath, claimed that he was certain, despite the pitch darkness of the night and the fact that the figure's face was blackened with coal, that the culprit was an eighteen-year-old boy named Thomas Helliker. It was Helliker, Heath insisted, who held him at gunpoint while his cohort torched the factory, destroying the machinery inside and causing £8,000 in damages — a massive sum, roughly $780,000 in today's dollars.

Helliker was a "colt," an apprentice in the cloth finishing trade. Cloth finishers, also known as croppers, were responsible for smoothing, or finishing, woven products. In Wiltshire, where Littleton lay, they were in the thick of a contentious strike. The authorities tried to get the Helliker boy to talk, but he refused. Despite evidence quickly surfacing that he had not been there at all, he faced the gallows.

The chief investor in the factory system and automated machinery in the region, and the workmen's main antagonist, was a man named John Jones. He'd built the Staverton Superfine Woollen Manufactory, a high-tech, six-story factory "filled with every sort of newly invented machinery," despite vocal public opposition, and now led an effort to cut the cloth finishers' wages. When the workers went on strike, he invested in still more automatic machinery in a pointed effort to replace them. That was the final straw for the croppers and their allies; they smashed windows of the Staverton factory and organized a raid on the mill at Littleton and set it ablaze.

This act of sabotage, in turn, broke the factory owners' will. They agreed not to lower wages and abandoned the gig mill technologies. Jones, meanwhile, swore revenge.

Incidents like the torching of the factory in Littleton mark a part of story of the Industrial Revolution that often goes untold: alongside every major labor-saving innovation, a spasm of protest burst out from the workers whose lives it disrupted. And because laws that regulated their trade or protected their rights were often lacking or unenforced, cloth workers and artisans resisted conditions they found unfavorable by breaking the machines used to exploit them. If merchants or shop owners refused to pay established rates or tried to bypass legal regulations with new technology, workers might smash the machinery they deemed "obnoxious." At a time when organizing unions was illegal, it was a strategy embraced by workmen whose jobs were on the line with no other recourse: "collective bargaining by riot," as one historian termed it. "The eighteenth-century master was constantly aware that an intolerable demand would produce, not a temporary loss of profits, but the destruction" of his machinery.

These workers did not view technology as inherently progressive; they had not been taught to lionize disruption. To them, devices that would degrade their working conditions, or harm their ability to earn a living, were a moral violation, plain and simple. The artisans and craftsmen knew exactly what these machines foretold for their trade — they considered technology as a force taking shape not in the future but in "the present tense," as the historian of technology David Noble put it.

This is why the weavers in Manchester didn't want to help Cartwright with his loom. They were all too aware of the kind of innovations that were accelerating the Industrial Revolution. In 1710, cloth merchants had taken to using unpaid apprentices on their own stocking frames instead of paying skilled workers fair wages. After weeks of fruitless negotiation, the stockingers smashed a hundred of the hosiers' machines. In 1768, after James Hargreaves unveiled his spinning jenny, a band of cotton spinners broke into his shop and destroyed it. He'd developed the device in secret, fearing precisely this outcome, and was forced to leave town before he tried to commercialize it. In 1779, angry workers burned one of Arkwright's factories to the ground while town officials looked on in sympathy — with the workers. In 1792, cotton weavers destroyed the first two dozen power looms that Edmund Cartwright had managed to sell. Now the same fate befell Jones's gig mill, which had automated the work carried out by the croppers, some of the most powerful of the cloth artisans. The spirit of Ned Ludd was in the air.

The trial of Thomas Helliker was a farce, according to the locals who helplessly watched it unfold. A friend of Thomas's told the authorities that they'd gotten drunk together the night of the attack, and that he'd actually locked Helliker inside a room to sober up. That meant it was impossible for him to have been present at the mill at the time of the arson. Meanwhile, Heath, the factory manager, had a motive to blame Helliker — Heath's employer offered him a hefty reward for identifying troublemakers. Heath had no doubt seen Helliker around town, could easily ID him, and might have thought he would make an easy mark. Finally, it was unlikely that Helliker would even have been *allowed* to join in an attack. As a mere colt, he was excluded from major planning sessions or actions undertaken by more senior craftsmen.

Still, Jones pressed on with the trial, threatening Helliker if he didn't name names. If the boy did know the men who had destroyed the factory, he never said a word. He was hanged on his nineteenth birthday. The outpouring of rage and grief was overwhelming, and reverberated far beyond Littleton. Workers carried his coffin across the Salisbury Plain in a grand procession, with "thousands of women dressed in white" walking alongside it, to a tomb in his hometown that commemorates him as a martyr. It reads, THOMAS HELLIKER: THE THREAD OF WHOSE LIFE WAS CUT IN THE BLOOM OF HIS YOUTH.

By the end of the first decade of the nineteenth century, the Industrial Revolution was in full swing. The most successful entrepreneurs and industrialists — the first tech titans, we might call them — had accumulated impressive stores of wealth and power, and more followed in their footsteps. Entrepreneurs and businessmen were mechanizing work in England's largest industry, automating jobs that had been bound by tradition for generations, and building new monuments of power and production. Political unrest was rampant, labor organizing and radical speech were forbidden, and an unpopular war was about to enter its third decade.

Those factories and machines now confronted the younger generations of workers and artisans with a dark vision of the future. And those workers, those craftsmen, apprentices, machinists, and croppers, had few tools at their disposal to push back. But they had each other, they had the pen, and they had the hammer.

PART I

THE GREAT COMET

GEORGE MELLOR

Spring 1811

Night after night, an unusual celestial body hung in the sky. It looked like a cross, people said, or a dagger. All told, it would be visible across Northern England for 260 days, the better part of a year. It was widely believed to be a bad omen. The comet had appeared in the spring and would burn brightest in the fall.

The Great Comet of 1811 became so famous that it was recorded in major works of popular culture. In Leo Tolstoy's *War and Peace,* the prodigal son Pierre gazes up at the "enormous and brilliant comet…that was said to portend all kinds of woes and the end of the world."

Bad omens are common when times themselves are bad; it's intuitive to predict the coming of more misery when you're already surrounded by it. And in 1811, for the working people of England, misery was everywhere.

The Crown was levying heavy taxes on its middle class to pay for its seemingly endless war with France. Worse, it had imposed trade sanctions, known as the Orders of Council, against French allies including the United States, cutting off huge, crucial markets for England's enormous textile industry, and plunging the cloth-producing regions into economic crisis. Across the nation, looms and stocking frames sat idle. Two consecutive years of bad harvests left food prices painfully high. And while the cloth trade was at a point of weakness, a growing number of businessmen had introduced more automated machinery into their shops and factories.

Twenty-two-year-old George Mellor would have been able to take in the sight of the Great Comet from his hometown of Huddersfield, smack in the middle of England.

Born in 1789, the year the French Revolution caught fire, George had a staunch moral code, a sharp tongue, and charisma to spare. He stood "six feet by the stick, and with shoulders well back, and strong, firm, warm hands that gripped you to make you tingle," as one local historian described him. "His eyes were brown and full of fire, and dark auburn hair curled close upon a rounded head. He had a temper, if you like, but he never bore malice."

George's strong, square jaw accentuated his "resolute, determined appearance," as another writer put it. He was righteous, impatient, and blunt, a man of "iron will and reckless daring." His father had died when he was three, and his mother married a man named John Wood, who owned a small cloth-finishing shop. George sparred with him constantly.

Wood grew ornery as his business dried up and orders that used to be sent to his shop were sent instead to larger operations outside town, to entrepreneurs who used more advanced machinery. George, meanwhile, chafed at being bossed around. Once, in a fit of anger, he ran away from Wood's home and returned only after he'd drawn up a list of conditions his stepfather would have to meet for George to continue living there.

George shipped off for war to fight "Boney," like so many young workingmen throughout the 1790s and 1800s. He was said to be a veteran of Britain's far-flung campaign against Napoleon in Europe. Also like many veteran workingmen, he came to detest the war as it dragged on.

When he was a teenager, George embarked on a seven-year apprenticeship to become a cropper. The cloth-finishing job required managing a pair of massive shears that weighed up to sixty pounds. A cropper used teasles to raise the nap, or the rough surface of a woven piece of cloth, then ran the great blades over the wool to smooth it. It was exceedingly hard, skilled work.

A good cropper was crucial to producing top-quality garments. Croppers earned more than most workers in the industry, and wielded more power, since well-finished cloth enhanced its value, and a shoddy finish could tank it. Croppers were also known to be unruly, free-spirited, strong-willed — and organized. Despite the Combination Acts, they had been known to strike in protest over low pay or to resist the introduction of machinery that might impact their jobs.

But the work left the body callused, bent, and molded. You could tell a cropper by his enormous forearms and by the "hoof" of callused skin that built up on his wrist. In the spring of 1811, George was in his early twenties, and he'd spent his post-adolescent life learning the trade. Seven years of hard, exacting labor;

seven years of paying his dues. That led to pride and attachment to the work, to a brotherhood, to an identity.

And George had learned with gusto, making friends and allies along the way. He'd come to be the most charismatic of the croppers at John Wood's shop, where he was respected by his colleagues, and was well-liked across town. As the Huddersfield writer D. F .E. Sykes put it, "A right proper man George Mellor was."

With its population of fewer than ten thousand souls, Huddersfield was small, but it convulsed with change. It sat twenty miles southwest of Leeds and seventy miles northeast of Nottingham, in the West Riding region in the county of Yorkshire, whose capital was the medieval city of York. Like so much of Britain, the West Riding was industrializing fast — thanks to those machines and factories that Wood and George eyed warily — against the will and comfort of many. The region was famous for its dissent-prone denizens, and would come to be known even more widely as the "metropolis of discontent."

But just decades earlier, George's homeland, draped in fog and surrounded by steep, moss-green hills, had been hailed as beautiful and idyllic. When the writer Daniel Defoe traveled to the West Riding in the 1720s, he was stunned. He praised not just the region's beauty, but its high quality of life, its humming industry, and the unusually egalitarian society it was home to. "The sides of the hills, which were very steep every way, were spread with houses," he wrote. "This whole country, however mountainous, and that no sooner we were down one hill but we mounted another, is yet infinitely full of people.... Those people were all full of business; not a beggar, not an idle person to be seen." Defoe liked the region so much that he decamped there to finish writing the novel that would secure his fame. He even had his hero, Robinson Crusoe, hail from the same county.

The business that kept Huddersfield residents busy, of course, was the clothing trade. By the time Defoe had visited, locals had been weaving wool for hundreds of years — almost everyone who lived there was connected to the industry. George was a cropper, his stepfather owned a cloth-finishing shop, his uncle William was a master handloom weaver, and William's son, Ben, who was close with George, was following in his father's footsteps. With the whole family pitching in,

William ran the weaving business from their home, along with a small farm. It'd been like this for as long as anyone could remember.

In *Ben o'Bills*, by D. F. E. Sykes and George H. Walker,* Ben describes the family business, offering a window into the daily life of the tens of thousands of English weavers who worked in a similar way:

> [My father] bought his wool of the stapler at Huddersfield — old Abe Hirst; — it was scoured and dyed in the vats in the farmyard; my mother and my cousin Mary, and Martha, the servant lass, that cleaned the house and milked the cows [and] helped with the spinning. The warping and the weaving we did at home in the long upper chamber. We had four looms at home, and, moreover, we put our work out to the neighbors. It was a busy house you may be sure.

With those four looms to work, journeymen weavers would join Ben and his father. They sold the woven cloth to a local merchant, who had it finished at a shop like John Wood's. Defoe noted that the people of the West Riding, like George and his stepfather, and Ben and his father, "generally live to a great age, a certain testimony to the goodness and wholesomeness of the country," and are "constantly employed, and…working hard."

This was the original "cottage industry," and it made for a flexible and family-oriented lifestyle. There was demanding work to be done, and everyone was expected to pitch in, but it was common to work just thirty hours a week, on one's own schedule, and take long weekends. The London-born novelist Elizabeth Gas-

---

* *Ben O' Bill's, The Luddite: A Yorkshire Tale* (London: Simpkin, Marshall, Hamilton, Kent, 1898) was researched and written by Huddersfield historian D. F. E. Sykes, with George H. Walker. They interviewed people directly familiar with the events in question, and the book recounts actual events and players, but they chose to publish the work as fiction, with the disclaimer that "the story, *Ben O' Bill's*, is mostly true, and the authors have not felt called upon to vary in any material respects the story as it was gleaned in part from the lips and in part from the papers of the narrator." (The narrator is Ben Bamforth, George Mellor's cousin.) Its depiction of major events, as corroborated by other histories, newspapers, and court records of the time, and the book's record of regional dialogue, family politics, and Luddite organization are invaluable. But, as the introduction to one printing noted, "It is interesting that, as a historian, Sykes chose to embellish the facts, that were available to him at the time, with fiction, and his purpose must have been literary."

kell wrote of being shocked by Yorkshire's egalitarian community, where "nearly every dwelling seems devoted to some branch of commerce. In passing hastily through the town, one hardly perceived where the necessary doctor or lawyer can live, so little appearance is there of any dwellings of the professional middle-class." She was "repulsed" by how comfortably these working people addressed those who were "above their station."

But in 1811, those small businesses were struggling. New operations had moved in, and the local blacksmiths' shops were busy building machinery. In fact, all George's life, change had seemed to consume his homeland, and, as he saw it, little for the better. The public land across Huddersfield had been steadily walled off and enclosed into private property, and was now controlled by a handful of elite landowners. Factories blotted the landscape, and coal mines and steam engines blighted it. Huddersfield was on the front lines of the Industrial Revolution, though in 1811 no one was calling it that yet.

For George, it was personal. Two new machines, the gig mill and the shearing frame, were turning up around Huddersfield, and they could do with an unskilled operator, even a child, what croppers like George had taken years to perfect. The gig mill could raise the nap of an unfinished piece of cloth automatically, while the shearing frames were mechanized to cut it. The Wiltshire croppers had stymied the gig mill's advance back in 1803, with a strike and with arson. But now trade was bad, the croppers had less leverage, and the entrepreneurs had brought the machines back. So the smiths kept cranking them out, and the bigger businessmen were adopting the technology for use in their shops and factories.

George Mellor had finished seven years of punishing service just in time for these machine owners to move to erase his future. He knew as well as anyone that the working people of England had plenty of reason to see bad omens in the night sky.

GRAVENER HENSON

March 1811

To a twenty-six-year-old framework knitter named Gravener Henson, the Combination Acts were a joke. And he intended to prove it.

Every worker knew that the laws, passed a decade earlier amid the Crown's swelling paranoia over reform movements, made it illegal for workers to form unions. But the acts were, technically, a two-way street. Business owners were barred from "combining" as well. That meant no price fixing, no colluding to set wages.

So it seemed absurd to Henson that a group of Nottingham hosiers, the business owners who bought and sold stockings, felt comfortable publicly announcing their joint decision to slash wages. Four of the city's biggest hosiers had decided to reduce the rate that they paid knitters, or stockingers, like him — those who worked a stocking frame to make lace, hose, and knitted goods — and they published an article in the *Nottingham Journal* documenting their reasoning.

That letter was a new volley in an escalating conflict between the knitters and the hosiers, the workers and the bosses. The hosiers' timing in colluding against the knitters could not have been worse. As in Huddersfield, the Nottinghamshire cloth trade had cratered, thanks to sanctions, war taxes, and poor harvests. Nottingham was the epicenter of the stocking industry, where knitting and lacemaking were about half of the region's economy.

Stocking-making had long been a respectable way to earn a living, even if framework knitters were never as prosperous as croppers like George Mellor. They were skilled, dedicated workers who enjoyed flexibility and autonomy in their workdays, if not extravagant pay. The cloth seller William Gardiner, in his 1838 memoir *Music and Friends,* described the life of a stockinger before the depression:

> What contributed to their solid comforts was the common and open
> field. The stocking maker had peas and beans in his snug garden, and a

good barrel of humming ale. To these comforts were added two suits of clothes, a working suit and a Sunday suit; but, more than all, he had leisure, which in the summer-time was a blessing and delight. The year was chequered with holidays, wakes, and fairs; it was not one dull round of labor. Those who had their frames at home seldom worked more than three days in a week.

Now countless framework knitters were out of work. With lowered demand, hosiers had cut wages, were using apprentices and children for cheap labor, and were churning out low-quality knockoff goods. The war was on, they said, and times were tight. They couldn't help it if markets were restricted. Besides, stockings were out of fashion that year, they said. And they weren't wrong. One ominous indicator of the widening divide between rich and poor was that trousers and boots — which better covered up the corpulent forms of well-fed aristocrats — were in style in London, not form-fitting stockings.

But a number of hosiers were also engaging in unscrupulous behavior, like forcing the knitters to accept payment in "truck," or goods, rather than money. Worse, most knitters had to rent the frames they worked on, and speculators outside the industry had realized they could buy the machines in bulk and jack up the prices. It had become common for these businessmen to charge a knitter 30 percent of the revenue they earned.

If the workers were to gather and demand the hosiers raise wages or address any of their concerns, however, they could be thrown in jail; that would be an illegal "combination" under the acts. But when those four Nottingham hosiers jointly announced that they would "reduce the price to our own Workmen, leaving it entirely at their option, whether they would take the prices we offered them," they did so with a brazen confidence that no such fate would meet them.

So Gravener Henson decided to do something about it. He was going to march to the magistrate's office and prosecute them for violating the Combination Acts.

Born in Nottingham in 1775, Henson had little to no formal education. He took to reading for personal pleasure, however, and started sitting in on free Wesleyan Sunday school sessions. He taught himself to write, and then to write well. He apprenticed as a framework knitter, and spent years in the trade before emerging as an unlikely activist and advocate. He was puckish yet dedicated, passionate but odd, and he could be both churlish and startlingly eloquent. One of his contemporaries described him as "thick set, with short neck, keen small eyes, and a head very broad at the base, rising up angularly to an unusual height."

That broad, angular head contained an almost unparalleled store of knowledge about the clothing trade and the laws that governed it. "He possessed an extraordinary memory, and delighted in the histories of manufactures and commerce," a colleague wrote. "He knew most of the laws of his own country and France regulating these matters."

Word of his unique expertise had spread, and in 1809, some frame workers enlisted him to help them resolve a dispute they were having with their bosses. "They compelled me to act, in a good-natured way," he later quipped. "They stopped my machine."

Now, in 1811, with his textile knowledge and experience in dealing with authorities in tow, Henson went to the Nottingham town hall, where he charged the manufacturers with combining to lower wages. The officials were not moved.

"The magistrates told me that I had not got sufficient evidence," he recounted, so "I produced the newspaper. I told them, I thought they might summon the printer, under the Combination Act, and he would give evidence." When asked further what that evidence was, Henson was blunt: "They published them in the public papers, week after week, signed by four distinct persons, to reduce the wages."

None of this inspired the local investigators to pursue a case against the largest business interests in town. "They said, their opinion was, that I had better not go on," Henson explained. They showed him the door. He was undaunted, though, and when he came back, the magistrates grudgingly admitted that he had a case. This time, they agreed to issue a warrant against the owners. Henson was sent to the town clerk, where the trail went cold again.

"He refused to grant me a warrant," Henson said, "because I could not prove the parish where they had met in." He gave up the case, but the point was made: the knitters and the hosiers profiting from their knitting were subject to very different rules. From there, the bitterness between the workers and the hosiers deepened. The workers continued to plead their case with increasing desperation; the hosiers had driven wages too low, raised rents on the devices too high, and finally pushed many to the brink.

"The manufacturers would not listen," Henson later told the authorities, "and [the stockingers] at last came in crowds."

"Very soon it was supposed," he said, "that the men would begin to break the machinery."

THE MACHINE BREAKERS

About halfway through the same month that the Great Comet first appeared, on March 11, 1811, a desperate, angry crowd filled the grime-encrusted medieval square in the center of Nottingham, swelling to a thousand strong.

The city smelled like human waste. It was overcrowded, poor, and suffered from sanitation problems. The population in many of England's cities was exploding, and local governments couldn't keep up.

Between 1650 and 1750, England's populace had held relatively steady at just over 5 million people. Fifteen percent lived in cities, the rest in rural areas like George Mellor's West Riding. But by 1811, there were nearly 9 million people; by 1820, 10 million. The urban-rural script was flipped on its head. Eighty-five percent lived in cities, and just 15 percent in rural areas. In Nottingham, "the poor lived on each other's backs" as development struggled to keep pace.

"For twelve months past many working men had swept the streets, paid a pittance by the city, with no other work to do," recalled William Felkin, a teenaged framework knitter's apprentice. "Threats of vengeance had been loudly uttered against hosiers paying reduced wages....Many of these men came in from all parts of the county."

Fiery speeches broke out, and orators aired grievances. Some blasted the Crown for the war against Napoleon and the Orders of Council. Others yelled out against shop owners setting food prices too high; whole families, including children, were going for days without eating. Still others called out for political reform: minimum wages, worker protections, representation.

The speakers also took aim at another pressing trend—automation. They railed against the master hosiers and businessmen who had taken to using a device known as a wide frame to jack up the rate of production of garments and to justify lowering wages further still. Those machines, the stockingers yelled, were putting craftsmen out of work. Wide frame technology had been around for a while but had chiefly been used to make cheaper, more disposable goods. Now, in leaner times, the clothing business owners started using it to allow children, colts, and less skilled workers to make stockings, too — but badly. Someone previously unfamiliar with the trade could use a wide frame to churn out stockings roughly six times as fast as a framework knitter, though the product was of obviously inferior quality. Skilled knitters made the entire stocking in one continuous piece, while wide frames churned out "cut-ups" in two pieces, to be stitched together.

The stockingers didn't have any qualms with the use of the wide frame back when it was making simple goods, like hose, that they did not compete with. They didn't hate or fear it for being new technology, either. "If workmen disliked certain machines, it was because of the use that they were being put, not because they were machines or because they were new," as one historian noted. There were plenty of cases where workers were staunchly in favor of new tech: they had asked hosiers to adopt a machine that would more thoroughly tally the thread count in a garment, to attest to the quality of their work. But the hosiers largely refused to adopt it, preferring to retain the unilateral power to determine the quality of a garment themselves, and to offer workers the prices they approved of, regardless of quality.

But as the shop owners used the wide frames to slash costs — and the knitters' wages — the machines came to embody the growing list of exploitations the men felt they were suffering, and the displacement of workingmen caused an outrage. Framework knitters may not have been wealthy, but they were proud. Now, on top of saddling them with high rents and payments in goods, merchants and entrepreneurs were using machines and children to take over their work, to increase their own profit — and in a time of crisis, no less. That year, fully half of the town's thirty thousand citizens were listed on the poor rolls, each desperate to receive charitable relief. The economic situation was so dire that hopeless framework knitters were committing suicide.

Before the crisis had reached such heights, the cloth workers of Nottingham, energized by men like Gravener Henson, had tried to make their case to the hosiers and magistrates. They argued that substandard products would hurt the reputation of the industry, that using the wide frames violated regulations governing the use of apprentices put in place by Charles II, and that it was unethical to put men out of work for the sake of directing profit into the hands of a few when times were as bad as they were for so many.

They made some good points.

But their public argument had fallen on deaf ears. So had lobbying. For a decade, the trade group representing weavers, knitters, and croppers had petitioned Parliament to enforce the laws and regulations that the manufacturers violated. But time and again, Parliament sided against the workers. In 1809, the government, led by prime minister Spencer Perceval, had overturned the trade regulations altogether. The year before, they had thwarted a popular mass effort calling for a minimum wage.

Peaceful acts of protest had not fared much better. Just weeks before the workers gathered in Nottingham, a group of framework knitters had waited until dark, snuck into a factory that deployed the wide frames, and removed the jack wires they needed to function. This mild act of rebellion did not sway the hosiers, however; they still refused to restore rates paid to knitters to pre–wide frame levels.

All this was carried in the undercurrent of that day in March of 1811, as the crowd swelled in Nottingham's square.

"A number of individuals from the adjacent villages," the *Nottingham Journal* reported, had made their way into the city "with a view of representing to their employers the hardships they were subject to and of intimidating others into compliance with their demands by which alone they can be enabled to obtain a subsistence."

Tempers flared, and nervous city officials called in the dragoons—sword-wielding, horse-mounted infantry—to suppress the crowds. The assembly was a peaceful, albeit electric and volatile, show of protest until nine o'clock at night, when the crowds dispersed and the military let down its guard.

After nightfall, the crowd reassembled just outside the city limits, joined with other protesters from the neighboring county, and grew to a throng of two, even three thousand people.

The teeming, haphazardly organized legion of workers made its way to the nearby city of Arnold, where a new textile factory stood. In the dark, the men covered their faces with black masks and slipped inside. Gravener Henson had been right: they were ready to break the machines. With a massive blacksmith's hammer, they destroyed the implements of automation one by one. Before sunrise, sixty of the "obnoxious machines" had been smashed to pieces.

Small, organized attacks, carried out by bands of masked men, continued in the following days. More wide frames — and only the wide frames — were left in ruins.

The disturbances convinced Nottingham hosiers to call a meeting two weeks later, on March 26, where the first order of business was to condemn "the illegal measures which have been resorted to by some of the framework knitters."

The second order was to issue a proclamation that resolved to "recommend to the Trade in general to give for all full-fashioned work the OLD PRICES."

These machine breakers had won. They had, for the moment, preserved their ability to earn a living.

THE ENTREPRENEURS

1800s

"Good God, Francis, it is no use your pretending to make cloth or any but those who have such places as me," the factory owner John Jones scoffed at a smaller rival in the early 1800s. His meaning: unless you automate your business and undertake mass production, you won't be able to compete.

Despite the protests of local working people, Jones had built the largest and most mechanized cloth-making operations in the West of England. It was his gig mill operation that the croppers had fought in 1803, and his insistence that the teenaged apprentice Thomas Helliker be hanged for the mill's destruction. And it was the ardent commitment to building high-tech factories evinced by men like him that drove other business owners to adopt machinery, even when their communities, or they themselves, opposed it.

"For several years the more enterprising merchants and some large master clothiers had glimpsed the riches that might accrue if they organized all stages of production under one roof, making full use of the machines that were already transforming the cotton industry," one Yorkshire historian noted. It was relatively easy, on a logistical level, to take the plunge into organizing production this way — essentially to develop small factories. The textile industry at the time was "the land of opportunity for the energetic and ambitious man with little capital," according to one economic historian of the Industrial Revolution. In other words: it was an ideal time to launch a start-up.

In the late 1700s and early 1800s, many entrepreneurs who embraced new technology did so in humble, rented spaces: old grain storage rooms or empty halls — the cluttered Silicon Valley garages of the day. The equipment necessary to jump-start an operation was not all that expensive — with a modest loan or some saved money, one could afford a few used looms, enough factory space, and even a monthly payment on a steam engine, if your business required it. (Though then, as now, loans were granted on the basis of many factors, including one's public standing, background, and reputation, limiting who was eligible from the outset.)

But as entrepreneurs invested in machinery, experimented with the division of labor, and increased their profit margins, it became clear that their gains were coming at a cost — they were sowing inequalities and resentment. They knew that working people, who were often their neighbors, opposed their actions, and had heard stories of protests, threats, and smashed equipment.

There were not a lot of such audacious entrepreneurs in those days, but their ranks were growing. One particularly ambitious man named William Horsfall ran an enormous wool factory outside Huddersfield that employed four hundred workers; he had rows and rows of shearing machines. His operation, per the *Leeds Mercury,* had "attained considerable perfection." But most cloth was still manufactured by men like the cloth maker whom Jones was talking down to, and those men still owned most of the machines, too.

And as often as not, *those* men sympathized with the weavers, stockingers, and croppers when they protested that machinery would cost them their jobs. Yet many small business owners found themselves pulled toward the tech titans of their day, not toward the workers they sympathized with. Why?

Imagine you run a healthy business manufacturing wool in the preindustrial age. You buy the raw materials, or the yarn that's already been spun, and deliver it to a weaver, who turns it into a rough piece of cloth. You take that cloth to be finished — all the rough parts sheared off, and the garment smoothed and softened — at a cropping shop. You then take the "finished" cloth and sell that to your clients, or at a market.

Your customers are a mix of local buyers and larger distributors to markets in farther-flung places like London or even the United States. You inherited the business and most of your contacts from your father, who was a cloth merchant like his father before him. You do not make an exorbitant amount of money, but you do well, and have close relationships with the weavers and croppers you do business with. You know their families, are friends with some of them, and are happy to pay a fair price for their work. You're satisfied with the arrangement; obviously, it isn't perfect, and you get into arguments over the quality of a garment here or a delayed order there, but most parties feel respected, compensated, and content with the level of autonomy they enjoy in their work. It upholds tradition, and it upholds the community.

But lately, business has been tough. The trade sanctions — which nearly everyone you know, from weavers to your competitors, hates bitterly — have dampened demand for your goods. There are new men getting into the business, and some of the old men have taken on a new attitude. These entrepreneurs often come from outside the community, new to their part of the trade, and don't have the same commitment to the people who live here. Instead of buying from weavers who work at home, they employ apprentices and journeymen directly in their own shops. And they have bank loans to invest in machinery that automates the process you've done with real workers for years. Most of these operations are small, but there are more and more of them.

Then, some of the old masters who are a little more steely-eyed than you, along with a handful of new competitors, have invested even more grandly in new technology. They have built larger operations to gather workers under one massive roof and economize on labor. They've built factories. They know that people won't like them, but they stand to make a *lot* of money. Pointing to the doctrine of laissez-faire or gesturing toward progress, they announce themselves as the future. (You might blame this trend on the teachings of Adam Smith; *Wealth of Nations* was published in 1776, and has exploded in popularity among businessmen in recent decades.) Some now seem to feel that innovation in and of itself is a virtue.

Now these competitors can sell their cloth more cheaply than you can. Some of your clients, whom you have met every week at the cloth hall for the last ten years, have told you they must buy the less expensive option instead. *Sorry, it's nothing personal, times are tough.* Your cloth business is not large enough to take many hits like this, and the writing is on the wall. Maybe the future *is* the gigging machine that allows a child apprentice to do multiple croppers' jobs at once, or the wide frame that can churn out cut-up stockings faster and for a fraction of the price.

So you face a choice: Do you buy the machines to keep up, so you too can lower your prices — and apologize to these men and women whom you've known all your life, whom you're now helping to put out of work? Or do you stand with the cloth workers as your income ticks down?

Some merchants, master clothiers, and small manufacturers did back the workers against the large entrepreneurs, and refused to buy new frames. *Ben o' Bill's* recounts what this dilemma looked like in practice: Ben's father, the master weaver William Bamforth, is weighing whether to invest in automated machinery, as a few of his competitors are doing, and the prospect troubles him so much he visits his pastor, Mr. Webster, to seek advice. Webster quotes him a parable, saying "it is written that the laborer is worthy of his hire, and ye may not muzzle the ox that treadeth out the corn." Using this machinery would stamp out the weavers' chance to earn an honest living, in other words. Webster warns Mr. Bamforth against "making haste to be rich," as this new generation of businessmen seems inclined to use new tech to do.

"You speak of building a mill for these new methods." He continues:

And now I ask you, brother Bamforth, can you be rich toward God, if you build up your fortune on the ruin of your fellow men? You say one of these new finishing frames will do the work of four, maybe of six men. Aye, also is there talk of looms that shall need neither skill nor care. It may be true, I know not. But oh! it will be a sore day for this hillside, and all the country round when that day shall be. What is to become of those who now keep a decent roof over their heads?

The times may be bad, he said, but men could still earn a little bit, to feed their families, their babies. "You may make new machines but you cannot make new men to order." In the story, Bamforth, who employs journeymen weavers himself, decides against automating his business.

One powerful cloth merchant, the father of future conservative politician and child labor reformer Richard Oastler, also refused to invest in automation, believing it to be "a means of oppression on the part of the rich and of corresponding degradation and misery to the poor." Instead, he gave up the trade altogether. Others stuck with the domestic system, even when it risked driving them to bankruptcy.

Many, of course, did not.

Plenty were conflicted as they joined the technological race, though. They knew the social compact that workers had observed for generations was being trampled by every machine added to a shop floor. But they felt pushed toward the machines by the strongest-willed entrepreneurs. And those entrepreneurs, like Jones, Gott, and Horsfall, had no such compunctions about the workers they displaced, and were driven instead by ambition, a deeply held belief in

Smith's fashionable free-market philosophy, sheer greed, or a cocktail of all of the above.

This was the scattershot process that set the Industrial Revolution turning: minor, staggered investments in machinery by small manufacturers, all revolving around the examples set by large factory owners. Incidentally, it is what set later industrial revolutions turning, too. As David Noble, a historian of twentieth-century automation, wrote, "Managers feel they must automate because 'everyone's doing it,' out of fear that they will be undone by more up-to-date competitors (a paranoia encouraged by equipment vendors). There is this vague belief that the drive to automate is inevitable, unavoidable, and this belief becomes a self-fulfilling prophecy."

Enough of the entrepreneurs displayed a callousness toward the working people's situation that disaffection grew throughout their ranks. "Now the masters are for makin' bad worse with this new machinery," a man named Soldier Jack says of the Huddersfield businessmen in *Ben o' Bill's*. "They're crying 'Every man for himself an' devil take the hindmost.'"

The working people of the 1800s knew who was on which side in their community. They knew who refused the frames, who had worked to uphold tradition and to pay fair wages, and who believed it was their God-given right to buy machinery that would put their neighbors out of work in the middle of a deep recession.

THE PRINCE REGENT

Summer 1811

The so-called Mad King had finally been pronounced too ill to rule. After spending long stretches of his reign battling bouts of mania, George III lapsed into a permanent state of derangement in 1811. His eldest son, George, the Prince of Wales, officially became the Prince Regent and ruler of England, though there would be restrictions on his power for a year, in case the king recovered. Among the Regent's first orders of business was to throw himself the most lavish party London had seen in ages. No expense was spared. The guest list sprawled past fifteen hundred names, including Louis XVIII, the self-proclaimed king of France, now entering his third decade in exile after the French Revolution swept away the monarchy. The highest ranks of British society and Parliament were invited, as well as the A-list celebrities of the day: writers, artists, and aristocrats.

The ball was all anyone in London society talked about. It eclipsed all other news, from the war against Napoleon to the spasms of desperation from the nation's poor.

On the evening of June 19, 1811, there were so many carriages bringing the British nobility to the prince's fete at Carlton House that the traffic choked the streets, and most were drawn to a complete stop. Crowds gathered to watch the guests arrive. "The remarks of the people on the occupants of the carriages, as the latter crawled or jolted on at a snail's pace, were sometimes very droll and apt, though not always complimentary," George Jackson, a British diplomat, remarked, rather diplomatically.

At least the aristocrats had the glory of the Carlton House itself to take in as they waited. The prince had been renovating the mansion for years, and some proclaimed it "the equal of Versailles" in splendor.

When those guests did finally arrive, there was indeed no shortage of luxury on display. They would enter the Blue Velvet Room, under a painted ceiling where a three-tiered crystal chandelier glittered. Then they would move into the Throne Room, with its "curtained Romanesque bays flanked by gilded Corinthian columns." Priceless paintings hung in the silk-draped audience rooms; Rembrandt's masterwork, *A Shipbuilder and His Wife,* was the prince's latest acquisition. A silver fountain fed water into an indoor canal that ran across the sprawling Gothic conservatory. "The value of the diamonds," Jackson wrote, "no doubt exceeded what was ever before seen in any assembly."

Attendees would finally shuffle into the overstuffed ballroom, where the guests of honor, including the French exiles — the Prince Regent adored royalty and considered Louis to be the rightful monarch of France — were congregating. The prince, dressed ceremonially as a colonel (a rank he had failed to obtain in his actual military service), greeted guests with a star splayed across his chest and a saber protruding from his belt.

Removed from his immediate environment, the Prince Regent cut an absurd figure against the dark mood and starved strife pervasive in 1810s England. He was self-absorbed, obese, and derided by his own constituents for his ignorance of domestic policy matters and his excessively lavish tastes.

"There was something puckish in his face, an air of mischief and even of innocence that went oddly with his almost indecently corpulent body," one historian wrote. "His features had always been more pretty than handsome, round and cherubic, set off by a fine head of light brown hair that had not yet begun to thin. When at his best his grace of manner was extraordinary, though he was often not at his best, or even civil. His boyishness had its petulant side; he sulked, he was capable of malice. He could be cruel."

He was an inexperienced playboy, born into immense wealth and elevated to power. "The enduring image of George IV is that of 'Prinny,' the overweight, overdressed, and oversexed buffoon waiting for his periodically deranged father to be declared unfit to rule," wrote one of his biographers. He was a spendthrift, too: by 1795, the prince had amassed an astonishing £630,000 in debt — the equivalent of nearly $76 million today. His father refused to settle the debt unless he married, so George did, and the Crown picked up the tab. The grudging marriage did not stop Prinny from having countless mistresses and liaisons (and rumors of nearly as many children), or being followed by scandal during his entire tenure in power. At one point, he even tried to assign one of his known mistresses to oversee his wife's bedchamber, but was talked out of the idea, on grounds that it would be too spiteful.

George III had been an attentive, even obsessive ruler. He kept meticulous notes about how ministers and Parliamentarians voted and about their political opinions and positions. He was closely and fervently involved in matters of domestic policy and military operations. The king had drawn out the American War of Independence for years after it was clear England was losing, motivated by personal, aggrieved fury, and aimed to starve the opposition's will to resist.

The Prince Regent had no such stomach for policy, nor military matters; he preferred to rubber-stamp the documents and legislation put in front of him by his aides. As one lord described it, "He sits at a table with General Turner on one hand, and Colonel McMahon on the other, the one placing a paper before him for signature, and the other drawing it away." He was confirmed as Regent in February of 1811; by August, he had signed thousands of documents this way.

"Playing at King is no sinecure," Prince George quipped to a lord present for one rubber-stamping session.

This Regency period in England, from 1811 to 1820, is what historians like to describe as "a time of contrasts." While farmers, artisans, and what would now be called middle-class workers were enduring levels of suffering and uncertainty they'd never experienced before, elites enjoyed unprecedented luxury. Cosmopolitan art, fashion, and architecture thrived, and a thirst for elegance and high culture was the order of the day — all embodied in, and often sponsored by, a new ruler whose tastes were unrepentantly sumptuous. (Regency England captures the pop cultural imagination like few other periods; it was the dashing aristocratic setting for *Pride and Prejudice* and *Sense and Sensibility,* and, two hundred years later, for the hip opulence of Netflix's *Bridgerton.*)

"Night and day, it seems, the English think only of pleasure," remarked the visiting ambassador from Iran, Mirza Abul Hassan Khan, after spending some time with the Prince Regent and his court.

Hence the Regent's ball, which was packed, hot, and, to onlookers, dizzyingly grandiose. Dinner was served at two thirty in the morning, booze flowed, and droves of servants catered to the attendees' every whim. While the upper crust declared the event a glorious success, it left a bitter taste for just about everyone left watching the proceedings outside the Carlton gates.

"It is said that this entertainment will cost £120,000," a young Percy Bysshe Shelley wrote in disgust. "Nor will it be the last bauble which the nation must buy to amuse this overgrown bantling of Regency." That amounts to almost ten million dollars in 2023, spent on a party at a moment when hundreds of thousands of English workers and farmers could not afford to buy bread. The lawyer, politician, and reformer Samuel Romilly, who attended the fete, later wrote, "It does not seem likely…to gain the Regent much popularity. The great expense of this entertainment has been contrasted with the misery of the starving weavers."

It had been mere months since those weavers had revolted. Yet the riots, as well as the workers themselves, were often derided in London circles, especially among the conservative establishment: if strained trade and poor harvests had dealt them bad luck, that was unfortunate, but going around breaking productive machinery simply revealed that the workers couldn't see the bigger picture. The historian Frank Peel, mimicking the elite mood of the day, tongue very much in cheek, wrote that "those weavers were too ignorant to understand that if they were miserable and starving, their masters were waging a great and glorious war; they knew only that their children were crying pitifully for bread.…Had they been better instructed, they would have known that it was their duty to lie down in the nearest ditch and die." The poor would have to recognize, as the upper classes did, that this was simply part of the march of progress.

Then again, it surprised few that the prince would go ahead with such a gaudy affair in troubled times. If a matter did not regard the arts, leisure, or appeal to the prince's own vanity, then it likely did not interest him much. As with most issues of domestic policy, he left them to his Tory prime minister, Spencer Perceval, who had responded to the unrest in March by sending in British troops.

A rank of precarious, angry, and impoverished workers rapidly growing across the country; new forms of technology, control, and production bringing advantages to a few at the expense of the many; a detached, vain, and despised leader at the helm: the comet had dimmed for the summer, but it was about to burn brighter than ever.

GEORGE MELLOR

Summer 1811

"There's hundreds of men and women and helpless babes that's just starving to death," George Mellor said to his cousin Ben, as they walked together through the damp, green fields of Huddersfield one evening, in a scene described in *Ben o' Bill's*. "And now the masters are for doing the work of men and women too with cunning contrivances that will make arms and legs of no use."

They were discussing the poverty that seemed to be everywhere, that touched everyone they knew, and they were discussing automation.

"Water and steam in time will do the work that nature intended to be done by good honest muscle," George said to Ben.

He had good reason to predict as much. Machines that could effectively replace key parts of his trade were indeed proliferating. Between the 1790s and the 1820s, the amount of labor required to produce a piece of broadcloth would drop by 75 percent. What previously took four workers to make would just need one. With a million workers employed at the peak of the trade, that meant hundreds of thousands of lost jobs. And those thrown out of work often had nowhere to turn. That's why so many in Nottingham were put to work by the city as street sweepers before the unrest in March: there was nothing else for them to do.

"A little saving of manual labor here an' there's one thing, the displacement of human agency altogether's what you prophesy," Ben shot back. If a machine could save men a little work, it wasn't *that* big a deal. Despite his youth — Ben was seventeen — he was even taller than George, and "strong with a strength that frightened me," as he described himself in *Ben o' Bill's*. We might imagine Ben shaking his head, waving away the visions of this steampunk robot jobs apocalypse; calm down, George.

George Mellor would not. He understood that the problem lay not with the machines themselves, but with the men who owned them.

"I've no patience with thee, Ben," George said. "Unless the toilers of England rise and strike for their rights, there'll soon be neither rights nor toilers. I've looked into this thing further nor you, an' I can see the signs o' the times. The tendency's all one way."

George's worry was not that machines might become all-powerful themselves, but that those machines' owners would use them to accumulate power and wealth at his expense. There's a big difference. (Ben shouldn't feel bad for conflating the two; people still do this two hundred years later.) George recognized that those who deploy automation can use it to erode the leverage and earning power of others, to capture for themselves the former earnings of a worker.

For decades, their trade had been governed by the notion of "fair profits"—a compact about what was morally reasonable for a merchant or businessman to earn by selling a worker's wares. The entrepreneurs and factory owners were now using machinery and laissez-faire philosophy to unspool that compact. Now George feared that this "tendency" that's "all one way" would result in a small number of people—our first tech titans—accumulating a large amount of wealth, forcing workers to submit to their terms and conditions. And he felt that no one in power—not the state nor the magistrates, nor the businessmen—was looking out for the workers, or the poor.

"What I've seen has burned into my soul," George said. "The natural rights of man are not thought of in this country, the unnatural rights of property have swallowed 'em up. It's all property, property."

It makes sense that George would link laments of rising automation with lost property rights, too. He had grown up in the era of "enclosure," with the national government enforcing a major transfer of the public commons—land once used by whole communities for sheep grazing, farming, and recreation—into land possessed and controlled by a much smaller group of owners, the landed gentry. Enclosure hit George's town four years before he was born. He would have seen stone walls being erected between homes and around pastures for the first time, the nebulous concept of private property stumbling into being. It would have helped convince George that the spree of mechanization was funneling its gains into an economy increasingly marked by rising privatization and inequality, that excluded more and more men like himself and his peers.

"For two centuries there had been a steady concentration of economic power in the hands of a small class," J. L. Hammond and Barbara Hammond wrote in

The Skilled Labourer: 1760–1832. The historians trace the growth of this power through

> the appropriation of the monastic lands, the decay and disappearance of the guilds, the enclosures, the changes in school and university, the rise in one trade after another of capitalism in a form that enables the few to control the productive energy of the many....In the medieval village all over Europe, here as elsewhere, the medieval man had certain rights. On the dissolution of that old village society in England, those rights were lost.

This inequality expressed itself in yet another way: In eighteenth-century England, communities lived under the constant threat of a virus. Records show that in the 1700s, up to 10 percent of all burials were the result of smallpox. "Smallpox was probably the single most lethal disease in eighteenth-century Britain," according to historical demographers. A vaccine was discovered in 1796, when George Mellor was still a boy, but inequality and the deadly disease combined to begin a pernicious trend. The late eighteenth century witnessed "the emergence of class differences in mortality," where rich, urban residents had "significantly higher survival chances than their rural and poorer peers."

For a young person, the future must have looked bleak, ruined by ambiguous authoritarian forces — power that was driving the foundations under George's feet to shift. These new forces were stamping out centuries of tradition to rearrange who controlled what, like the factory masters adopting machinery to erode standards and cut into the croppers' business.

"The French have more sense than us," George said. They saw the threat and toppled the ruling class in their revolution. His statement flirted with treason, even if it was not an uncommon opinion. Another Yorkshire resident echoed that sentiment, stating that, given how fed up people were with "bribery, corruption, luxury, tyranny, oppression, and poverty," if Napoleon had invaded England, "they would have joined Bonaparte, for they were so much oppressed, they were ready to proceed to any act of violence and desperation, for none could get bread enough, and many died for want of it."

The French, George said, "saw all the good things of this life were grasped by the nobles and the priests....They saw that the people made wealth by their toil;

and the seigneurs, that's lords, and the church enjoyed the wealth they made, only leaving them bare enough to keep body and soul together.... They sent their proud lords and ladies packing."

Moral outrages are most acute when they are personal. And George's was certainly personal. The breakneck evolutions of his time, the relentlessly accelerating technologies, the harsh politics of inequality, the privatization, and the pandemic: he was in the process of suffering the injustice of seeing his working identity stripped from him — seven years of training pronounced worthless.

"There's a conspiracy on foot to improve and improve," he said, "till the working man that has nothing but his hands and his craft to feed him and his children will be *improved* off the face of creation."

THE MACHINERY QUESTION

1800s

The notion that textile workers like Ben Bamforth and George Mellor would dis-cuss the prospect of losing their agency to machines in the early 1800s, when the machinery in question was a wooden, water-powered loom, may strike us as implausible. Our twenty-first-century visions of displacement by machines, after all, concern omniscient artificial intelligences and sleek autonomous robotics. But George and Ben's argument was colored by deeper anxieties, too — what is the value of human labor, they wondered, or even human life, in a world where technologies seem constantly poised to replace us? In fact, it's remarkably similar to debates we're still having two hundred years later, in books like *Rise of the Robots,* film franchises like *Terminator,* and cable news segments about the loom-ing AI takeover.

Our fascination with the idea that the machines we build will one day do our jobs for us, become as intelligent as us, or even replace us entirely and move among us as humans, dates to some of our earliest popular culture. Jewish folk-lore of the golem, an anthropomorphic mass of mud or clay brought to "life" by divine magic, predates the Bible. AI is older than Jesus. And the first widely known story of job-replacing machinery is even older.

The word *automaton* first appears in Western literature in Homer's *Iliad,* where it's used to describe the "self-moving and intelligent machines fabricated by Hephaestus," the blacksmith god of technology, according to the Stanford folklorist and historian of ancient science Adrienne Mayor. Around 700 BCE, Homer wrote about Hephaestus's various automated inventions, which included "a fleet of driverless three-wheeled carts that delivered nectar and ambrosia to the god's banquets," automatic gates, bellows that self-adjusted their trumpet blasts as needed, and a crew of artificially intelligent golden female androids that could anticipate the blacksmith god's every need. And the Greeks were drawing on even older oral traditions. "That means that more than 2,700 years ago, people were able to imagine automatons and self-moving devices long before technological inventions made them feasible," Mayor explained.

These imagined innovations were forged in service of accomplishing wondrous stuff. Greek myths about automation, as well as the semiautomated devices they actually tried to build, presented a hopeful, rather than ominous, vision for the future of machinery, as the AI historian Kanta Dihal notes. They signal an understanding of the power that lay in automation's potential—if a machine could assist the *gods,* it would be extremely powerful indeed.

Powerful enough to end the need for slavery, the philosophers argued. In one of the most odious segments of his *Politics,* Aristotle lays out his defense of "natural slavery"—that some men are simply better suited to be servile, essentially—and muses that enslaved persons would not be necessary if only the labor they carry out could be automated. "If every instrument could accomplish its own work, obeying or anticipating the will of others, like the statues of Daedalus, or the tripods of Hephaestus," he wrote, and "if, in like manner, the shuttle would weave and the plectrum touch the lyre without a hand to guide them, chief workmen would not want servants, nor masters slaves."

Or: automation might abolish slavery. Aristotle owned slaves himself, and this helped enshrine the idea that a certain class of people are interchangeable with machines, an attitude that would undergird the actions of elites in coming millennia. The philosopher also helped inaugurate the tradition of insisting that exploiting the human laborer is a necessary evil on the path to full automation, which is always just around the bend. Notably, by the time Ben and George were having their debate, the weaving shuttle *had* already been automated, via a device called the flying shuttle, and yet there were plenty of servants and poor workers still to be found in England.

Still, for nearly a millennium, before there were many meaningful real-world examples to draw on, dreams of automation presented in prose and poetry were optimistic. The concept was admired by elites, who may prefer not to own slaves or pay servants, and perhaps even by workers, who would prefer not to be either. These dreams were a source of hope and excitement—until the Industrial Revolution. In the Middle Ages, humans began to widely fear automated machines as a force that could run dangerously amok, and it wasn't until the Industrial Revolution, when automation began to intrude on the livelihoods of the middle, educated, and even upper classes — the classes that tend to produce culture — that automation was seen as a threat to our humanity.

At the end of the eighteenth and the beginning of the nineteenth century, the factories that sent foreboding plumes of smoke and pollution into the skies offended the sensibilities of some of the aristocrats and all of the poets, and a discomfort with mechanization spread. Machinery and mass production drove the rise of the "factory city" and the industrialization of the countryside. They were aesthetically unpleasant and endangered human health. They were the heart of "the dark satanic mills," as the Romantic poet William Blake famously put it, converting pastoral idylls into polluted, mechanized blight. They were the engines that drove the exploitation of working people, of children — but would they bring progress, too? Would the mass adoption of technology be worth it?

This was articulated as the so-called machinery question: reductively, the "Are robots coming for our jobs?" of the Industrial Revolution. "The ultimate, the most exciting, and the most threatening development of all," historian Maxine Berg wrote, was "the replacement of man by machine." It towered over all other inquiries about the impacts of technology. And in the 1800s, it really was an open question: it was not obvious which path history would take, or what the outcomes of adopting mass automation would be. The moral possibility of reversing the entire "tendency," as George Mellor called it, was very real. "It was far from clear whether it was a portent of inevitable economic revolution, or but one course of development among several, which might be adopted or rejected," Berg wrote. Would the machine "bring wealth only to those who owned it, or to society as a whole? Would it make work or create unemployment? Would it unite society or foment class conflict?"

One's answer to the machinery question was often split, as it is today, along class lines. The middle class, the Romantics, and small landowners tended to oppose the new technology, while entrepreneurs and elites embraced it. And for the workers directly affected, then, as now, the answer was clearer-cut — it would not be helpful, to them, if machinery was used to degrade or take their jobs.

The writer and critic Thomas Carlyle speculated that, should the workmen lose their faith in God, presumably a last remaining comfort against the crushing factory system — a "blind No-God, of Necessity and Mechanism, that held them like a hideous World Steam-engine" — then the outcome seemed clear. "Their only resource," he wrote, "would be, with or without hope, — revolt."

NED LUDD

Fall 1811

After dimming in the summer months, the body in the sky was burning brightly again. Now it could be seen around the world. Napoleon worried it would bring him misfortune as he prepared for his campaign into Russia. The crew of the *New Orleans,* the first steamboat to make a journey through the western rivers of the United States, could see the comet from the deck of the ship.* Astronomers would later determine that the Great Comet's tail was likely over a hundred million miles long, with a coma larger than our sun.

Soldiers had been garrisoned in Nottingham since the disturbance in the spring of 1811, but the months had passed quietly, if miserably, since then. Hundreds more cloth workers had been put out of work, according to a parliamentary inquiry, and harvests and trade conditions had not improved in the slightest. Food prices had climbed to about twice what they were before the war. Thousands of weavers, artisans, and machinists had petitioned Parliament for help, and were denied outright.

In late October 1811, an official notice was put up in Nottingham offering a reward for information about an incident in Arnold, the site of the previous spring's machine-breaking. Someone had snuck into a shop and "did completely Destroy" five frames owned by a major hosier. In the first week of November, six wide frames were smashed in Bulwell. The men materialized under the cover of night, forcing their way in through the windows and doors. They left the older machines intact. Shortly after, they disbanded, firing a lone pistol shot into the air.

These incidents were merely a prelude.

---

* Once, when the ship docked at a port along the river, locals ran up, startled by the mechanical noise — they at first thought it the sound of the Great Comet falling into the river.

On November 10, a band of seventy cloth workers carrying hammers, axes, pistols, and swords marched to a shop that belonged to a master hosier named Edward Hollingsworth. Inside was an array of Hollingsworth's wide frames, which he used to make cheap cut-ups.

Hollingsworth had been warned that trouble was coming. He had barricaded his house and posted half a dozen of his workers, his brother-in-law, and neighbors to stand watch with muskets. When the artisan army approached and saw the defenses manned, they demanded Hollingsworth offer up the frames willingly. He refused.

A gunshot rang out, then another, and the exchange erupted into a furious shootout. The men rushed the building and began to smash the door with a hammer. In the chaos, a young framework knitter named John Westley was shot as he tried to tear away the window shutters. He fell back, knowing he'd suffered a fatal wound.

"Proceed, my brave fellows, I die with a willing heart," he called out, and collapsed. He was the first to be killed in the growing revolt against the automators of the Industrial Revolution.

His companions carried his body out of firing range and returned to Hollingsworth's shop "with a fury irresistible by the force opposed them." The master and his guards fled upstairs, exchanging fire with the machine breakers, and the enraged band smashed every one of the wide frames in the house, as well as the furniture and doors.

The men left Westley's body by the road through Bulwell forest, where it was discovered the next day. Westley was a single parent; his two young children were orphans now.

With that, the dam burst.

The same night of the assault on Bulwell, another group infiltrated a factory in Kimberley, where a hosier had been abusing underage, unapprenticed labor. The men smashed ten of his frames. As the hosiers and manufacturers got word that attacks on machinery were rampant, many quickly organized convoys to move their investments into Nottingham, where they thought their technology would be better protected. The machine breakers intercepted the hosiers' wagons and smashed the devices in broad daylight. They crushed the metal parts so they could not be salvaged and set the wooden rubble ablaze.

Near Sutton-in-Ashfield, which had a reputation as a manufacturing center for cut-ups, the use of wide frames, and employers who paid in truck, well over a hundred men assembled at a local alehouse. Many were armed, and someone passed out gunpowder. The garrison then marched to the hosier Francis Betts's house and demanded he turn over his wide frames. When he refused, they hauled the machines out of his factory one by one and smashed them in the street — seventy in all. A local militia caught up to the frame-breakers, and managed to capture a handful; four men were arrested and placed into custody.

That made four major attacks within two days. After forty-eight hours of machine-breaking fury had reigned across the county, magistrates called an emergency meeting to organize a local military response and appoint constables to give watch. Meanwhile, news of the uprising against entrepreneurs spread fast, and in neighboring counties, concessions from hosiers eager to avoid the same fate came quickly: they agreed to raise their wages in order to prevent another outbreak of sabotage.

One seventeen-year-old apprentice, William Felkin, was ordered by his boss, a major hosier, to ride his horse as fast as he could to every nearby village and spread the word that there would be an advance paid to all cloth workers if they spared his and his partners' three thousand machines. Felkin rode hard through "a dreary afternoon with heavy rain and winter sleet" to deliver the news to the framework knitters across the region. The frames, which "had been undoubtedly doomed...as belonging to one of the most influential houses in the trade," were spared. "Not one of their frames was injured, and no further fears were excited as to the safety of their property."

However, while some manufacturers were willing to raise their prices to stop the carnage, others were not. Some refused in principle; some feared they wouldn't be able to compete.

The militia that local magistrates assembled to investigate the "disturbances" sought witnesses for the crimes, and information about the perpetrators. But a funny thing happened, a thing that set weavers and workers and croppers like George Mellor excitedly discussing the news amongst themselves: no one named a soul. No evidence surfaced. No one talked.

On November 14, three days after the attack on Hollingsworth, John Westley's funeral was held. A thousand people showed up to pay their respects to the fallen cloth worker. At the time, the entire population of Arnold was three thousand people. Ben Bamforth described being there in *Ben o' Bill's:*

I saw his body brought into the town on a stretcher by two constables. I can see his eyes and open mouth, with the yellow teeth, and the tongue thrust out between them, and blood trickling down the sides of his chin and his hands, the fingers of one wide outspread, the other gripping tight some grass and sand he had clutched, and his right knee drawn up so rigid they could not stretch the body.

His body was carried through the streets as if he were a felled saint; a local reporter noted "looks of stern and savage defiance...on every countenance."

Sides had been chosen.

With many machine owners still unwilling to pay a living wage, halt practices like paying in truck, or slow the use of wide frames, the attacks accelerated. Factories and large workshops were infiltrated, and the machines were destroyed. Trade caravans importing new machines were ambushed by figures with faces painted black, the frames smashed.

Soon, fifty machines a week were being destroyed around the region. There was an operation nearly every night across the Midlands. And there were obvious patterns to the attacks. The most hated hosiers and manufacturers in town, those most aggressive in deploying the wide frames, were the most targeted. The frames made a loud, distinct noise when they were being used to make cut-ups, so the insurgents had a good idea of where the devices were, even if the owners had not publicly admitted to using them. As a rule, they smashed only these wide frames, used as they were for slipshod automation, and left the entrepreneurs' other machinery alone.

The machine breakers had prepared for this uprising; they were organized, strategic, and intentional in their displays of power. Those quiet months over the summer had evidently been put to productive use. "The practice of these men was to assemble in parties of from six to sixty, according as circumstances required, under a supposed leader...who had the absolute command of them, and directed their operations; placing the guards, who were armed with swords, firelocks, etc. in their proper places, while those armed with hammers, axes, &c. were ordered to enter the house and demolish the frames," wrote the *Nottingham Review* journalist John Blackner, "and when the work of mischief was completed, he called over the list of his men, who answered to a particular number, and he then gave a signal for their departure, by discharging a pistol, which implied that all was right."

The leader of each of these parties went by the same name: General Ludd.

Local magistrates desperately roused volunteer militias and yeoman cavalries and called on the state to send military aid. Nottingham and the surrounding towns were soon occupied by troops, patrolled by constables, and heavily surveilled — and still the assaults carried on.

The attacks were often prefaced with a note, or a letter, pinned to the door of a factory, like the one sent to Hollingsworth two days before his operation was besieged.

Mr H
At Bulwell

Sir,
if you do not pull down the Frames
or stop pay in Goods onely for
work or make in Full fashon
my Companey will visit y machines
for execution against you —
Mr Bolton the Forfeit
I visit him —
Ned Lud
Kings

The demand was clear: Remove your frames — or we will destroy them. The rise of the Luddites had begun.

LORD BYRON

Fall 1811

The place was a mess. His crumbling estate, his personal affairs, his entire hometown, the region at large; it all seemed to be in chaos.

After two years of travel, sailing across the Mediterranean, and visiting Portugal and Spain, the twenty-three-year-old George Gordon, Lord Byron, had finally decided to come home. He was not looking forward to it.

"My prospects are not very pleasant," Byron wrote in a letter to his friend Francis Hodgson during his voyage from Greece, as he sailed home on a ship full of British soldiers returning from war. Rumors of sexual impropriety had been trailing him, he was for all intents and purposes broke, and his poetry had not yet found the success he felt it deserved.

> Embarrassed in my private affairs, indifferent to public, solitary without the wish to be social, with a body a little enfeebled by a succession of fevers, but a spirit, I trust, yet unbroken, I am returning home without a hope, and almost without a desire.... The first thing I shall have to encounter will be a lawyer, the next a creditor, then colliers, farmers, surveyors, and all the agreeable attachments to estates out of repair.

Debt collectors were indeed haunting his home, the sprawling Newstead Abbey, just north of Nottingham, where his mother had been managing their affairs. Unbeknownst to him, she was in ailing health. "You will be good enough to get my apartments ready at Newstead," Byron wrote to her. He had not seen her for those two years abroad. "I must only inform you that for a long time I have been restricted to an entire vegetable diet, neither fish nor flesh coming within my regimen; so I expect a powerful stock of potatoes, greens, and biscuit: I drink no wine." He continued: "I hope you govern my little *empire* and its sad load of national debt with a wary hand."

When he landed in the summer, he had posted up in London, visiting friends and arranging for the publication of a manuscript he'd written during his travels. Byron wasn't entirely convinced that *Childe Harold's Pilgrimage,* a romantic, politically charged epic that was obviously inspired by his own life, was suited for publication. A business partner convinced him otherwise. He sold the manuscript to an enterprising Fleet Street bookseller named John Murray, who was just embarking on his literary publishing career; he would soon publish Jane Austen and Thomas Malthus, too.

Byron was still in London when he received word that his mother had fallen gravely ill. He rushed home for Newstead but did not make it in time. "I heard one day of her illness, the next of her death," Byron wrote. His grief would only compound, as a season of death set in. In a matter of months, three of his friends would pass away, two of consumption and one of drowning.

Many days, Byron wouldn't leave the grounds. He loved Newstead, which he would later immortalize, in his epic *Don Juan,* as "embosom'd in a happy valley, / Crown'd by high Woodlands, where the Druid Oak / Stood."

The natural beauty and childhood nostalgia of the place were deeply resonant to Byron. But the tranquility it once offered had been disturbed, first by the loss of his mother and friends, and now by a workers' uprising. Byron had returned as the upheaval consumed Nottinghamshire, as shops and factories suffered what seemed like nightly assaults. It felt like anarchy, and it left the Romantic dazed. He was blindsided by the rampant poverty of his home district, the gaping discontent, and the government's near-total ignorance of his countrymen's plight. He might well have felt like smashing something, too.

Byron had taken his place in the House of Lords only two years earlier, in 1809, before departing for Italy and Greece. He was from an aristocratic family, but he wasn't a direct successor to the lordship; a series of family deaths and a circuitous bloodline left him at the helm of Newstead. In keeping with the Romantic ethos of championing the individual spirit over the systems that restrained it, Byron was sympathetic to the artisans. They had good cause to revolt, he thought, and felt the state should intervene to address their plight, not with soldiers, but with alms. He worried that as tensions continued to rise, all-out rebellion, even bloody revolution, was on the horizon.

"Nightly outrage and daily depredation are already at their height," he wrote to a friend. "And not only the masters of frames, who are obnoxious on account of their occupation, but persons in no degree connected with the malcontents or their oppressors, are liable to insult and pillage."

He spent much of the fall languishing in depression, proofing the pages of *Childe Harold,* trying to get the estate in order, and seducing — or rather, preying on, as powerful landowners in his position often did — the young maids who worked the grounds. Byron's looks and appetites were already bordering on infamy, though he tightly controlled his public image. He would only allow dramatic, flattering portraits to be circulated; in reality, he often looked pallid, boyish, even chubby. He slept with curl-papers in his hair at night.

Plagued by his debts, Byron set in motion a plan to renew the rights to land he held in Lancashire that had fallen into abeyance while he was gone. He hoped either to sell the land, or to start mining its coal-rich deposits for profit. On a short trip to survey the estate, he stayed with a friend. Another visitor captured a snapshot of the poet: "He is a pale, languid-looking young man who seems as if he could not walk upright from sheer weakness. He has a fine large blue eye, but it is so wild and odd that I have doubt, if he lives, he will be mad," she wrote. Byron wore "long white linen pantaloons and a long gold chain around his neck, and his embroidered shirts had 'a foreign look.'" One of his legs seemed shorter than the other, owing to a deformity in his right foot, and he walked with a slight limp. He was extremely self-conscious about what many described as a clubbed foot. The boarder noted that "his fidgeting manner sets my heart beating."

The fidgeting was understandable. Since he'd arrived back in England, his friends and family seemed to be dying en masse. He was freshly concerned about a new crop of rumors, most of them true, that he'd had sexual relationships with younger men, and an affair with his half-sister. His business partners were trying to censor the political content of *Childe Harold,* while assuring him the book would become a sensation. These anxieties were no doubt amplified by the fact that he'd soon be courting public opinion directly: nobles were expected to participate in politics, and he was overdue to give his debut speech in the House of Lords. It had no doubt occurred to him that his maiden speech would prove a fine opportunity to raise his profile before the release of *Harold* in early 1812.

The plight of the frame breakers might make for a worthy subject. If this was the season of tragedy, perhaps he could channel some of his sorrow and anger on behalf of his fellow countrymen, bereaved as they were for very different reasons.

GEORGE MELLOR

Fall 1811

NOTTINGHAM, DEC 1.

Three troops of the 15th light dragoons are here, and detached in many small parties in the adjacent villages, in parties of eight or ten men. The yeomen calvary from Bundy are ordered in by the County Magistrates. Two companies of the 2nd Nottingham local militia are here also, under the direction of the Magistrates.

Last night parties of Ned Lud's men (the nick-name for the frame breakers), to the amount of 250 men, broke several frames in the villages near this town…

They broke only the frames of such as have reduced the price of the men's wages; those who have not lowered the price, have their frames untouched; in one house, last night, they broke four frames out of six; the other two which belonged to masters who had not lowered their wages, they did not meddle with.

It is supposed that the masters will come to the old prices and there the business will end.

— Leeds Mercury, *December 7, 1811*

When George Mellor heard the news about the uprisings, his reaction was simple. "I wish I was there."

The men called themselves Luddites. They'd taken the machine-breaking tradition, the bargaining tool of last resort, vitalized it with the myth of young Ned Ludd, who turned on his master, and organized a fearsome, even awe-inspiring rebellion.

Throughout the fall and into the winter, as the machine breakers continued their attacks and notched one stirring victory after another, word quickly reached weaving towns like Huddersfield and rapidly industrializing cities like Manchester. Wherever men were watching factories built and lined with new machines, wherever wages for weavers were in freefall, wherever the work-from-home cottage industries and tradition-bound jobs seemed to be on the verge of extinction, these rousing reports of brazen — and apparently *successful* — attacks on hosiers, factory bosses, and machine owners caught fire.

"Those Nottingham lads are showing them specials a bit of real good sport," George said loudly at John Wood's finishing shop one day, to a room full of friends and fellow croppers, who shouted in agreement.*

These men included Will Thorpe, a fellow cropper at another finishing shop across the street, whose "stolid aspect" and "half dreamy" eyes occluded a conviction that seemed to run as deep as Mellor's, and Thomas Smith, George's dependable right-hand man at John Wood's.

They were each in their early twenties. Like George, they'd all finished the long slog of apprenticeship just in time for the occupation they'd studied to confront an existential threat. Other workers in shops like this had been croppers their entire lives and had no other prospects to speak of.

They admired these followers of Ned Ludd, who were organized, surgical, and smart in their machine-breaking campaign. They marveled that teams were so organized that they could slip into the entrepreneurs' shops, with guards stationed a stone's throw away, smash the offending machines, and slip out. At night, they moved stealthily, in military formation, not unlike an efficient machine themselves. By day, they were bold, and had won the hearts and minds of the poor and the working class.

They'd heard stories of daring feats, like the one about the lone Luddite who snuck inside a hosier's shop and smashed the wide frames on the third story before he was spotted by a constable with a small force, who surrounded the building. The Luddite noticed some freshly dug ground in the backyard garden of a neighbor's house, and leapt out the window. On the first floor, a family was gathered for a meal; they watched him crash down, let himself into their back door, and escape through the front — all to the cheers and hollers of the crowd out on the street.

---

* The dialogue in this chapter appears here as it was recorded in Frank Peel's oral history, *The Risings of the Luddites, Chartists, and Plug Drawers* (1880), 19.

"It would be glorious to dash them cursed frames into a thousand pieces," George said.

"Aye," said Thorpe. "But wishing's all naught. It strikes me we've had rather too much of that."

"You're right, Will," George said, "but what can two or three do? What are we doing here?"

One thing they weren't doing as much was working. The shop's orders had slowed to a trickle. "John Wood's cake's baked," as George would say, and it followed that his was too.

It's not clear how much direct contact those "Nottingham lads" had with workers in the neighboring regions. But because Ned Ludd had become a mythic symbol, a figurehead, there didn't *need* to be any meticulous long-distance coordination for new cells to crop up and declare war on the implements enabling their exploitation. After all, the machine breakers were not ultimately after the machines themselves but rather the men who were using them to transform social relations and gain power. The Luddites were technologists themselves; they did not hate the machines, though they did not hold any undue respect for them, either.

The uprising had inspired copycat actions in cloth-manufacturing regions like Leicestershire and Derbyshire. The fever was spreading, and why was no mystery. By now, every weaver, cropper, and tradesman in town knew a family, or several, that was not just out of work but starving outright.

"You should have gone with me yesterday to Tom Sykes's," Ben Walker, another cropper, said. Sykes had been unable to find work for a month, and had become so poor he couldn't buy bread. His wife, a "poor delicate creature," had fallen sick and died of starvation.

"Pined to death, they say," Walker said, "and I believe she was. When I got to his house he was just opening the door and ordering a person out, and he called out after him, 'I want none of your sympathy; if it hadn't been for such as you, she'd have been alive still.' And there she lay on the bed, poor thing, skin and bone, nothing else."

Half of all the laborers in the region had not worked for months. Poor Law relief payments had reached a record high of £6.6 million, but many of the newly out-of-work weavers did not qualify. And there was no sign that any other kind of help was coming.

The croppers were not yet starving but saw the injustice making its way to their doorstep. There was no *reason* for anyone to starve; the factory owners were still turning *profits*. If they stopped running new machines, there would be more work to go around. Besides, to the artisans who had taken seriously the traditions and contracts of their trade, using machinery to take market share at the cost of the worker was neither moral nor fair. It infuriated them that the entrepreneurs and factory owners claimed the right to automate people's work, when it so obviously led to suffering.

Imagine dedicating years of your life to learn a difficult job that was supposed to guarantee you a good living — playing by the rules, you might say; going to school, learning a trade, investing incredible volumes of time and resources to obtain a modest level of station and security — only to realize that the deal was suddenly void. Your faith in systems working as intended would be as broken as a hammer-smashed frame.

Now, you do not know if, when, or how your job will vanish altogether, along with your identity. Imagine not having any other realistic options. What you *did* know was that someone else, with advantages you did not have, could suddenly acquire new technologies that would allow them to reorganize those rules for their benefit.

What are your options, really? You might go to work in the factory that has absorbed and degraded your job. It will pay less, the conditions will be miserable, the freedoms you are accustomed to will be gone, and you will now have bosses controlling every aspect of your daily routine, critiquing and denigrating your work. They would prefer a child apprentice anyway; they're more pliable and don't talk back. Or, retrain for a new job? You have built a life around the one you thought you had, and given years to it; you are proud of it. You would be far behind, paid less, and unconditioned to a job in a new line of work. *If* there is a new job available at all — in these towns and regions dominated by one or two industries, there often simply is not. Start a business or mount a challenge to the entrepreneurs and industrialists? With what money? Perhaps the option is there for those with the capital to compete with the men who have threatened your livelihood, or those who know how to win the favor of the bankers for a loan. But it is not there for most.

Maybe you don't even need to imagine this, because you live in the twenty-first century and have seen a corporation, a platform, or a manager use technology to rewrite the social contract that once defined your own job. Maybe you are a cab driver who saved up for years to own your own car and medallion, who knows miles of city streets as if they were your own backyard, only to have Uber show up on the scene, undercut prices with its store of venture capital and its algorithmic management system, and render your investment worthless. Or maybe you have worked for years as a salesperson, acquiring contacts and relationships with vendors, only to see your company introduce an automated portal that performs most of your role. Or maybe you are a writer or an artist who was let go from the digital media company that published your work, just as the outlet announced it would begin using AI to generate content. Or maybe you are a warehouse worker dependent on the job to pay rent for your family, and you are hustling to keep up with the array of robots on the shop floor. Maybe you are a truck driver who has trained yourself to weather the endless stretches of highway, and now you are reading about an autonomous big rig that drove itself from Los Angeles to Las Vegas.

Maybe you've just spent years in high school and college, studying and working around the clock to get good grades and to pad out your resume, yet the only jobs you seem to be able to find after graduation are benefits-free contract work through online temp agencies or gig apps.

Whether an eighteenth-century power loom or a twenty-first-century gig app platform, new technologies are often the "form that enables the few to control the productive energy of the many," as the historians J. L. Hammond and Barbara Hammond put it. The rising cloth workers saw that novel technologies promised owners not just an opportunity to improve efficiency, but an excuse to disrupt a previous standard of work — even when the technology is not capable of performance superior to that of the human worker. It's a useful vector through which to justify disturbing a norm, shocking a system, or evading rules and regulations. And it is not necessary for the job to be *automated* away entirely; a worker might feel the impacts of new labor-saving technologies secondhand, through a loss of wages, increased surveillance and control over their performance, or higher productivity demands. Automation is only the simplest and most straightforward manifestation of what has been a defining dynamic of technological capitalism since its earliest iteration, and it was in its most stark configuration in the days of poor weavers like Tom Sykes.

That configuration looked like this: Tom Sykes had a job, a machine was built that could do that job, and someone across town decided they could increase their profits by purchasing the machine to do it instead of Tom. Now that machine does some or all of Tom Sykes's job on behalf of its owner, and Tom has less work. The owner profits and Tom is out of luck. At no point does Tom have a say.

Workers have been told for centuries that this process is normal, even morally good, because it will benefit society in the long run by making some products cheaper and more plentiful. But men like Tom Sykes and George Mellor did not accept this justification. In George's day, a craftsman could walk by the building where his work had been relocated to be done by a machine, while he might be hungry and the factory owner was getting richer. (In our day, that job might be shipped overseas or routed through a gig app, and the anger may be more diffuse.) There was no negotiation involved in this transfer of earning power, and to the worker it looked an awful lot like theft.

No wonder so many of the working young were filled with righteous fury. Their traditions were being voided, the world they had been promised was being degraded, its physical shape was being scarred and polluted. The world was undergoing a paradigm shift, and there, at the helm of the ship, was the new automation technology used by profiteers. There was no promise of jobs, of stability, of a future. You can imagine all six feet of George Mellor, masculine frustration surging through that massive frame.

And then, one more time:

"I wish I was there."

NED LUDD

November 1811

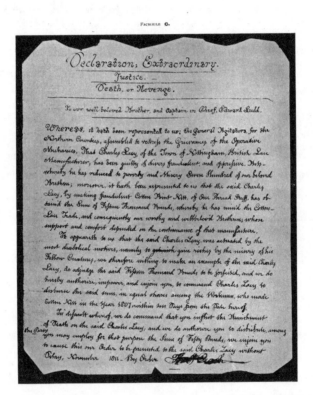

One of the first known Luddite letters, this proclamation threatens a major Nottingham investor in automation. [Image courtesy of the Thoroton Society of Nottinghamshire.]

The men who called themselves Luddites turned out to be prodigious writers.

DECLARATION: EXTRAORDINARY.

Justice.

Death, or Revenge.

The first warning letter, delivered to a local factory chief, Charles Lacy of Nottingham, was blunt and threatening. The missive was posted on a handbill and soon reprinted by the newspapers.

To our well-beloved Brother, and Captain in Chief, Edward Ludd.
Whereas, it hath been represented to us, the General Agitators, for the Northern Counties, assembled to redress the Grievances of Operative Mechanics, That Charles Lacy, of the Town of Nottingham, British Lace Manufacturer, has been guilty of diverse fraudulent, and oppressive, Acts — whereby he has reduced to poverty and Misery Seven Hundred of our beloved Brethren.

The complaint went on, outlining grievances and promising vengeance unless the machines were shut down.

It was signed, "General Ludd."

The missives became the machine breakers' calling card. A torrent of subsequent letters, written in various hands, bore that name: Ned Ludd. Some called him General Ludd, some Captain, others King. This one accused Lacy of profiting off the lost jobs to the tune of fifteen thousand pounds, and of giving local tradesmen a bad reputation with his factory's shoddy workmanship.

The turmoil that followed the letters could not be confused for random acts of vandalism. This was a movement, and it had a name, a symbol, and an animating principle.

Its aim was to destroy, specifically, "machinery hurtful to commonality" — machinery that tore at the social fabric, unduly benefiting a single party at the expense of the rest of the community. If factory bosses continued to use machines to do artisans' work, and do it sloppier and more cheaply, their recourse was simple: the Luddites would break the machines that were "stealing their bread."

Reports of the first attacks on Bulwell, where John Westley had been killed, noted that a man calling himself Ned Ludd had led the charge. Yet no one knew who he was.

This General Ludd quickly became a kind of legendary figure, one that allowed an anonymous movement to rise up in his name, staying underground while mobilizing behind a common cause. He was a permutation of Nottingham's last mythic hero, Robin Hood — a recognizable icon that the downtrodden could rally around and emulate. It proved an ingenious, decentralized way to organize: distributing power, maintaining secrecy, building a legend, all through this nineteenth-century meme.

But why "Ludd"? There was the legend of Ned Ludd, sure; but that was just that, a legend. There *could* have been a real Ned Ludd in Leicester (or elsewhere) who smashed his master's frame and caused enough of a stir that insurgent workmen might adopt his name as their own. Or, the legend could have been retrofitted to suit the machine breakers' cause. Lludd Llaw Eraint, for instance, was a Welsh hero who lost an arm in battle and was exiled as god-king, only to receive a new silver replacement that allowed him to return as a sort of cyborg warrior deity; Lludd of the Silver Hand would evolve when adopted by British folklore, be drafted into a fairy tale in which he must thwart three plagues, two dragons, and one miscarriage epidemic, and go on to found the township of Lluodon, or Loudon, or London. As one newspaper observed,

> The word *Uuyd,* in Welch, signifies an army or camp; *Lluydda,* in the same language, is make war; and *Lluyddar* is a soldier. The rioters Nottinghamshire Cheshire call themselves Luddites; perhaps some Welchman among may have originated the name; and Cheshire borders upon Wales.

There was also an ancient British king named Lud, who, as the writer Brian Bailey notes, "according to tradition, built the original Lud Gate in London in 66 BCE. He was buried 'nigh unto that gate, which even yet is called Porthlud in British but in Saxon Ludgate,'" according to Geoffrey of Monmouth. Bailey continues:

> A posthumous statue of King Lud from the rebuilt medieval gateway survives in the church of St Dunstan-in-the-West in Fleet Street. The hall of the Framework Knitters' Company, long demolished, once stood only half a mile away in Redcross Street.

Whatever its provenance, the Ludd name proved invaluable: everyone, cloth workers and factory bosses alike, read those letters, reprinted in the press. The uprising was "a potent and forceful movement across regions because it was generated... by 'the creation and appropriation of the eponym Ned Ludd,'" as the historian Katrina Navickas put it.

And while the authorities attempted to call the participants by other names — "rioters," "Jacobins," and "depredators" were all trotted out to counter their popularity — the Luddites kept posting their own letters, and signing them General Ludd. The term stuck; they controlled the narrative, to powerful effect.

"Fear and panic fostered rumors of unrest, which temporarily paralyzed local communities, even if no machines were fired or no marauding mobs ever appeared," according to Navickas. The looming threat of General Ludd also helped amplify routine complaints into menacing grievances. It gave the workers leverage.

The attacks continued with ferocity. Nightly raids, more frames smashed, and some in broad daylight, too, with who knows how many witnesses but nobody leaking. It was all anyone was talking about.

THE PRINCE REGENT

November 1811

It had been weeks, and the Prince Regent would not get out of bed. He had twisted his ankle demonstrating a popular dance step, the Highland Fling, to his daughter, and now Prince George said he was too injured to conduct any governance at all. Instead, he lay on his stomach, imbibing a self-prescribed dose of laudanum, a mixture of alcohol and opium tincture: one hundred drops every three hours.

"He will sign nothing, and converse with no one on business, . . . and you may imagine, therefore, the distress and difficulty in which the Ministers are placed," William Fremantle wrote to the Marquess of Buckingham on November 28. "The Duke of Cumberland is going about saying it is all a sham, and that he could get up, and would be perfectly well if he pleased."

The duke was not alone in accusing the Prince Regent of faking to get out of doing his job. Much as he enjoyed the parties and the pomp, the Prince Regent had not taken to the politics. Parliament had anointed him as acting ruler with some major restrictions; he could not create peers — the process of ennobling lords, and giving them membership in Parliament — or appoint public ministers for a term longer than a year. These restrictions, and the pretense of not wanting to upset his ailing father, gave him an excuse to refrain from politics, to preserve George III's unpopular right-wing Tory cabinet rather than make any potentially divisive changes, to keep out of the messy business of governing, and to avoid making enemies.

But every month brought his restricted regency closer to an end, and the prospect that the prince would have to weigh in on the formation of a new government grew closer. At the heart of the issue was that he had always fancied himself a liberal and had publicly professed his loyalty to the more progressive Whigs, many of whom were old friends. The Tories in power, facilitators of an unpopular war and holding court over mass misery, were deeply disliked by the public. But Prince George was also heavily influenced by his father, a staunch, conservative Tory. Thus, he was in favor of continuing to fund the war against Napoleon, and stood against Irish Catholic emancipation — two positions anathema to the Whigs' platform. As for the Luddites, it's not clear the prince cared much about them at all.

The Tory prime minister, Spencer Perceval, had sent the armies north to quell the machine-breaker riots in the spring, and was doing so again now, as unrest stirred once more in Nottingham.

Shortly after George was made Regent, the Whigs, believing themselves to be his allies, pressed upon him to form a new government. But he was allergic to confrontation and caved under pressure. He punted, basically, citing the notion that his father might recover, and kept the Tory configuration in power, saying he'd have to wait until his restricted regency was over, and *then* he could assume full powers over England.

He had always been indecisive, even lazy. Immediately celebrated as the first male heir to a British monarch in seventy-five years, it was all disappointment from there in his father's eyes. He was raised in the gated luxury compound known as the White House at Kews. Unlike his siblings, he was fickle and lethargic; he would give up on sports if he fell down, preferring to avoid the risk of injury. When he was inoculated against smallpox, leaving him ill and bedridden for weeks, the prince stated he was quite content just to lie there, while his peers complained and chomped at the bit to get out and play. One of the prince's courtly educators fretted that George might fail not only to make a suitable member of a royal court but to be "a gentleman at all."

His insulation from the harsh realities of the age, his acclimation to a life of pleasure and luxury, his ambivalence to action, had all carried over into his reign, as he was charged with overseeing what had suddenly become the largest domestic military occupation in England's history.

Richard Ryder, Perceval's appointee as secretary of the Home Office, the agency in charge of national security and law enforcement, had shipped a thousand troops of infantry to Nottingham, and nine hundred cavalry. Combined with the local militia, there were thousands of soldiers now assembled to quell the attacks. The military and local magistrates set about signing in new constables and pushing a sweeping set of rules and restrictions — pubs were to close at 10 p.m., for instance — and martial law was instated over the district.

Armed dragoons and militiamen patrolled the streets, crowded the inns, and became an unignorable fact of life for the citizenry. Worse, they were entirely unaccountable.

The occupation was met with anger, even protest, among the populace, who feared that the move would solve nothing, and only stoke more violence. Moreover, it wasn't working. An editorial in the *Leeds Mercury* captured the popular angst and trepidation:

> MILITARY coercion has not been so operative in Nottinghamshire as we were led last week to suppose. The depredations of the Frame-Breakers are still continued, and on occasion they seem to have had a "method in their madness" that cannot fail to create alarm...To what an alarming crisis is this country brought when its Military force, instead of being employed against our foreign enemies, is obliged to act, reluctantly indeed, against our own subjects, made blind and desperate by privation and want. Nor is the cause of this riotous disposition confined to Nottinghamshire — it exists in almost all the manufacturing districts of the kingdom.

The editorial ends by suggesting that the Crown is trying to paint the frame breakers as ignorant savages to justify the use of force, but that no one is being fooled. Everyone knows, it insists, that the problem rests with the government still taking its cues from the free-market principles of former prime minister William Pitt the Younger, and refusing to implement measures to regulate industry or to assist the working class:

> These riots owe not their rise and progress to the gold of Napoleon, but to the politics of PITT and his adherents; and so long as the system acted upon by that Minister [is] pursued, we may be distressed, but we cannot be surprised either at disaster abroad or commotion at home.

Nottingham had become an occupied territory. The people were calling for aid. The frames continued breaking.

The prince remained on his stomach in an opium daze.

"The prince is still very unwell, and it is much believed that the attack in his arm is paralytic," wrote Lord Grenville, a progressive Whig who was eager to get the regent to act on Catholic emancipation and other reforms. "The language of his Court is that he has taken no decision."

Until then, when Parliament would be brought back into session, the Prince would lay in a stupor, preferring bed to governing what was becoming the world's first great industrial power.

You can almost hear Grenville sighing, his hopes collapsing, in his letter: "This is only to put off the evil day."

WILLIAM HORSFALL

Fall 1811

Every Tuesday afternoon, William Horsfall made the trip by horseback from his factory in the rural hamlet of Marsden to the cloth market in Huddersfield, one of the major fixtures of the crowded downtown. Clad in a top hat, single-breasted vest, and fine tailcoat, the entrepreneurial factory owner cast a familiar and imposing figure over the verdant pastures and small villages he passed on the coach road. Horsfall's business employed four hundred workers, who operated the automation technologies the artisans hated most: the shearing frame and the gig mill.

The ride took about an hour, and along the way, Horsfall was often taunted by the local children, some of whose parents were put out of work by his mechanized factory.

"I'm General Lud!" the kids would say, jumping out at him along his path. He would respond by lashing them with his horsewhip.

Horsfall knew he was not well-liked — he'd loudly championed adding more automated machinery to his factories, and openly derided the struggling weavers — and he did not seem to care. As the cloth workers rose up, and many entrepreneurs approached their business with newfound caution, Horsfall only vowed to redouble his efforts.

"I'm going to build more frames, even if it means I have to ride up to my saddle in Luddite's blood," he had boasted in town to other manufacturers.

His belligerence was rooted in a personal grievance. Around the turn of the century, angry croppers had burned down his family's mill, Ottiwells, after his father and his uncle first installed shearing frames, which were illegal at the time.

But Horsfall believed, above all, in the factory. He rebuilt Ottiwells, took over its operations, fitted it with gig mills and more shearing frames, and encouraged a pair of enterprising blacksmith brothers, Enoch and James Taylor, to set up shop next door. He commissioned tools and new devices, at specification, from their business — a major advantage when it was difficult or risky to transport machines and parts over long distances. Horsfall had become perhaps the most aggressive entrepreneur in the region in adopting the use of automation. He succeeded where his father had not, becoming among the first to successfully install the shearing frames that mechanized the croppers' work, back in 1807, driving prices down and forcing his competitors to follow suit.

The only other man in the region who could seriously compete with the scale of Horsfall's ambition was William Cartwright, a more taciturn and less bombastic entrepreneur. Cartwright had arrived in the West Riding from parts unknown in 1809, and installed a fleet of shearing frames at his mill at Rawfolds a few miles away (he had no relation to the power loom inventor of the same surname). He too was aggressive in expanding his operations.

Cartwright was enigmatic. He spoke French, which was enough to render him a person of interest among the locals, and he was an acquaintance of Patrick Brontë, an eccentric reverend in Yorkshire. When Cartwright met Patrick's daughter, Charlotte, he left a lasting impression. She would later base a major character on him in her second novel, *Shirley*, the follow-up to the blockbuster *Jane Eyre*. Brontë described Cartwright to her biographer, Elizabeth Gaskell, as "a very remarkable man, having, as I have been told, some foreign blood in him, the traces of which are very apparent in his tall figure, dark eyes and complexion, and singular, though gentlemanly bearing."

Charlotte Brontë, who used her acquaintance of Cartwright, along with local lore and newspaper accounts when researching and writing *Shirley*, introduces his fictional counterpart, Robert Gérard Moore, as an entrepreneur deplored by the locals for driving progress.

"Misery generates hate," Brontë wrote.

These sufferers hated the machines which they believed took their bread from them; they hated the buildings which contained those machines; they hated the manufacturers who owned those buildings.... Hollow's Mill [the stand-in for Rawfolds] was the place held most abominable; Gérard Moore, in his double character of semiforeigner and thorough going progressist, the man most abominated.

The hate for the machines is exaggerated; the hate for the buildings and the men who owned them is not. Cartwright's massive operation at Rawfolds did indeed, like Ottiwells, make for an imposing symbol of the changing times, and just whose power and capital were changing them. "The intensity and extent of hostility to machinery cannot be explained simply by fears of redundancy," the historian Adrian Randall wrote. "Innovation was frequently met with resistance not merely from those in immediate danger of losing work nor even from those under longer-term threat but also from other trades and groups not threatened in any way by the machine. In many ways the fear which united this opposition was the fear of the factory."

Workers were rising up not just to smash the machinery degrading their working conditions or to win short-term concessions, but in an effort to dismantle the emerging factory system itself. And men like Cartwright and Horsfall were the ones driving factorization. Manchester, in the neighboring county of Lancashire, as home to the power loom and the water frame, saw a particularly notable economic explosion. A market town of ten thousand residents at the beginning of the eighteenth century, it expanded into one of the first truly industrial hubs, a city of nearly two hundred thousand by 1820 — and nearly, it seemed, as many factories. The conditions grew notoriously brutal. One typical weaving factory was described in 1810 by inspectors as "overcrowded," with "a great want of ventilation." The workers' "diet was inadequate," "the privies were too offensive to be approached by us," and the "apprentices complained of being overworked." Physical "abuse and exploitation" were routine.

By the end of the nineteenth century, textile production would increase fifteenfold, and an entire class of workers — skilled, proud artisans who had plied their trade for centuries — was facing extinction. Many of England's cities, in fact, were packed with the displaced and the impoverished. And they were forced to work in the factory, that symbol of brutal conformity, individual submission, and the erosion of traditions.

"The image of the factory was amazingly powerful and fear of the factory widespread," Randall wrote, as it "symbolized a new power, a new centralized force in the industry." These new factories were despised more by the worker for the domination they stood to bring to bear than they were by any poet decrying the blight on the countryside. They reminded the public of forced workhouses — literally of prisons, which helped enshrine "the popular association of the factory with places of confinement and loss of liberty."

No wonder that workers hated them and the men who were building them. But entrepreneurs like Horsfall and Cartwright were at peace with that hatred; in fact, it may have animated them further.

"Faced with resistance," the historian Maxine Berg wrote the factory owners and their associates "adopted an attitude of increasingly aggressive determinism and optimism. For them, it became axiomatic that mechanical change was natural and evolutionary, the very motor of progress itself." This attitude remains predominant in the modern tech sector, two hundred years later. The historian of Silicon Valley Margaret O'Mara describes that region's "near-maniacal" focus on building new commercial technologies, "and as a consequence [tech entrepreneurs] often paid little attention to the rest of the world. Why care too much about the way government institutions or old-line industries worked, when your purpose was to disrupt them in favor of something far better? Why care about history when you were building the future?"

The entrepreneurs' faith in "progress" was rooted in the trendy philosophy of the Scottish economist Adam Smith. "Laissez-faire," the Luddite historian Brian Bailey wrote, had become "the political dogma of the English bourgeoisie. In fact, it represented freedom for the employers and intolerable repression of the workers." Smith had published *The Wealth of Nations* in 1776, and while initial reaction was limited, it became influential in the following decades. The book, a synthesis of recent strands of economic thought, presented the doctrine of laissez-faire, or "let people do" (as they see fit), as a science. It was profoundly pro–free trade, pro-entrepreneur, anti-regulation, and anti-monopoly.

"It was a deliberate attack on the existing economic policies and dominant economic assumptions," the economic historian Kirk Willis wrote. Monopolistic mercantilism, the order of the day, à la the East India Company, was out. Competition, free-market economics, and deregulation were in. So was the factory, and the factory owner, as the chief engines of economic prosperity. "Adam Smith's discussion in *The Wealth of Nations* united two key concepts: division of labor as a motor for generating prosperity, and market systems based on self-interest as a fuel for that motor," the political scientist Mike Munger explained.

Two of Smith's earliest and most influential advocates were the British prime ministers guiding England's post–Revolutionary War economic policy: the Earl of Shelburne and William Pitt the Younger. During a debate in the Commons on the fiscal health of the nation on February 17, 1792, Pitt offered a paean to Adam

Smith: "The writings of an author of our own times…of a celebrated treatise on the Wealth of Nations…will, I believe, furnish the best solution to every question connected with the history of commerce, or with the systems of political economy." Those theories imbued Pitt's "New Tories" with a sheen of modernity and lent an ideological fuel to the ascendent entrepreneurial class.* Factory owners began invoking Adam Smith, and workers began despairing of him. (And newspaper columnists would continue to lambast Pitt, years after he was in office, for turning Smith's ideas into public policy.)

When Horsfall and Cartwright did complete their weekly journeys to the busy Huddersfield market, heads would turn as they made their transactions and moved more and more sums of money. There was already a sense, even then, that they were doing more than enriching themselves at the expense of the workingmen whom their devices were making redundant; they were using those machines to impose an entirely new mode of work onto the populace.

Theirs were the guiding hands on new technologies that were helping to forge the very shape of industrial capitalism.

---

* Also crucial were the ideas of the economist Thomas Malthus, who held that periods of mass suffering among the working poor were inevitable in any prospering economy as the population grew too fast for food production to keep pace.

B

November 1811

In a secret meeting at the Falstaff Hotel in Manchester, in the middle of a black November night, a man who dressed and presented as a cloth worker listened in as a delegate from the town of Royton spoke up. His people, he said, were "worne out with petitioning." In fact, they were "Ready at a days warning to Come forward if they had a Leader." As delegates from other cities stepped forward one by one to declare the same, the man took further note.

When he got home, he recorded the details of this clandestine conference of a fast-growing group that called itself the Manchester Committee; he quickly dispatched a letter to the magistrate Colonel Ralph Fletcher in Bolton, who paid him by the day and reimbursed his expenses. He signed his name, simply, "B."

B was a spy.

Since October, B had attended the secret gatherings in the lantern-lit back rooms of public houses, in sodden fields outside Manchester, and all across Lancashire and Cheshire. He sat among artisans, weavers, and small-shop owners, some of whom he insinuated he had known for years, and listened as they discussed what they called "the Business." He would nod and cheer as calls for an uprising filled the halls, and as the Luddites cursed the machine owners and mill bosses. B would propose seditious and traitorous schemes of his own.

In a rendezvous with a contact from Cheshire, B was told there were a thousand workers there who were ready to rise. The men were armed and knew combat; the "general wish was a revolution."

B was an agent in a vast state-funded network organized by magistrates like Fletcher and funded by the highest levels of government. They were freelance informers, paid to infiltrate underground meetings and to report back with intel on radical and putatively insurrectionary movements. The spy system was informal, for-profit, and unaccountable. Their messages must, as a result, be read skeptically. Many spies were loyal to the Crown and relayed relatively accurate information, though many would act as provocateurs to get it. Others surely intuited that tales of impending revolution and reckoning fed the imaginations and biases of their employers — and hence the demand for further use of their paid services — and elaborated accordingly. There was more than one way to make a living in turbulent times.

But spies like B soon discovered that infiltrating Luddite cells was easier said than done. Agents like him had been tapped for decades to subvert radical political clubs, where most members didn't know each other outside the weekly gatherings, and where an unfamiliar face didn't arouse much notice. Spies in such circles were highly successful; they not only obtained information used to arrest and convict dissidents, but sowed distrust and discord among the groups organizing for reform. The Luddites, on the other hand, were close-knit; the men lived and worked together, often for many years. Outsiders were treated with suspicion.

The task may have been a little easier in Manchester, where B was based. Lancashire, Cheshire, and Derbyshire were all cotton-weaving districts, and together they comprised the third major hotspot for Luddite activity. Unlike the West Riding of Yorkshire, which was home to wool production, and Nottingham, which specialized in silk and lace, the cotton industry in Manchester was comparatively new. It did not have deeply entrenched customs and centuries-long traditions defining the trades, and had fewer strong trade groups to protect it.

Cotton was not grown natively in England; for decades, it had been imported from India. It was only when the rising demand drove the American states and the West Indies colonies to improve their own technology — and radically expand slavery and cotton plantations — that the industry boomed in Lancashire. Much of that cotton flowed in through Manchester, where mechanization and the factory transformed the city into Cottonopolis.

The number of spindles in cotton mills rose from 1.7 million in the 1780s to between 4 million and 5 million in 1812, essentially tripling in output, making cotton "the most important British manufacturing industry." These factories had a major demand for cheap labor, so they turned to children. The kids were shipped in, like the young apprentice Robert Blincoe, from orphanages to work, for free, on early spinning machinery, alongside migrant workers from

Ireland and women who had been displaced en masse by the last wave of machines.

In patriarchal, industrializing England, the loudest outrage was over the degradation of mens' jobs, but the hand spinners, who had also worked at home, also with their families, had just seen their livelihoods wiped out en masse by automation technologies like the spinning jenny and the water frame. "Hand spinners were the most numerous industrial employees right across Europe," one economic historian noted. "They've been forgotten, because they were women. And their livelihoods were demolished in the space of fifteen years."

Now, the Lancashire weavers and artisans stood opposed to one of the most potentially disruptive automated technologies of all. Nottingham knitters rioted against wide frames and cut-ups, West Riding croppers eyed the shearing frame and the gig mill, and the Lancashire weavers had in their sights Edmund Cartwright's power loom.

It had taken a decade or two, and had required some further innovation and adjustments to make it profitable, but Cartwright's device had finally caught on. In Manchester, the looms were often gathered in a factory powered by the steam engine, too. There were not yet a large number of power looms in operation, but, as in the other Midland and Yorkshire regions, trade was depressed, poverty was rampant, and families were suffering. Any entrepreneur who sought to install a steam-power loom, or who already had one in operation, now faced the wrath of thousands of cotton weavers.

As production boomed and more factories turned to child labor or adopted the looms, and as wages plummeted — the weekly pay of a Lancashire weaver declined from 25 shillings in 1800 to 14 shillings in 1811 — the weavers organized to push for a minimum wage. When they met resistance, they undertook a peaceful and tightly orchestrated strike that brought wages back up, though the rates did not last. So the Lancashire weavers brought the matter to their magistrate, who in turn told them to write the secretary of the Home Office. Richard Ryder replied bluntly that "the Steam Looms were a Great Service to the State."

Anger was percolating here, as it was just about everywhere else. The cotton weavers tried to meet with the factory bosses, the owners of the power looms, to see if a deal could be struck. But the options were running out.

The most recent letter from B relayed that the last meeting he'd attended had a relatively simple aim: to send a Luddite delegate on a mission to several nearby counties, gathering intel, testing "the political, economic, and social waters," and gauging their readiness for uprising. They planned to meet back in two weeks "with reports from their local areas about the numbers of people who wished to be involved in the Business."

There would no doubt be many.

B was nothing if not productive. He had already provided a slew of detailed accounts of seditious-sounding meetings; the delegates from other counties who had passed through them; talk of impending, full-blown insurrection—his "specialty"—as well as information about how many men were preparing to revolt, and how many weapons they'd use.

Soldiers, the royal dragoons, and organized militias had been the face of the Crown's opposition to the Luddites in Nottingham. The spies kept quietly to the shadows, as spies do, though the workers were well aware they were out there.

The Nottingham Luddites had, after all, begun to win some victories and concessions. Not only were they popular with the working poor, but they had driven some shop owners to abandon the new machines, raise wages, and return work to the craftsmen. The Luddites were nothing less than a threat to the ruling class's economic power.

Magistrates like Ralph Fletcher were convinced that, whether by rifle in broad daylight or by arrest at night, these threats had to be rooted out immediately—or else they'd spread.

MASS UNEMPLOYMENT, INSURRECTION & MILITARY OCCUPATION

There are 20,000 stocking-makers out of employment. Six regiments of soldiers from different parts of the country have been sent into this town; and 300 new constables have been sworn to keep the peace. But all this is of no avail as the practice of setting fire to corn and hay stacks, and breaking open houses still continues.

Nine Hundred Lace Frames have been broken, which cost £140 each; from twenty to thirty of them are destroyed in a night. The whole country, for twenty miles round, is full of these ruinous proceedings, nor can they be checked. Nottingham jail is full of debtors; and the country is equally distressed. No trade; no money.

This has been the case for two months. This town is now a garrison, and strictly under martial law.

God only knows what will be the end of it; nothing but ruin.

— The London Statesman, *December 15, 1811*

NED LUDD AND THE
PRINCE REGENT

December 1811

The towns around Sherwood Forest were full of smashed machinery. The assaults on the wide frames had continued unabated for over a month, the toll of lost property rose by the day, and the letters threatening more attacks piled up on the doorsteps of manufacturers and magistrates. The raids had spread beyond Nottingham, and into the neighboring counties of Leicester and Derby.

The Nottingham City Council put up a bounty of £500 for anyone who offered information about the "Authors, Writers, Publishers, or Senders" of letters "under the fictitious name of Ned Lud." The Duke of Newcastle, meanwhile, put up a £2,000 reward for information about the "illegal & felonious proceedings in the Town."

Some suggested that a mischievous and well-spoken framework knitter, Gravener Henson, who had helped cloth workers organize against the hosiers a few years back, must be involved somehow. But if anyone knew more than that, no one told the authorities.

The bounties offered in response, content posted on doors of factories and city buildings, fueled the viral spread of the Luddites' narrative. The story of the uprising was, to the working classes, resonant and irresistible—the surgical nature of the strikes, the disguises and secret identities taken on by the crusaders, the blunt moralizing of the threats, the name General Ludd in the land of Robin Hood. It was a legend unfolding in real time. The letters, sent to factory bosses, magistrates, and machine owners at an increasing clip, gave the sense that the Luddites were everywhere, united.

Whether it was the media—maybe thinking it would make the movement look juvenile or foolish—that invoked the story of Ned Ludd, the young apprentice who angrily lashed out at his master, or Luddites themselves—aiming to stamp their rebellion with an origin story — a more detailed explanation of the Luddites' beginnings started to circulate.

On December 20, 1811, a story ran in the *Nottingham Review:*

There are few persons in this part of England, who know any thing of the History of the Stocking Frame, who, probably, have not heard it was the invention of William Lee, of Notts, student in the University of Cambridge. This Gentleman, it is said, being in love with a young Lady, found that her incessant occupation in knitting, left no leisure to receive his addresses; and resentment for slighted love prompted him to invent a machine, which should supersede the necessity of knitting.

At present, a person named Ned is become more famous, by the destruction of this machine than William Lee, by its invention. Ned Ludd is not, as many people suppose, an ideal personage; but is, or lately was, an inhabitant of Antsley, near Leicester, where he was apprenticed to learn the art of frame work knitting.

Ned being rather adverse to the satisfaction of his master, who complained of him to the Magistrate. As a remedy for Ned's disorder, the Magistrate, it is said, recommended a little whipping.

This, however, was far from curing the patient, that he took the first opportunity of getting a great hammer, and entirely demolishing the machine, which he considered as the occasion of his punishment. Hence the persons who have lately repeated Ned's operation, on a very extended

This news brief, published in *The Nottingham Review* on December 20, 1811, is the earliest known explanation of the Luddites' origin story.

scale, in this and the neighboring counties, have thought it proper to assume his name and conceal their own.

The story was reprinted in newspapers across England.

On the day the story appeared in the *Nottingham Review,* the Crown issued a royal proclamation: "A considerable Number of disorderly Persons, chiefly composed of Stockingers," it began, "have for some Time past assembled themselves together in a riotous and tumultuous Manner..."

These "Persons" were attempting to "compel their Employers to comply with certain Regulations prescribed by themselves," using "measures of Force and Violence." The florid proclamation declares that the state would use all its power to punish the guilty parties and apprehend the Luddites, whom the Crown did not refer to by name.

The Prince Regent's office promised that any Briton who offered information leading to the arrest of these rioters "shall be entitled to the Sum of FIFTY POUNDS for each and every Person who shall be so convicted," as well as a pardon, in case the informer might be liable for prosecution themselves.

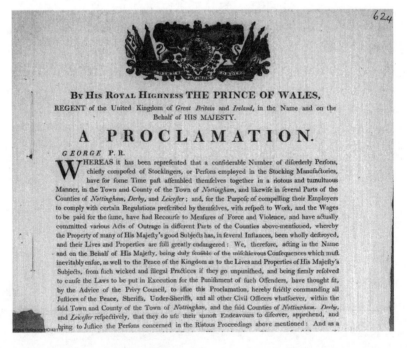

The text of Prince George's proclamation was carried in the same Dec. 20 edition of *The Nottingham Review.* It was also printed as a handbill, pictured above.

There were rumors of a proposal to render machine-breaking a capital crime, as a number of powerful Tory legislators felt that severe punishment was the only way to deter further outbreaks. But popular sentiment differed.

The poet Percy Bysshe Shelley wrote to a friend on December 26:

This proposal will be (if made) a proof of the imbecility of aristocracy. I have been led into reasonings which make me hate more and more the existing establishment, of every kind. I have beheld scenes of misery. The manufacturers are reduced to starvation. My friends the military are gone to Nottingham.... Curses light on them for their motives, if they destroy one of its famine-wasted inhabitants. [Robert] Southey thinks that a revolution is *inevitable.*

Shelley had a stark vision for what would happen if the ruling class continued to turn the screws: "The groans of the wretched may pass unheeded till the latest moment of this infamous revelry, till the storm burst upon them, and the oppressed take ruinous vengeance on the oppressors."

The state had placed a bounty on the Luddites' heads, if anyone dared take it up.

As soon as a copy of the proclamation was nailed to the church door in Sheepshead, someone posted a proclamation of their own alongside it: As the government had offered a reward of £50 for the conviction of offenders, there were 50 bullets ready for the body of the first man who should give information.

It was signed, "Ned Lud."

GEORGE MELLOR

Christmas Day, 1811

Church bells rang for midnight as George Mellor trudged up the icy path to his aunt's farmhouse in Lower Helm, his coat flecked with snow. Some way to ring in Christmas. It was a frigid walk, and he'd practically fled the home of his stepfather, who was in a terrible mood over the state of his shop's business. The snow was falling, and some of the lights of the neighbors' houses were still flickering up on the hillside. There should have been more. At the door, he cleared his throat.

"Christians awake," George bellowed, "and salute the happy morn!" In a scene described in *Ben o' Bill's,* a commotion stirred inside the modest but spacious four-room house where his uncle, the weaver William Bamforth, ran his looms on the second floor.

William opened the door and George leaned in, taking off the huge red scarf and the soggy overcoat wrapped around his formidable frame. His auburn hair and brown beard were soaked. As he swept inside, he gave his aunt a "smacking" kiss, and another to Martha, the housekeeper, who kissed him back so enthusiastically the whole room erupted in laughter. And there was his affable mountain of a younger cousin, Ben, and their charming, sly-eyed cousin Mary — whom he sent a lingering smile — and 'Siah, the stablemate, who was well past drunk. Dirty plates and the detritus of a rare hearty meal were piled off to the side.

Since the Bamforths ran a farm as well as the four looms upstairs, there were eggs, milk, and vegetables; a decent buffer against hunger, and means for a small feast in trying times. Better than a lot of the neighbors could say. That was the unspoken truth that haunted any move to merriment here, George knew, even if it was Christmas. For many, there was little to celebrate.

"It seemed an age since I saw you all; and our house's none too cheerful just now," George said, explaining his nocturnal arrival. "Trade's fearful bad, and John Wood's as sore as a boil. I brought this sprig of mistletoe."

He pulled out the festive plant, the family laughed again, and Ben went for more ale, just as carolers arrived at the porch.

The next evening, George proposed that he and Ben take a walk to nearby Marsden for a pint at the Red Lion. They took off across the darkened pasture toward Huddersfield, animated by jokes and gossip. As they made their way into town, a voice called out and surprised them both.

"Ben, lad?" It belonged to William Horsfall, the factory owner, who was a polite acquaintance of Ben's father.

"Hello, sir," Ben replied.

"A right good Christmas to you, and my compliments to my good friends at Helm," Horsfall said. "And who's your friend, Ben?"

Ben introduced George. Horsfall said he recalled knowing George's mother, years back, and offered his hand in greeting, but George froze over.

"Come along to Ottiwells and taste our spiced ale," Horsfall said. "My wife will be glad to have a crack with you." Ben was ready to accept, but George, who'd been so mirthful a minute earlier, didn't budge. He said nothing. Horsfall wasn't used to having his offers rejected, and he turned away to the soundtrack of Ben's apologies. All George could manage was a curt nod and a limp handshake.

As soon as Horsfall was out of earshot, George sprung back to life.

"I wonder you can speak civil to a man like that," he erupted. "Don't you know that Horsfall is foremost of all in pressing on the use of the new machines? Don't you know that he has put them into Ottiwells? Don't you know he is sacking the old hands and will have none but young 'uns that will and can learn . . . how to work the new frames? Don't you know that there's many a family in Marsden now, this very merry Christmas that we're wishing each other like prating parrots, that has scarce a fire in the grate or a scrap of meat on the table, or warm clothing to the back, just because of Horsfall and such as he?"

By now, Horsfall's hatred for the Luddites was well known, and it was common knowledge that he'd said he'd ride up to his saddle in their blood. "Don't you know that in Huddersfield Market Horsfall has sworn hanging isn't good enough for the Nottingham lads? If you don't know, you live with your eyes shut, Ben, and your ears waxed, for I'll never believe that your heart's shut, lad. And then you ask me why I couldn't take him hearty by the hand."

Ben still wasn't satisfied. "But what has *Horsfall* to do with all this?" he asked.

"He has this to do with it, Ben. Ever since the bad times began, Englishmen have been told to stand together shoulder to shoulder against a common enemy," to do their patriotic duty.

As long as everyone suffered together, rich and poor alike, George mused, all was fine. The rich have no monopoly on patriotism, he said. *He* had gone to war for England, after all. But then a funny thing started happening. Amid this calamity, the rich started getting richer.

Those new machines, George told Ben, cut costs to an impossible degree. A new wide frame in Nottingham cost £120, and did the work of four weavers — the owner made the cost back in a year. "He saves it, but who loses it?" George said. "Why the wage earners to be sure." Hence the twenty thousand stockingers out of work in Nottingham right now. They stopped in the street; fire was in George's eyes, worry in Ben's.

"But what can you do, George?" Ben asked. Ben was always the cautious one, much as he loved his cousin. "You cannot fight against the law of the land. The masters have the law at their backs. It's not worth kickin' against the pricks. You surely will not have ought to do with machine-breaking."

George frowned. "I'm groping in the dark just now. Frame-breaking and rick-burning seems but spiteful work, but it is action, and action of some sort seems called for. If we submit like dumb cattle, our rulers say we are content and have no grievances; if we assemble in great numbers and proclaim our wrongs, they hang us for sedition. What can we do, where shall we turn?"

And that was when George told him.

"Steps are to be taken," he put it, carefully, euphemistically, but any cloth worker in the region would know exactly what he meant. These "steps" were intended to "dissuade" manufacturers from killing jobs with automated machinery. The first of these steps was to take an oath, and to bind oneself to never work the machines hurtful to commonality, and never to work in a shop or a factory where such machines were deployed. The second step was to push the manufacturers to stop using them, and George was careful to note that "no violence of any sort was to be employed either against man or machine, at least not if the masters proved amenable to reason." Of that George thought there could be little question.

Many of the masters, after all, were men that George and Ben would have known much of their lives. Men that knew their families, their history, their community. These social bonds were not easily broken; it was not easy to stare into the eyes of a friend and say, *I am taking your job.* (One benefit of machinery was that it could be used as a rhetorical tool as well, to muddy the moral clarity of the situation — a use it's been put to by owners ever since. It's the robots, not your boss, that's coming to take away your job.)

"They cannot stand against us, if we are united," said George; "our weakness lies in action unconcerted and without method."

It was to be an tightly disciplined campaign. Many craftsmen, like George, had military backgrounds, after all. Many had been involved in some form of organizing, though under the table, as it was illegal. "If we set our faces resolutely against the use of these newfangled substitutes for human labor, we can at least compel the masters to wait till times are better and trade mends," George said. It hardly seemed a radical demand. "It may be that when the wars are over and the market calls for a larger and a quicker output, machinery may be gradually introduced without hardship to those who have grown old in the old methods and who cannot use themselves to new ways."

In other words, what if, after their campaign clarified how urgent and serious they were, the artisans and factory owners sat down together and decided, democratically, how best the machines might be rolled out? How they might benefit the shop owners and trade workers alike? How they might reduce work and improve profits without needlessly tearing communities and lives apart? And if they could accomplish that, George wagered, they would in the process learn "the secret and the value of combination" — of forming a union — "and we may turn our organization to the protection and the improvement of the worker and to the wresting of those rights that are now withheld."

Still, George knew the risks. Ben pressed him again, and he demurred.

"I cannot see daylight which ever way I turn."

ROBERT BLINCOE

"I Shouted Aloud for Them to Stop the Wheels."

1811

Robert Blincoe, the orphan who left on a carriage ride for Nottingham over a decade before, had run four miles straight. He'd run across the jagged rocks and green glens that hid the factory from the outside world, his body covered in welts, bruises, and blood. By the time he arrived, out of breath, at the door of his former overseer Johnny Wild, Robert looked like "a wild creature." Wild was now a stocking weaver who worked out of his home, and he was shocked to see his former adept run up, half-dressed, drenched in sweat, with "terror in his looks."

Wild jumped up from his frame and let Blincoe in. It had been about a decade since he'd last seen Blincoe, when the boy had been a child apprentice at Lowdham Mill, outside Nottingham. Now, he learned that Blincoe, nearly twenty years old, was making his way to the local magistrate to inform him about the severe abuse he and other apprentices received, constantly, at the hands of his current employer, at Litton Mill, where he'd been moved after Lowdham. He was past his breaking point. He told Wild and his wife, as they rushed to feed him, the story of his predicament, gesturing toward the wounds that marred his face and chest.

Lowdham, his last factory, had been bad. The food was awful, the hours endless, the rate of work relentless. Blincoe had started out forced to do dangerous jobs like picking stray cotton out of the rumbling machinery so it wouldn't jam up the works, and had then been moved on to roving the threads to be fed into the great churning wheels. He was still little, just seven years old, and "was not able by any possible exertion, to keep pace with the machinery." When he fell behind, he was beaten by his overseers, most of whom were not as kind as Johnny Wild. The beatings happened so often that he looked "as spotted as a leopard" with bruises.

When he complained to the manager, the man retorted, "Do your work well, and you'll not be beaten." The overseers were paid for meeting their quotas, and risked being discharged if they did not. Abuse as a work incentive was tacitly tolerated throughout the factory system, which then employed twenty thousand children in the cotton industry alone. Blincoe realized that it was useless to complain. Fourteen hours of work a day amid the clank of machinery and the stink of oil, grease, and sweat, under the constant threat of the whip; it was too much. Blincoe ran away for the first time.

"I cannot deny that I feel a glow of pride, when I reflect that, at the age of seven years and a half, I had courage to resent and to resist oppression," Blincoe would say years later, "and generosity to feel for the sufferings of my helpless associates." He told his peers his plan, and invited them to join him, but no one was willing to risk the long road back to London. Which was well enough for them; Blincoe was spotted by an associate of the factory's management as soon as he slipped past its walls. There was a cash reward offered to anyone who caught a runaway apprentice, since escape attempts happened so frequently. His captor beat him and returned him to Lowdham.

Cowed, Blincoe saw no choice but to submit. He worked as hard as he could, accepting the beatings when he couldn't go fast enough. Aside from the threat of the managers, the machinery in cotton mills was extremely dangerous. By the time the first year was through at Lowdham, most of his cohort had been injured in one way or another. "Some had the skin scraped off the knuckles, clean to the bone, by the fliers; others a finger crushed, a joint or two nipped off in the cogs of the spinning-frame wheels," he recalled. One day while Blincoe was tending the machinery, he got the forefinger on his left hand stuck in one of the gears, which tore the finger off at the joint. While he screamed in pain, desperately "clapping the mangled joint, streaming with blood, to the finger," the overseers laughed in his face. Blincoe ran to the doctor, who dressed the wound and sent him back to work.

It could have been worse. While ten-year-old Mary Richards was getting ready to leave one night, her apron got caught in the horizontal shaft that turned the immense drawing frames, and her entire body was immediately pulled into the machinery. Blincoe was one of the few workers still in the factory that night. He called out for help, but could do little but watch as she "whirled round and round" the machine. He listened to the awful cracking of her bones. "I cannot describe my sensations at this appalling scene," Blincoe said. "I shouted out aloud for them to stop the wheels! When I saw her blood thrown about like water from a twirled mop, I fainted." Her head "appeared dashed to pieces," and her body was "mangled." Amazingly, she survived the incident, but she was disabled for life. Still, she too was sent back to the mill.

When he was eight years old, Blincoe recalled standing at a window on an upper story, contemplating throwing himself off the side.

It was horrific. It was also standard practice at most factories to work apprentices, children, women, and poor unskilled laborers precisely this hard. Conditions in the mills varied somewhat, but most defaulted to excessive demands on the workers: ruthless production rates, fourteen hours a day, six days a week.

Many factory owners, especially those who had built their operations in more remote places, could not find enough labor to run them — they did not want skilled workers, who cost more and were not pliable, and few wanted to work in the factories voluntarily — so they brokered deals with workhouses and orphanages like St Pancras, where Blincoe was signed up. Ironically, some of those factory owners, like Richard Arkwright, had built those operations in remote locations precisely because they were remote — away from population centers where workers intuited that factories signaled a hellish future of work and would try to burn them down. Hidden from view, these operations often turned into precisely the hells that workers feared.

Disgruntled, anti-factory working people, like those who became Luddites, had an array of disturbing contemporary examples assembled before them. The true scope of horrors unfolding at the mills was not yet widely known, but plenty of examples were.

After the politician and mill owner Robert Peel paid a visit to his operation, which was run by a subordinate, he was shocked to discover how bad the conditions were *at his own factory*. The children who worked there were maimed, their skin sallow, and they were clearly malnourished. They had no time for education or exercise. Peel was appalled. He lamented that he did not personally have the time to invest in overseeing affairs on the factory floor. So he pushed for legislative reform that would limit the number of hours children worked, and the conditions in which they lived, though he didn't stop employing child labor. He needed them to run the machines, after all.

In 1802, the reforms passed Parliament easily. Shortly after, an investigative committee arrived at Lowdham and conducted a lengthy report. Conditions improved dramatically after that—new housing was erected, working hours were limited, and the ventilation inside the factory was improved. Not long after, however, the operation closed. Blincoe was sold to another master to finish his term of indenture, which still had nearly a decade to go.

Once again, he was loaded into a wagon and transported to his new place of employment: a factory in a sequestered valley, far from any major city, in Litton, Derbyshire. Any faint hopes that Robert had that his situation might improve in a new environment, with new management, were quickly stamped out: conditions at Litton were, somehow, far worse than even Lowdham.

The work hours were longer, the food even less edible, the quarters more packed and squalid. In fact, everything was more squalid, from the factory floor to the dorms, where disease ravaged the workers. At Litton the abuse transcended mere brutal punishment; it was elevated by overseers there to a kind of sadistic sport. The children were tied up and beaten, or forced to jump on one foot around the deadly machinery, or placed on top of a cylinder set in the way of the rotating spokes of the spinning machine, so that it would knock them down, and risk pulling them into the gears if they weren't quick to duck out of the way. The apprentices' teeth were filed down, their ears pinched until they bled, and they were dragged by their hair across the floor.

Bad as Lowdham was, none of the children were killed while Robert was indentured there. At Litton, so many children had died that the factory's management distributed the bodies to churches across the region so that the sheer volume would not arouse the suspicions of the ministers who buried them.

Blincoe had survived this torment for almost ten years, and now, at nearly twenty years old, with his freedom on the horizon, he was overcome with outrage at all of it. He'd skipped a few hours in his sixteen-hour workday with his friends, and when his absence was noticed, he was thrown out of the apprentice house for the night with nothing to eat. When he was let back in, the factory's owner, Ellis Needham, beat him with a walking stick until he was covered in welts and oozing with blood. Blincoe stopped only to eat on his way out and took off running—he was finally going to tell the magistrate about what was going on at Litton Mill.

At Lowdham, Blincoe didn't even know that so much of what went on in the factory was against the law — the problem with Peel's child labor reform act was that it lacked any kind of enforcement mechanism. After 1802, and an initial round of interest in improving conditions and inquiries into known offenders, the law was essentially recognized as a "dead letter." Operations like Litton had not changed at all. If anything, they had only grown worse. So the young bodies piled up.

Life expectancy wasn't long in the eighteenth and nineteenth centuries, and most working people were fortunate if they lived past their forties. And progress was beginning to cut its forked path. There was a vaccine for smallpox now — even Blincoe got a dose — and yet conditions in factories for the working poor left them worse off, physically stunted from malnourishment, or killed prematurely in the machinery. And it was a certain kind of death — to be killed by the cutting-edge technology, for a profit — that cast so dark a shadow. When these deaths were reported, they made the news, plainly underscoring the dangers of the factory and how commonplace this kind of carnage risked becoming.

Imagine the stocking frame workers, croppers, and weavers, already concerned that the factory was coming for them, reading or hearing headlines detailing the abuses of the factory in gatherings and watercooler talk. Imagine George, Ben, and Gravener reacting to each of these stories, all of which ran in 1811, the year the Great Comet burned and the Luddites rose up to push back against automation, exploitation, and the factory:

> *Thursday, a child playing in one of the foundries on Leith Walk, got entangled with the machinery, and was killed on the spot.*
> — Perthshire Courier, *February 28, 1811*

> *Andrew Bordie, employed in the lint mill of Buxburn, near Aberdeen, walking across the loft, slipped his foot, and falling backwards upon the water wheel, then in motion, was instantly killed. The body, which had stopped the motion of the machinery, was immediately taken up.*
> — Belfast Commercial Chronicle, *Monday, March 4, 1811*

> *A dreadful accident happened on Sunday morning the 4th instant, in the carding-mill, at Beckfoot, near Skinburness, in Cumberland. Mr. Isaac Saul, the proprietor of the concern, was caught by some part of the machin-*

ery, and crushed to death.... The body was found broken and lacerated in a manner too shocking to describe.
— Westmorland Advertiser and Kendal Chronicle, *Saturday, August 17, 1811*

On Wednesday morning a boy working at a factory in Fisherton was caught by part of the machinery, was dreadfully crushed in that part and the head, had a part [of] the scalp torn off, and arm and thigh broken, before he could [be] extricated.
— Salisbury and Winchester Journal, *Monday, September 23, 1811*

Saturday evening, a laborer employed at a mustard mill in the neighborhood of Mile End, having approached too near the machinery, was suddenly drawn into the midst of it. [He] had his skull fractured, both his thighs broken, one of them in two places, and four of his ribs on the left side also fractured.
— Oxford University and City Herald, *Saturday, October 12, 1811*

But many more of these stories went unreported at places where factory entrepreneurs had rushed to adopt automation — places like Litton Mill, where no reporter knew to look, and no manager reported the dead.

Fueled by rage over being beaten, on the brink of becoming an invisible statistic for too long, Blincoe wrapped up his story. He won the sympathies of his old superior Johnny Wild, whose wife said they'd do what they could and gave him some food and shelter. He was aware enough now to know that the abuses he suffered weren't merely unjust, but illegal. He did not yet know that his sufferings would eventually help expose the barbarity of the entire factory system.

After he rested, Robert Blincoe sprinted off again to find the magistrate, to try, if in vain, to hold one of the first tech titans accountable.

THE FIRST TECH TITANS

The first tech titans were not building global information networks or commercial space rockets. They were making yarn and cloth. A *lot* of yarn, and a *lot* of cloth.

Like our modern-day titans, they started out as entrepreneurs. But until the nineteenth century, entrepreneurship was not a cultural phenomenon. Businessmen took risks, of course, and undertook novel efforts to increase their profits. Yet there was not a popular conception of the heroic entrepreneur, of the adventuring businessman, until after the birth of industrial capitalism. The term itself was popularized by Jean-Baptiste Say, in his 1803 work *A Treatise on Political Economy*. An admirer of Adam Smith's, Say thought that *The Wealth of Nations* was missing an account of the individuals who bore the risk of starting new business; he called this figure the *entrepreneur*, which translated from the French as "adventurer" or "undertaker."

For a worker, aspiring to entrepreneurship was different than merely seeking upward mobility. The standard path an ambitious, skilled weaver might pursue was to graduate from apprentice to journeyman weaver, who rented a loom or worked in a shop, to owning his own loom, to becoming a master weaver and running a small shop of his own that employed other journeymen. This was customary.

In the eighteenth and nineteenth centuries, as now in the twenty-first century, entrepreneurs saw the opportunity to use technology to disrupt longstanding customs in order to increase efficiencies, output, and personal profit. There were few opportunities for entrepreneurship without some form of automation; control of technologies of production grants its owner a chance to gain advantage or take pay or market share from others. In the past, like now, entrepreneurs started small businesses at some personal financial risk, whether by taking out a loan to purchase used handlooms and rent a small factory space, or by using inherited capital to procure a steam engine and a host of power looms.

The most ambitious entrepreneurs tapped untested technologies and novel working arrangements, and the most successful irrevocably changed the structure and nature of our daily lives, setting standards that still exist today. The least successful would go bankrupt, then as now.

In the first century of the Industrial Revolution, one entrepreneur looms above the others, and has a strong claim on the mantle of the first of what we'd call a tech titan today. Richard Arkwright was born to a middle-class tailor's family and originally apprenticed as a barber and wigmaker. He opened a shop in the Lancashire city of Bolton in the 1760s. There, he invented a waterproof dye for the wigs that were in fashion at the time, and traveled the country collecting hair to make them. In his travels across the Midlands, he met spinners and weavers, and became familiar with the machinery they used to make cotton garments. Bolton was right in the middle of the Industrial Revolution's cotton hub hotspot.

Arkwright took the money he made from the wigs, plus the dowry from his second marriage, and invested it in upgraded spinning machinery. "The improvement of spinning was much in the air, and many men up and down Lancashire were working at it," Arkwright's biographer notes. James Hargreaves had invented the spinning jenny, a machine that allowed a single worker to create eight threads of yarn simultaneously—though they were not very strong—in 1767. Working with one of his employees, John Kay, Arkwright tweaked the designs to spin much stronger threads using water or steam power. Without crediting Kay, Arkwright patented his water frame in 1769 and a carding engine in 1775, and attracted investment from wealthy hosiers in Nottingham to build out his operation. He built his famous water-powered factory in Cromford in 1771.

His real innovation was not the technology itself; several similar machines had been patented, some before his. His true innovation was creating and successfully implementing the system of modern factory work.

"Arkwright was not the great inventor, nor the technical genius," as the Oxford economic historian Peter Mathias explains, "but he was the first man to make the new technology of massive machinery and power source work as a system — technical, organizational, commercial — and, as a proof, created the first great personal fortune and received the accolade of a knighthood in the textile industry as an industrialist." Richard Arkwright Jr., who inherited his business, became the richest commoner in England.

Arkwright was the first start-up founder to launch a unicorn company, we might say, and the first tech entrepreneur to strike it wildly rich. He did so by marrying the emergent technologies that automated the making of yarn with a relentless new work regime. His legacy is alive today in companies like Amazon, which strive to automate as much of their operations as is financially viable, and to introduce surveillance-intensive worker-productivity programs.

Often called the grandfather of the factory, Arkwright did not invent the idea of organizing workers into strict shifts to produce goods with maximal efficiency. But he pursued the "manufactory" formation most ruthlessly, and most vividly demonstrated the practice could generate huge profits. Arkwright's factory system, which was quickly and widely emulated, divided his hundreds of workers into two overlapping thirteen-hour shifts. A bell was rung twice a day, at 5 a.m. and 5 p.m. The gates would shut and work would start an hour later. If a worker was late, they sat the shift out, forfeiting that day's pay. (Employers of the era touted this practice as a positive for workers; it was a more flexible schedule, they said, since employees no longer needed to "give notice" if they couldn't work. This reasoning is reminiscent of that offered by twenty-first-century on-demand app companies.) For the first twenty-two years of its operation, the factory was worked around the clock, mostly by boys like Robert Blincoe, some as young as seven years old. At its peak, two-thirds of the 1,100-strong workforce were children. Richard Arkwright Jr. admitted in later testimony that they looked "extremely dissipated, and many of them had seldom more than a few hours of sleep," though he maintained they were well paid.

The industrialist also built on-site housing, luring whole families from around the country to come work his frames. He gave them one week's worth of vacation a year, "but on condition that they could not leave the village." Today, even some of our most cutting-edge consumer products are still manufactured in similar conditions, in imposing factories with on-site dormitories and strictly regimented production processes, by workers who have left home for the job. Companies like Foxconn operate factories where the regimen can be so grueling it has led to suicide epidemics among the workforce.

The strict work schedule and a raft of rules instilled a sense of discipline among the laborers; long, miserable shifts inside the factory walls were the new standard. Previously, of course, similar work was done at home or in small shops, where shifts were not so rigid or enforced.

Arkwright's "main difficulty," according to the early business theorist

Andrew Ure, did not "lie so much in the invention of a proper mechanism for drawing out and twisting cotton into a continuous thread, as in…training human beings to renounce their desultory habits of work and to identify themselves with the unvarying regularity of the complex automaton." This was his legacy. "To devise and administer a successful code of factory discipline, suited to the necessities of factory diligence, was the Herculean enterprise, the noble achievement of Arkwright," Ure continued. "It required, in fact, a man of a Napoleon nerve and ambition to subdue the refractory tempers of workpeople."

Ure was hardly exaggerating, as many workers did in fact view Arkwright as akin to an invading enemy. When he opened a factory in Chorley, Lancashire, in 1779, a crowd of hundreds of cloth workers broke in, smashed the machines, and burned the place to the ground. Arkwright did not try to open another mill in Lancashire.

Arkwright also vigorously defended his patents in the legal system. He collected royalties on his water frame and carding engine until 1785, when the court decided that he had not actually invented the machines but had instead copied their parts from other inventors, and threw the patents out. By then, he was astronomically wealthy. Before he died, he would be worth £500,000, or around $425 million in today's dollars, and his son would expand and entrench his factory empire.

The success apparently went to his head — he was considered arrogant, even among his admirers. In fact, arrogance was a key ingredient in his success: he had what Ure described as "fortitude in the face of public opposition." He was unyielding with critics when they pointed out, say, that he was employing hundreds of children in machine-filled rooms for thirteen hours straight. That for all his innovation, the secret sauce in his groundbreaking success was labor exploitation.

In Arkwright, we see the DNA of those who would attain tech titanhood in the ensuing decades and centuries. Arkwright's brashness rhymes with that of bullheaded modern tech executives who see virtue in a willingness to ignore regulations and push their workforces to extremes, or who, like Elon Musk, would gleefully wage war with perceived foes on Twitter rather than engage any criticism of how they run their businesses. Like Steve Jobs, who famously said, "We've always been shameless about stealing great ideas," Arkwright surveyed the technologies of the day, recognized what worked and could be profitable, lifted the ideas, and then put them into action with an unmatched aggression. Like Jeff Bezos, Arkwright hyper-charged a new mode of factory work by finding ways to impose discipline and rigidity on his workers, and adapting them to the rhythms of the machine and the dictates of capital — not the other way around.

We can look back at the Industrial Revolution and lament the working conditions, but popular culture still lionizes entrepreneurs cut in the mold of Arkwright, who made a choice to employ thousands of child laborers and to institute a dehumanizing system of factory work to increase revenue and lower costs. We have acclimated to the idea that such exploitation was somehow inevitable, even natural, while casting aspersions on movements like the Luddites as being technophobic for trying to stop it. We forget that working people vehemently opposed such exploitation from the beginning.

Arkwright's imprint feels familiar to us, in our own era where entrepreneurs loom large. So might a litany of other first-wave tech titans. Take James Watt, the inventor of the steam engine that powered countless factories in industrial England. Once he was confident in his product, much like a latter-day Bill Gates, Watts sold subscriptions for its use. With his partner, Matthew Boulton, Watts installed the engine and then collected annual payments that were structured around how much the customer would save on fuel costs compared to the previous engine. Then, like Gates, Watts would sue anyone he thought had violated his patent, effectively winning himself a monopoly on the trade. The Mises Institute, a libertarian think tank, argues that this had the effect of constraining innovation on the steam engine for thirty years.

Or take William Horsfall or William Cartwright. These were men who were less innovative than relentless in their pursuit of disrupting a previous mode of work as they strove to monopolize a market. (The word *innovation*, it's worth noting, carried negative connotations until the mid-twentieth century or so; Edmund Burke famously called the French Revolution "a revolt of innovation.") They can perhaps be seen as precursors to the likes of Travis Kalanick, the founder of Uber, the pugnacious trampler of the taxi industry. Kalanick's business idea — that it would be convenient to hail a taxi from your smartphone — was not remarkably inventive. But he had intense levels of self-determination and pugnacity, which helped him overrun the taxi cartels and dozens of cities' regulatory codes. His attitude was reflected in Uber's treatment of its drivers, who, the company insists, are not employees but independent contractors, and in the endemic culture of harassment and mistreatment of the women on staff.

These are extreme examples, perhaps. But extremity is often needed to break down long-held norms, and the potential rewards are extreme, too. Like the mill bosses who shattered nineteenth-century standards and traditions by automating cloth-making, today's start-up founders aim to disrupt one job category after another with gig work platforms or artificial intelligence, and encourage others to follow their lead. There's a reason Arkwright and his factories were both emulated and feared. Even two centuries later, the most successful tech titans typically are.

PART II

METROPOLIS OF DISCONTENT

[The year] 1812 opens with a gloom altogether so frigid and cheerless, that hope itself is almost lost and frozen in the prospect.

— The Manchester Gazette, *January 1, 1812*

GEORGE MELLOR

Winter 1812

The moon hung low in the night sky as the men organized themselves in an empty field outside Huddersfield. Soon there were a dozen, then twenty, and finally forty-five. They were dressed in black, and some were disguised, their faces painted with coal. Some carried pistols, others gripped hatchets and knives.

In a series of uncertain movements, the men, most of them in their twenties, formed into three companies. They were nervous and exhilarated. For some, their movements were guided by military experience. For others, the drills were still new. It was eleven o'clock on a wintry Saturday night, and the regiment had a simple objective: destroy the machinery that undercut their jobs and tore at the social fabric of their community.

The Luddites anxiously debated their target. Some wanted to assault the massive Bradley factory, but the majority decided, ultimately, they weren't ready for an operation of that magnitude yet. "The system was yet in its infancy and they had but eleven guns and pistols that night," according to one account.

The men marched to the shop of Joseph Hirst. They found the mill boss and two of the boys who worked for him still running the machines. The gunmen approached first. They demanded that the boys let them in, which they did. While two sentries kept watch, the crew split into two squads; one stood guard while the other methodically destroyed the shop's seven gig machines and twenty-four pairs of shears.

"Now for the windows," a voice called out, and the men started firing guns at the house. Three windows were broken before someone the crew referred to as General shouted out for them to stop. They did.

"Now for the house and master!" someone else yelled.

Again, the man they called General ordered them to desist. This would do for now, he said. Do not hurt a hair on their heads. The order would stand, the leader warned, unless Hirst resumed automating cloth production and they were forced to return. In that case, he told him, "We'll do it to the bottom."

The disguised figure, carrying a scythe-bladed sword in one hand and a pistol in the other, called roll, and each man answered to a number. A shot was fired into the air, and the General gave a shout.

The Luddites of the West Riding had completed their first mission.

The deputy nodded to their General, the man they knew from work and around town as George Mellor, and they were off to the next target.

ANNA LÆTITIA BARBAULD

February 1812

In a year that opened with newspapers printing stories about the pervading sense of hopelessness itself, it would have been a dereliction of duty if Britain's Romantic poets didn't weigh in on the darkening future. In the frozen early months of 1812, Anna Lætitia Barbauld, a widely respected poet whose work had influenced writers like Coleridge and Blake, published what would come to stand as one the most unrelentingly grim, pessimistic, and ultimately accurate poems of the day.

Anna Lætitia Barbauld's epic "Eighteen Hundred and Eleven" is an apocalyptic vision that castigates England's despotic rulers for dedicating the nation to war while its people starve:

> *Colossal Power with overwhelming force*
> *Bears down each fort of Freedom in its course;*
> *Prostrate she lies beneath the Despot's sway,*
> *While the hushed nations curse him — and obey [. . .]*
> *Man calls to Famine, nor invokes in vain,*
> *Disease and Rapine follow in her train;*
> *The tramp of marching hosts disturbs the plough,*
> *The sword, not sickle, reaps the harvest now,*
> *And where the Soldier gleans the scant supply,*
> *The helpless Peasant but retires to die; [. . .]*
> *So sing thy flatterers; but, Britain, know,*
> *Thou who hast shared the guilt must share the woe.*
> *Nor distant is the hour; low murmurs spread,*
> *And whispered fears, creating what they dread;*
> *Ruin, as with an earthquake shock, is here.*

The poem broadcast openly what many, especially the poor, were feeling but lacked the platform to articulate: that thanks to Britain's forever war with France and its blind eye to its own citizenry's well-being, the nation faced disaster; that ruin lay ahead for the farmer and the worker; and that even England's former colony, the United States, would soon eclipse its power.

Barbauld's prophesy was not only controversial, but a genuine shock to polite society. After the poem's debut, critics and commentators responded with such virulent revulsion that Barbauld was browbeaten into never publishing another poem again. "Eighteen Hundred and Eleven" had slipped a toe across a line that the aristocracy couldn't stomach. This level of solidarity with the working class, the antiwar sentiment, the grappling with the abuse of state power, and the notion that all of the above might foment decline — these were too radical for the culture-making men of Barbauld's day. Folks like George Mellor, however, would probably have been inclined to agree with her.

The man wielding — supposedly wielding, anyway — that "Colossal Power with overwhelming force," the "Despot" himself, knew the war he'd inherited was unpopular. So the Prince Regent was cheered when, on January 19, a rising star in Britain's military campaign against Napoleon on the Spanish peninsula pulled off a rare decisive victory. In a quick, unexpected attack, the Duke of Wellington besieged the fortress at Ciudad Rodrigo and captured the Spanish city from Napoleon's forces.

The victory, however modest, landed with an outsized impact. It opened up a channel for the British to push into Spain, then held by the French, and served as justification for the Prince Regent to continue backing the war. And just in time. The guardrails were finally coming off: Prince George's restricted regency was set to expire the next month. It had been nearly a year since he'd been appointed Regent, and he'd mostly left the government as his Tory father, who remained sidelined by mental illness, had shaped it. But now, finally out of bed after his ill-fated dance step, the Prince Regent would at last be forced to govern. Some of his old allies in the more liberal Whig party, who believed the war to be disastrous, held out hope that George would make good on his earlier word and put their leaders, Lord Grey and Lord Grenville, in charge.

The bond of that word was fading fast. Over the last year they had seen no movement toward placing them in power, only a string of excuses, evasions, and deferences to George III and his Tory allies.

The Whigs weren't the only ones pressing for a change in government. The Luddites were, too. On February 8, the *Leeds Mercury* reported that "no depredations have been committed in Nottinghamshire, by the Frame-breakers for several days past," and that the reason why could be explained by a letter the Mayor of Nottingham said he had received; signed by General Ludd. It read:

No more frames will be destroyed while the Restrictions on the Prince Regent remain in force, which expire on the 18th...when it is hoped his Royal Highness will take the distressed situation of trade into his most serious consideration; and if he fails in this, the last hope of redress, the Luddites will destroy every frame they can find in the county.

The government, led by Spencer Perceval, however, had no intention to assist the distressed working men and women. On the contrary, as the machine-breakers had been destroying the wide frames across Nottingham over the winter, Perceval's Tories, led by the Home Office secretary, Richard Ryder, had begun drafting a law that aimed to punish the Luddites with the full brunt of the state's power.

Five days before his restricted regency was set to expire, the prince wrote a letter to the Duke of York, remarking that, while he had maintained the executive government out of a "sense of duty" to his father, "a new era is now arrived and I cannot but reflect with satisfaction on the events which have distinguished the short period of my restricted Regency." The victory in Spain had "added most important acquisitions" to Britain's empire, and, to the risk-averse prince, that was enough. He asked the Duke to convey his message to Lords Grey and Grenville. The writing was on the wall.

"I shall be most anxious," George wrote, "to avoid any measure which can lead my allies to suppose that I mean to depart from the present system."

NED LUDD

On January 1, 1812, a handbill was posted in Nottingham titled simply, "By the Frameworck Knitters. A Declaration." The document explained that the knitters "are Impowre'd to breake and Distroy all Frames or Engines that fabricate Articles in a fraudulent and Deceitful manner," seeing as how such machinery violated regulations put into place long ago by King Charles. That charter, while old, did in fact state that it was legal to dismantle machinery producing counterfeit goods. The new declaration held that on these grounds, machine-breaking was not just morally justified, or strategically advantageous — but legal, too.

The bill was signed, "Ned Lud's Office, Sherwood Forest."

Newspapers across England continued to break stories about the Luddites' nightly raids against the machine owners. "It is 'impossible' to convey a proper idea of the state of the public mind in this town," as the *Nottingham Register* described the state of the city on January 29. "The constant parading of the military in the night, and their movements in various directions during both night and day, give us the appearance of a state of warfare."

The machine breakers continued to demonstrate a boldness much of the public found heroic. A squadron of Luddites infiltrated a factory within "a few hundred yards" of a major military barracks and destroyed twenty frames there. A man wearing a goatskin mask and a beard down to his waist intercepted a convoy of wide-frame machines bound for Nottingham and smashed them all. One Saturday night, a band of Luddites crossed the Trent River and smashed dozens of machines in a factory in a town called Clifton. Suddenly they were surrounded by mounted cavalry, summoned from a nearby garrison. The Crown's soldiers thought they had the Luddites as good as caught, but the Luddites seized a boat that was left unattended, split into two divisions, and escaped across the river "in perfect safety."

The Luddites worked to control their popular image. They donned disguises, even costumes, masquerading as local folk heroes or women, writing more colorful threatening letters, and even trying to correct the public record when necessary. After a newspaper item accused the Luddites of a theft unrelated to machine-breaking, a letter soon surfaced with an explanation: that had not been the work of Luddites, it said, and they had in fact found the man responsible, and punished him: he'd been "hang'd for 3 Menet and then Let downe againe" and instructed not to repeat his crime.

Each of those actions had taken place in the Midlands, around Nottinghamshire, where the Luddite campaign had been raging, and was no longer novel. But it certainly would have snapped plenty of heads to attention when they read the late February edition of the *Leeds Mercury,* which carried an article headlined simply: MACHINERY DESTROYED.

"It is with deep regret we have to state that outrages of a most alarming description and extent, have been recently committed in the neighborhood of Huddersfield," the story began. A band of men forced their way into Mr. Joseph Hirst's factory, and "proceeded to destroy all the machinery used in the dressing of cloth such as dressing frames, shears, and other implements used in what is commonly called Gig Mills, the whole of which they completely demolished."

That party, or one just like it, then moved on to another shop, owned by a man named James Balderson, who ran the same array of machinery. They smashed his operation to pieces, too.

As in the Nottingham press, the paper described the surgical nature of the attack:

> The depredators, or to use the cant terms, Luddites, assembled with as much privacy as possible, at the place marked out for attack, and *divided* themselves into two parties, the more daring and expert of which entered the premises, provided with proper implements for the work of destruction, which they accomplished with astonishing secrecy and dispatch.... They do not appear to have done any mischief besides breaking the machinery.

The Luddite movement had spread to Yorkshire, to the home turf of wool production and the proudly independent woollen worker, whose communities had for centuries been so industrious, lively, and egalitarian. Few there were eager to see it subsumed by the machinery of the entrepreneur.

GRAVENER HENSON

February 1812

Machine-breaking couldn't go on forever. No matter how cunning and popular the Luddites were, the core tactic was hardly sustainable. And at this point, in Nottingham and the surrounding area, where attacks had raged for three months, it seemed that many of the hosiers, factory owners, and shop owners were just as eager to work out a durable solution as were the textile workers, artisans, and laborers. A more *permanent* solution.

With that in mind, cloth workers across Nottingham initiated an effort to put a bill before Parliament that would ban the worst offenses of England's textile industry. Calling themselves the United Committee of the Framework Knitters, they planned to canvas local workers and owners about the current situation, and then campaign to push Parliament to enact legislation to regulate the trade. The effort was led by Gravener Henson.

First, Henson and a colleague wrote to the Earl of Waldgrave, who oversaw some of the troops occupying Nottingham. They assured him that the committee of workingmen had met with the master hosiers, many of whom were more than receptive to their cause. They had received "the Cordial support and cooperation of several of the most respectable," as Henson wrote, and had "taken up our Business in the warmest manner being fully sensible of the goodness of our intentions, and the beneficial result that will accrue from our endeavors." He added, "We humbly trust, my Lord, from your well known generosity, that you will nobly assist any endeavor that will conduce to the Public Good."

Two weeks later, Henson advertised a meeting in the *Nottingham Journal*, encouraging stockingers to join him and the committee to hash out their demands for Parliament.

GRAVENER HENSON

TO THE
FRAMEWORK KNITTERS
OF
NOTTINGHAM, THE COUNTY THEREOF,
AND THE
TOWNS AND VILLAGES ADJACENT

The troubled State to which the above Places are reduced, by the pressure of the Times and the operations of the Frame-Breakers, having at length excited the Attention of the Legislature; and as many Members thereof have expressed a desire to obtain every possible Information as to the probable Cause of these Disturbances, it has been thought prudent by many of the Workmen, that they, and their Fellow-Workmen at large, should contribute all their power toward furnishing such Information.

Henson requested that each town, village, and neighborhood send two representatives to the Sign of the Sir Isaac Newton, a pub in Nottingham, the next Tuesday at noon sharp. The meeting would be on the public record, and the intelligence gathered there sent on to Lord Holland, a powerful Whig leader, and other officials, including Samuel Whitbread, a sympathetic Whig MP who had been encouraging the workers to seek reform in Parliament.

Similar efforts were underway in Lancashire. There were meetings in Bolton, in fact; right in Colonel Fletcher's jurisdiction—precisely where they could be infiltrated by spies.

Henson's campaign to drum up support, both among textile workers and business owners, in the Midlands and beyond, would earn him the distinction of becoming what may have been the first paid union representative (though, of course, the committee was not *technically* a union — that would be illegal).

Yet Henson's well-written entreaties, and his advocacy on behalf of the textile workers, started to arouse some suspicions in many corners. There was another sort of letter-writing campaign underway, after all, where the targets were machine-shop owners who soon found themselves with considerably fewer machines. Rumors spread that Henson himself was involved in the Luddite uprisings. Some speculated that he was commanding the troops through the for-ests and the moors to smash the machinery hurtful to commonality by night, and using his careful hand and the Luddite leverage to ply for reform by day.

Some went further still, claiming that Henson was no less than King Ludd himself.

B

February 1812

"The meetings are packed now," the spy wrote, and he might know. B had been busy. He'd been attending those secret gatherings of weavers, spinners, and tradesmen, and penning feverish dispatches to his employer, the magistrate Colonel Fletcher. He anxiously wrote that a representative from Sheffield had shared at a meeting that, in his city, "children not forteen years of age Reals against this present Government."

And if his intel was to be believed, a common theme was developing.

A contact in Ireland told the group, B said, that two hundred thousand Catholics and Protestants stood "ready to rise in either April or May, with religion and sectarian division seemingly having as loose a grip on society than ever before." When those rebels did rise up, they planned on receiving support from Britain's arch military foe: "to receive help with arms and troops from the French."

B's reports, which were tailored to appeal to Fletcher's paranoia, suggested that the Luddites were just the beginning; that a more general uprising of disaffected workingmen was imminent, and that England's lower classes would join up with the Jacobins to overthrow the ruling elites (like him).

There was enough anger and desperation in the region to warrant the suspicion: conditions in Lancashire, just like everywhere else in Northern England, had not improved. On February 22, the *Leeds Mercury* reported that "nearly one sixth part of the whole inhabitants of Liverpool, and nearly one fifth of most of the other large towns in Lancashire, are now in a condition to require the aid of charitable relief.... This is a state of affairs which we believe is without example."

In the throes of the worst poverty in over a decade, the cotton weavers had not gained any traction in their years-long fight for a minimum wage, or even a temporary wage increase. In addition to hosting delegates to discuss Luddite activity or talk of uprisings that B was so eager to elevate, the meetings in the region were also focused on a less cloak-and-dagger matter: a last-ditch effort to secure a fair rate of pay for cotton-weaving work.

Years of petitioning the Home Office, entreaties to slow the rollout of automated power looms, and peaceful organizing had yielded no lasting results or protections for the Lancashire cotton weavers. Finally, in February 1812, they asked the local magistrate to organize a meeting between the weavers' representatives and the manufacturers. A conference was arranged, and the factory and shop owners agreed to a small but significant raise in baseline wages. Just days later, however, the master manufacturers met among themselves and promptly announced they would do nothing of the sort.

"They told me, that they were only making game of us; they said one thing in the presence of the magistrate, and another when they met themselves," the weaver Joseph Sherwin, the man appointed by the magistrate to represent the weavers, later explained. The factory owners had either lied to the workers to save face or changed their minds after meeting to discuss setting prices. The news was not well-received by the weavers. That night, the windows of a factory running the power loom were smashed in Stockport, and shots were fired into a similar operation owned by an entrepreneur named John Goodair.

But the weavers would not give up on the prospect of a peaceful resolution. Next on the agenda at this packed meeting, B wrote, was "petitioning the Prince Regent for reform, and for an end to the war." A meeting "of towns of Manchester" was called for. The organization among members, B seemed to be insisting, was improving. The committee was communicating with the other Luddite cells, and sympathetic groups and allies.

The efforts had grown expansive enough that other spies had reported that a code had been developed for acknowledging fellow Luddites. The code went as follows:

You must raise your right hand over your right eye if there be another Luddite in company he will raise his left hand over his left eye — then you must raise the forefinger of your right hand to the right side of your mouth — the other will raise the little finger of his left hand to the left side of his mouth and will say, "What are you?" The answer, "Determined" — he will say, "What for?" Your answer, "Free Liberty."

Such reports, true or not, strengthened the spies' implication that the machine-breaking campaign threatened to erupt into a greater insurrection. "Not only Nottingham," B wrote, "but also the surrounding area is 'ready' and waiting" for a leader to emerge, and "for a revelation [*sic*] and nothing Else as nothing Else would do."

The misspelling is notable. Many Luddites would probably have preferred a revelation — by the Crown, who would instate a minimum wage and protect their trade — to a revolution.

GEORGE MELLOR VERSUS
JOHN BOOTH

(Luddism versus Machines of Loving Grace)

Winter 1812

It was Saturday afternoon, and colder than usual. Most of the tools at John Wood's shop sat idle. George Mellor and the other croppers gathered around their visitor, a pale young apprentice with delicate features, as he read aloud the news from Nottingham.

The arrival of the newspaper, like the local *Leeds Mercury*, was "the great day of the week" for the men here; it was when news from the outside world reached Huddersfield, where it could be chewed over, debated, spat at. The croppers spent a good chunk of every Saturday, and sometimes Sunday, too, discussing "the exciting intelligence with which the columns of the newspapers were crowded," according to the historian Frank Peel. Many cloth workers were illiterate, so they listened carefully as a designated reader relayed the word.

The apprentice's cheeks flushed with excitement as he spoke. His name was John Booth, the son of the Reverend Booth, a clergyman with the Church of England and a minor celebrity in the area. The reverend had worked as a cropper himself, and kept a room on the premises; his son John had become friends with George and a number of the other croppers here. The dispatch he read aloud now was of great interest to everyone in the room: it detailed the latest in the uprising of disaffected English workingmen who were finally striking back against the bosses. The croppers shouted out approvingly between sentences. And one voice thundered above the others.*

---

* The dialogue and descriptions of the cropping shop come from Frank Peel's oral history; recall that this was not recorded contemporaneously, but was remembered years later and is

"That's right!" George Mellor yelled. "The Nottingham lambs are showing them clod-hopping soldiers a bit of real good sport."

The red on Booth's cheeks and the spike in his voice weren't due only to the drama he was describing, but because change was in the air.

Perhaps following in the footsteps of his father, Booth was drawn first to his studies, and then to radical politics. He was a follower of Robert Owen, the mill owner and collectivist who believed that workers should be organized into factory town cooperatives, and receive more equal pay and a good education. Booth had, according to Peel's account, "thoroughly imbibed the notion that the whole framework of society was out of joint and that the nations and governments of the earth required a thorough remodeling."

It had been less than a generation since the French Revolution swept the monarchy from power on the other side of the English Channel, leaving behind a fragile democracy, power vacuums, a bloodied guillotine, and Napoleon Bonaparte. And while reform clubs had sprung up in Britain in its wake, there had not been much revolutionary fervor in the country since. There had been no show of *force*. There had been nothing like the Luddites.

A report in the *Leeds Mercury,* perhaps the very one that Booth was reading now, noted that "since the commencement of the Luddite system in the neighborhood of Nottingham, 42 lace frames, and 544 plain silk and cotton stocking-frames have been destroyed; the value of the former, when new, is upon the average about [£]60 each; and that of the latter from [£]18 to [£]20 each." That amounted to around £13,400 total in 1811 prices — nearly £9 million, or $12 million in 2020 dollars.

That was serious damage.

Here in his stepfather's shop, George was "strongly moved by the news," so much so that the veins stood out on his temples "like a whip-cord, under the strong excitement which seems to agitate every fiber in his body." He was flanked by his right-hand man, Thomas, who worked alongside him at John Wood's, and his friend Will Thorpe, a peer at a nearby shop.

Recognizing George's charisma and leadership skills, Booth had tried to

---

likely embellished on some counts. It also contains an inaccuracy about the timeline of Robert Owen's rise to prominence.

push his friend into politics, with "indifferent success." George, meanwhile, tried to goad Booth into joining *their* cause.

George had proceeded with the plan he'd hinted at to Ben that night after Christmas. With his magnetic personality and military experience, he'd been marshaling the Huddersfield croppers, weavers, and artisans into a local troop of Luddites. They had undertaken at least two strikes, which Booth would have known about, both from his friendship with the men involved, and, well, by reading the newspaper.

The Nottingham revolt had reaped genuine successes and concessions from manufacturers; if that frightened men like Colonel Fletcher, it embolded men like George Mellor. And the new front for the uprising looked to be Huddersfield, where aggressive entrepreneurs like William Horsfall and William Cartwright were still expanding their factories and filling them with the gig mills and shearing frames that aimed to make the croppers' work redundant. The nexus of the struggle between machinist and machine owner, it seemed, was shifting to the metropolis of discontent.

To Booth, the expanding rebellion hinted at more than worker disputes. It heralded radical possibility.

To George, there was nothing theoretical about the situation.

"Look at Booth," George said,

he'll come here and talk about the evils under which we working men groan, by the hour together, and air all his newfangled notions that he's picked up through the Socialists; he knows very well that machinery is destroying us and nothing but the workhouse will be left for us soon, and yet he's never got farther than talk. Join us lad, join us, you've talked long enough, it's time for action now.... It's true these machines aren't taking Booth's trade out of his fingers, or he'd happen to see things in a different light.

"Come now George," a cropper in the back said, "let's have fair play. It's hardly the thing to set at Booth like that. He's never pretended to be one of us."

Rather than be insulted, Booth was moved by Mellor's appeal. "I quite agree with you, my friends, as some of you well know, respecting the harm you suffer from machinery, but it might be man's chief blessing instead of his curse if society was differently constituted," he said.

If society was differently constituted.

There it was, the machinery question again. The one that animated pub debates and intellectual salons, and would prove ubiquitous in England for much of the nineteenth century. (It would never go away, not really.) "The machine was not an impersonal achievement to those living through the Industrial Revolution; it was an *issue*,"* the historian Maxine Berg wrote. "The question was central to everyday relations between master and workman, but it was also of major theoretical and ideological interest."

The sight of looming factories enraged some onlookers and mystified others. The violent clangs of metal, the hiss of steam — to many, the unearthly soundscapes of industrialization inspired an aesthetic revulsion that dovetailed with their concerns about humans being replaced by machines. Capitalism itself was not yet fully established, much less accepted or celebrated. Whether automation was to be embraced or resisted, extolled or regulated, was hotly debated. George Mellor and Ben Bamforth had gotten into the question as they walked through the Pennine hills that night, and John Booth was wading into it again here in John Wood's idle cropping shop.

"It provoked the village cleric as much as it did the cosmopolitan intellectual; it concerned the politician as much as the workman and employer, the social reformer as much as the scientist and inventor. These groups contended over the costs and benefits of the new technology," Berg wrote; "they speculated on, and then welcomed or dreaded, the changes the machine would bring to social relations.... There was excitement and fear at this unknown force which swept relentlessly onward, casting the old society in its wake.... In the uncertainty of the times it still seemed possible to halt the process of rapid technical change." It was this possibility, of course, that ran hot in George's veins.

Barring that final conviction, these arguments are remarkably similar to the ones we have about automation today. Adam Smith's *Wealth of Nations* held that the division of labor, and attendant technological adoption, would eventually create as many jobs as it displaced; new, streamlined operations would produce more stuff,

---

* Emphasis mine.

creating more demand. So, much of Britain's elite political class was convinced it was all for the good (or at least a powerful instrument with which to bolster their self-interest). Machinery was transforming Britain into the world's dominant economic power, after all, and generating great wealth for elites in the process. If some jobs were lost in the short term, and if work became more precarious for the middle classes, it was unfortunate but unavoidable. That was the price of progress.

Workers like George, Thomas, and Will, meanwhile, recognized when a specific machine posed a risk to their employment; in such cases, the machinery question was clearer-cut, and their answer was "no." Some workers certainly hated the machines as a symbol of oppression, while others recognized their utility in different circumstances. But to many, it was quite obvious that machines were being used to accumulate wealth and power by those who could afford to collect, organize, and run them under one roof. That the owners would profit at the expense of the workers did not seem like bold entrepreneurship or inspired innovation. It seemed immoral.

"Where machinery augmented or facilitated employment within the existing work structure, it was rarely resisted," the historian Adrian Randall wrote. "It was when machinery threatened employment or relocated it into a new factory-based system that it generally encountered hostility." Like when it encountered the Luddites.

Yet then, as now, there was another response to the machinery question: that machines should be embraced, but our politics and societies transformed so that all men and women shared in their benefit. That machines of loving grace will do our work—if the focus is placed on reorienting machine politics, not their pistons.

This was the great project of Booth's hero, Robert Owen, who spent years, at significant personal expense, campaigning to convince England's leaders to reshape industrial communities into collectives patterned after his own factory system. There, working hours were limited and children had access to education; his hope was that everyone might benefit from the machines more equally. Owen had become wealthy by managing, then inheriting, a cotton factory, and he'd recognized that the system stood to accelerate unemployment and inequality (not to mention worker turnover).

"We and other countries are already so placed by [machinery] that a very large number of people are thrown idle greatly against their will and they must be supported or starve," Booth said at the shop. "We know this is so, but are we therefore to conclude that machinery is in itself an evil?"

He pressed his case to the skeptical, murmuring crowd on the shop floor. "Cropping by hand as you now practice it is by no means easy work," Booth said. "We all know that it is very painful for learners to handle the shears until the wrist has become hoofed up. Now look at one of these machines," he said, meaning the gig mill. "Observe how smoothly and how beautifully it works! How perfectly it does for the workman the arduous part of his task. By its aid, as we well know, your task has become chiefly one of care and watchfulness." A statement like this risked insulting the men who prided themselves on their ability to do this work well.

"To say that a machine that can do this for you is in itself an evil is manifestly absurd," Booth continued. "Under proper conditions it would be to you an almost unmixed blessing, but unfortunately the favorable conditions do not exist."

If such technology could be made to benefit society as a whole, it would be a great boon. The dream of automated luxury and a leisure society was being dreamt — and maybe, with this first bona fide surge of productive machinery, with the water frames and power looms and gig mills, even seemed plausible for the first time.

"If the capitalists and the millions of unemployed would abandon large towns and cities for communities of moderate size," Booth said, "and were all employed as economically as such a union would occasion, in agriculture, making and working machinery for the common benefit of the whole, these islands in the course of a few years would present an entirely different aspect, and poverty and starvation become utterly unknown." It was a radical solution to what Owen saw as rapidly escalating inequality and worker exploitation. (It also presaged the thinking of humanist entrepreneurs like Andrew Yang, who gained fame championing a universal basic income, another top-down social solution to the ills of automation and exploitation.)

"If! If! If!" Mellor said, almost yelling now. "What's the use of such sermons as this to starving men?"

This was the unanswerable retort to the machinery question. Such questions come down to who has the power and luxury to answer them in the first place, and who those answers apply to. Does labor-saving machinery benefit society? If you're a machine owner, or profiting from the success of the machine-owning class, it's obviously a lot easier to say yes. But if those technologies are degrading or eliminating your livelihood, the long view is irrelevant. You're worried about your next meal.

"If men would only do as you say, it would be better, we all know," George said. "But they won't. It's all for themselves with the masters. What do they care if a thousand or two of us are pined to death if they can make brass a bit faster?"

This criticism of the promise of automation, articulated by one of its early victims, holds true today. Today the promise is that robotics and algorithms will take on the "dull, dirty, and dangerous jobs" that no human wants to do. Automation *should* be a blessing, we are told. And have been told and told. In fact, the influential economist John Maynard Keynes predicted in 1930 that improvements in machinery and, subsequently, productivity would lead us to a fifteen-hour workweek at most.

If automation could be harnessed for the "common benefit," as Booth argues, that might be a plausible outcome. Instead, it has consistently played out as Mellor has feared; labor-saving technology has accelerated the accumulation of capital among an ever-shrinking pool of elites. Today, you might know them as the 1 percent. Advocates of automation technologies argue we should embrace them and their quality-of-life-improving capacities; critics counter that until there is a mechanism that allows workers to dependably share in the benefits of automation, it will only serve to displace them, make their work more precarious, be a threatened alternative to their services, and used as leverage against them.

"Hold!" protested Booth. "No man can feel for the poor, starving workmen more than I do, but I fear the course adopted by the Luddites to remedy it is not the right one. To confess the truth, I am in a strait. I am afraid your plan will never succeed, and I don't see much chance of reorganizing society on a better and sounder basis at present, workingmen being as a rule almost totally uneducated."

"Feel for them that's starving?" shouted Benjamin Walker, another cropper at John Wood's, and, according to Peel, a "violent" man. "You're either a liar or a coward. How can you feel for them when you won't lift up thy little finger to help 'em?"

"I am no coward, Walker," Booth said, "and again I say I do feel, from the bottom of my heart, for you.... It is hard for people to starve to death in their own houses in a Christian land, but would it not be better to lay these things before the masters and to reason with them, rather than to infuriate them by destroying their machines and —"

"Reason with them," Thorpe interjected, "reason with the stones I say, for their hearts are as hard as flint. What's the use of talking about reasoning with a man when his interest pulls the other way? They'll have these machines if we all clam to death. The only chap that can reason with them is Enoch — that chap is the best reasoner I know of, when he breaks them into a hundred pieces — they understand that!"

Enoch was one of the brothers and business associates of William Horsfall, who built, ironically, both the shearing frames and the giant hammers the Luddites used to smash them. The Luddites had a saying: "Enoch made them, Enoch shall break them."

Booth was at a loss — if there were any other viable trade in the region, he'd suggest they change jobs. "I know not what to advise or what to say," he managed, finally.

"Say you'll join us," Mellor said.

"I will join you," Booth said. "My head tells me you are wrong but my heart is too strong for it. Perhaps the masters, seeing you are driven to desperation, will after all be compelled to take your circumstances into consideration."

There was a murmur in the small crowd. A number of the men walked up to him and, one by one, shook his hand firmly.

"Then, as you have decided to join our cause at last and have avowed yourself a Luddite at heart, I will administer the oath of our society and enroll you as a member," Mellor said.

A couple of croppers moved to block the door; the rest formed a half-circle around Booth and Mellor, who had picked up a New Testament Bible.

Frame-breaking was a serious crime, and so was joining a combination or union, as was about every other activity associated with the Luddites, so secrecy was paramount. That meant members had to take an oath. Oath-taking was against the law, too, punishable by "transportation" to Australia or another British colony. Not only was oath-taking considered treasonous, it was also held deadly serious by those it bound; it was a weight, a shared burden, a brotherhood. Even when threatened by prison or worse, those who'd taken the oath tended to adhere to it — both out of deference to the cause, and out of fear of the fallout from breaking the silence.

The Oath of the Fraternity, as it was known, was used to bind new Luddites to the cause across England. George Mellor stepped forward to administer it. He took Booth's hand.

"What is your name?" Mellor asked.

"John Booth."

"Are you willing to become a member of our society and submit without demur or question to the commands of General Ludd?"

"I am."

"Then say after me: I, John Booth, of my own voluntary will, do declare and solemnly swear that I never will reveal to any person or persons under the canopy of heaven the names of the persons who comprise this secret committee, their proceedings, meetings, places of abode, dress, features, complexion, or anything else that might lead to a discovery of the same, either by word, deed, or sign, under the penalty of being sent out of the world by the first brother who shall meet me, and my name and character blotted out of existence and never to be remembered but with contempt and abhorrence. And I further do swear that I will use my best endeavors to punish by death any traitor or traitors, should any rise up among us, wherever I can find him or them, and though he should fly to the verge of nature I will pursue him with unceasing vengeance. So help me God and bless me, to keep this, my oath, inviolate."

Mellor held out a Bible to Booth, and the young man kissed it. That was that. He gave Booth a printed copy of the oath, and told him to commit it to memory, in case he came to a position where he might administer it to another conscript.

Punish by death any traitors. Name and character blotted out of existence. Unceasing vengeance. Booth was officially "twisted in."

He was a Luddite now.

MARY GODWIN

February 1812

The American vice president had come to dinner again. His life read like an epic poem — full of daring feats, bizarre turns, and ample drama — but he was also broke, which is probably why he kept showing up at the Godwins'. Aaron Burr was almost president of the United States, once; it was a tie, between him and Thomas Jefferson, but the politicians preferred Jefferson, so Burr had to settle for vice president. Then he killed Alexander Hamilton, fellow Founding Father, in a duel and was forced out of political life altogether. He bought up land out west in the US, was accused of being a traitor, and fled to Europe, where he spent his last years wandering, penniless.

Most fourteen-year-olds would probably find the appearance of a former vice president at their dinner table remarkable; Mary Godwin didn't.

There were always folks like that coming around. Bohemians, philosophers, statesmen, artists, writers — revolutionaries — Mary broke bread with all of them. She had a keen interest in poets (like Walter Scott, the sensation of the day) and the literary life. Her father, William Godwin, had helped advertise a series of talks that his acquaintance Samuel Taylor Coleridge was giving that winter, and she attended the last four of them with her little brother, William Jr. She was not impressed. Coleridge's delivery was too rambling and unfocused. Perhaps more of interest was a certain poet in attendance, Lord Byron, whose fame, work, and scandalous aura were all ascendent.

Life had always been like this; even more so, apparently, when her mother was still alive. Mary Wollstonecraft, after all, had attracted the attention of every variety of personality and pundit in her day. Her blistering broadside, 1790's *A Vindication of the Rights of Men*, argued for democracy and equality, and marked the beginnings of her yearslong association with the French Revolution. Her 1792 book, *A Vindication of the Rights of Woman*, turned her into a literary celebrity.

She would later be called a founding figure of feminism, and she drew visitors from around the globe, including Charles de Talleyrand-Périgord, whom she pressed to ensure that girls receive equal education in France's new monarch-free society. Wollstonecraft traveled to Paris just a month before Louis XIV was delivered to the guillotine, and she was there to see his bloody end.

All that was in an entirely different century, though, and young Mary had never known her mother. She did, however, harbor her mother's gifts for writing and political thought. By the early 1810s, she too became interested in radical politics. "Her rapidly growing powers of mind and observation were nourished and developed by the stimulating intellectual atmosphere around her," one biographer wrote; those powers were also being fed by the tumult and darkness of the days, of the unrest and inequalities, "of the anxieties and uncertainties which, like birds of ill-omen, hovered over the household and were never absent for long together."

But these were happy times in the cluttered family home on London's Skinner Street, and she passed them in high spirits. "These few months were, very likely, the brightest which Mary ever passed at home," the biographer wrote. This despite the fact that the household was itself on the verge of being broke.

Her father, William Godwin, was, as usual, in considerable debt, which caused him enough anxiety that it was obvious to visitors — Burr remarked on his host's financial woes in his diary entry about that evening spent at the Godwins'.

But mostly, the lonely and grateful Burr was delighted by the impressive little family.

Inspired by the lecture series, William Jr. decided to give a speech of his own, behind a little pulpit his father had built in the home. Little William stepped up before his audience and delivered his speech "with great gravity and decorum."

The talk's subject was "the influence of government on the character of a people," and it was written by his sister Mary.

RICHARD RYDER

February 14, 1812

The letters kept pouring in; the Home Office secretary, Richard Ryder, could hardly keep pace. So many were written in a familiar tone, alarmed and imploring, whether they were mailed by magistrates or manufacturers. The events they described sounded the same, too, and so did the nature of the complaints.

"The rioters assemble in the night, suddenly break the machines & depart," went one note, from Joseph Radcliffe, an exasperated and anxious magistrate of Huddersfield. "Guns are fired every night in various parts of the country, by the Luddites, to alarm & mislead." The letters all seemed to end with the same asks too—for military support. Send in the cavalry—restore order with a show of force. "Two Troops will be sufficient provided you will oblige us by ordering 100 Infantry," Radcliffe wrote, "as we are under the necessity to quarter men in all the neighboring Villages."

The Home Office had been criticized for being slow to address the unrest in Nottingham, other than to send platoons of soldiers north, to little effect. At the time, townships and cities did not have their own official police forces; magistrates would appoint special constables from the local populace to resolve disputes when necessary. An old law allowed officials to compel men of legal age to serve as those constables. In extreme circumstances, magistrates like Radcliffe would write to the Home Office in London with a request for assistance: a troop of armed dragoons, or authorization of special powers.

Richard Ryder was a dutiful factotum whom Spencer Perceval had appointed to head up the Home Office in 1810. He was a career politician who'd developed a reputation for serving the party and making as few ripples as possible. Now, with his small staff—he had just two dozen employees, including an undersecretary and some administrative workers—Ryder was overwhelmed by the scale and ferocity of the Luddite uprisings.

Letter after letter made it clear that the entrepreneurs, magistrates, and gentry were deeply rattled by this movement of machine breakers. So much so that the machine owners and their allies turned to the Home Office for armed backup whether or not their cities had seen any Luddite disturbances. (Though not all of them: some lords and officials revealed a tacit sympathy for the Luddite cause, through their silence.) But those officials were unsuccessful in capturing many Luddites, for at least one reason: the Luddites were popular. "Perhaps the most serious handicap of all which the authorities suffered was the fact that machine-breaking was carried out against a background of considerable public sympathy for the plight of the workman," the historian Malcolm Thomis wrote.

The handful of Luddites who were arrested in Nottingham at the onset of the uprising were awaiting trial. That hadn't slowed down their colleagues.

The magistrates, lords, and landed gentry feared that Luddite action was building to something more sinister: a conspiracy to topple the Crown. "It was widely said at the time that the Luddites had a further and more fundamental object," as the historian F. O. Darvall put it, "the object of overthrowing the existing system of government."

But what was Ryder supposed to do? Send dragoons to every nervous magistrate who wrote him a letter? He'd already dispatched a huge force to Nottingham, larger than the army currently fighting Napoleon in Spain, in fact. How many troops did these people think he could spare? There was a war on, after all.

The Home Office was broadly in charge of administering *all* domestic affairs, from issuing name changes to registering warrants for patents to overseeing church preferments and so on, *in addition* to supplying domestic military aid to keep the peace. It was an unusual organization that had been established just a few decades before, in 1782, in the wake of the war in America. The Home Office held great power; it was the means through which magistrates, Parliament, and officials could get a line to the king regarding domestic affairs. But it was also understaffed and devoid of a clear remit. As a result, Ryder's reaction to the Luddite uprising was muddled. Frustrated, he sent directives for magistrates to form committees and hire more informants.

Then, on February 14, 1812, Ryder stood to address the House of Lords and proposed a law that would make the consequences for destroying machines much more severe.

In his speech, Ryder painted the Luddites as terrorists, calling them "evil" perpetrators of "a system of riot [that] had existed for the last 3 months, a system bordering almost on insurrection." He did not mention that workers in Nottingham were desperate and impoverished, or that craftsmen had repeatedly petitioned the Crown and his administration for relief. Ryder said he "wouldn't comment on the substance of the dispute," though he did speculate that the trouble was caused by "unfortunate market fluctuations."

When Colonel Eyre, a Nottinghamshire representative, rose to second the motion, he added, with an elite condescension that would soon blossom into common sense, "when those deluded men, the frame breakers, were employed in destroying the machinery, they little thought that they were depriving themselves of the means of earning a livelihood."

In an effort to demonstrate the lengths his office had already gone to stop the Luddites, Ryder explained that by last December, he had ordered 900 cavalry and 1,000 infantry to Nottingham. This, he said, "was a larger force than had ever been found necessary in any period of our history to be employed in the quelling of any local disturbance."

Yet it wasn't enough. After all, some manufacturers had raised their prices, while others gave up the machines altogether. The Luddites' messy rebellion was gaining ground. To break it required a firmer punishment. The laws calling for a decade and a half of exile for breaking machinery "had proved completely insufficient," Ryder said.

He now proposed "that the offense should now be made capital."

Luddites who destroyed manufacturers' frames, he said, should face the death penalty for breaking the machinery of automation.

THE PRINCE

February 1812

The moment had finally arrived: the Prince Regent was officially and completely in control. The restricted regency expired on February 18, and the reins were off. The impatient Tories and expectant Whigs turned to the new ruler of England, the man as good as king.

So did the hundreds of thousands of working men and women who were on the brink of plunging into poverty, or who had already taken the fall. Many voices had joined the Luddites, whose Nottingham contingent had made good on its promise to hold off on further attacks if the prince introduced reforms at the end of the restricted regency. Here at last was a clear chance for Prince George to aid the suffering, to do something other than occupy their lands with troops and militias or to place bounties on the heads of machine breakers. He could abolish the Orders of Council to restore trade, regulate the use of machinery, or provide relief to the hungry through strengthened poor laws. He could at least signal a new direction by giving his old friends a chance to lead the government.

Instead, George announced that he would keep the arch-Tory Spencer Perceval on as prime minister. There would be no change in leadership after all. It was his "final betrayal" of the Whigs.

The inertia of power had proved too great; George may have play-acted as a progressive when he was a libertine prince, but now, with the full powers of king, he didn't care to perturb the machinery of governance his father had set into motion. Though his failure to serve Whig interests had been apparent for some time, his former friends and allies were furious. Some had still held out hope, believing his hands to be tied by circumstance and his father's illness. It was seen not merely as a divisive political move, but an act of irredeemable backstabbing. All but two of his former Whig allies would never forgive him.

"It grieves me to tell you, Sir, that the general astonishment at the step which you have taken is only equalled by a dreadful augury for your future security," wrote Lord Moira, who had served as one of his closest advisors for twenty years. "A disinterested public views with wonder your unqualified and unexplained departure from all those principles which you have so long professed."

Leigh Hunt, the editor of the London *Examiner*, published a missive openly deriding the Prince Regent as "a violator of his word, a libertine over head and ears in debt." Further embracing his rightward lurch and his new party's penchant for punishing dissent, the Prince Regent sued Hunt and his brother, the paper's publisher, for libel. The Hunts spent the next two years in prison.

Days after his retention of the Tories, the Prince Regent defended his actions during a dinner at Carlton House, ripping into the Whig leadership in a speech that Lord Buckingham described in his memoirs as "a furious and unmeasured attack." George's own daughter, the eleven-year-old Princess Charlotte, stood up at the table after the speech. Her father had raised her to admire the Whigs and their politics, and she counted many as close family friends. She began to weep.

"The Princess Charlotte rose to make her first appearance at the Opera, but rose *in tears,* and expressed herself strongly" to one of the Whig leaders in attendance, in defiance of her father, Buckingham wrote. Later, at the theater, Charlotte saw another prominent Whig in the box and stood up in front of the audience and "kissed her hand to him repeatedly, in sight of the whole Opera," in solidarity with the backstabbed Whigs.

It was common at the time for writers and poets to respond to major political events by publishing anonymous poems and lyrics in newspapers. After the spectacle at Carlton House and the opera, a poem entitled "Lines to a Weeping Lady" appeared in the *Morning Chronicle*.

Weep, daughter of a royal line,
A sire's disgrace, a realm's decay;
Ah, happy! if each tear of thine
Could wash a father's fault away

Weep — for thy tears are virtue's tears —
Auspicious to these suffering isles;
And be each drop in future years
Repaid by thy people's smiles.

The anonymous poet, it would later be revealed, was Lord Byron.

LORD BYRON

Inside the idyll of Newstead, the winter had passed in a melancholy fog. The mourning lord was marinating in loss, occasionally diverting himself by harassing his maidservants. He was writing and editing, finalizing the preparations for *Childe Harold,* and tending to estate affairs. He'd decided to sell Newstead.

Outside, Nottingham was racked by chaos, violence, and now an occupying force. The Luddite attacks had continued in every corner of Nottinghamshire and beyond. Militia and infantrymen sat menacingly in towns, cities, and pubs across the Midlands.

Byron had witnessed the effects of the occupation. He saw that the troops were quelling nothing, he saw the misery of the Nottingham workingmen, and he saw that Luddite raids continued undeterred. Byron felt the rage still percolating, and he sympathized with it. He believed it was all coming to a head.

"From all that fell under my own observation," Byron wrote to a friend, "I feel convinced that, if *conciliatory* measures are not very soon adopted, the most unhappy consequences may be apprehended." Now, as a newly engaged member of Parliament, he had a means to sound the alarm. Byron had decided to make his debut speech to the House of Lords a rousing defense of the Luddites.

He explained his reasoning to Lord Holland, a Whig leader who had become Byron's political mentor. (Holland saw in Byron a handsome, increasingly famous figure who might help further the Whigs' cause.)

"I consider the manufacturers as a much injured body of men, sacrificed to the views of certain individuals who have enriched themselves by those practices which have deprived the frame-workers of employment," Byron wrote in a letter to Holland on February 25. "For instance; — by the adoption of a certain kind of frame, one man performs the work of seven — six are thus thrown out of business.

But it is to be observed that the work thus done is far inferior in quality, hardly marketable at home, and hurried over with a view to exportation."

Byron continued:

Surely, my Lord, however we may rejoice in any improvement in the arts which may be beneficial to mankind, we must not allow mankind to be sacrificed to improvements in mechanism. The maintenance and well-doing of the industrious poor is an object of greater consequence to the community than the enrichment of a few monopolists by any improvement in the implements of trade, which deprives the workman of his bread, and renders the laborer "unworthy of his hire."

My own motive for opposing the bill is founded on its palpable injustice, and its certain inefficacy. I have seen the state of these miserable men, and it is a disgrace to a civilized country. Their excesses may be condemned, but cannot be subject of wonder. The effect of the present bill would be to drive them into actual rebellion.

The few words I shall venture to offer on Thursday will be founded upon these opinions formed from my own observations on the spot. By previous inquiry, I am convinced these men would have been restored to employment, and the county to tranquility. It is, perhaps, not yet too late, and is surely worth the trial.

A defense of the frame-breakers would be an unpopular but provocative position inside Parliament, which was surely part of Byron's calculation—he rarely missed an opportunity to make himself the topic of conversation.

Byron recognized the limits of acceptable empathy, however, and was sure to add in his letter, "Condemning, as every one must condemn, the conduct of these wretches, I believe in the existence of grievances which call rather for pity than punishment." Even as he readied himself to rally for their cause, Byron still regarded them as "pitiable wretches." Holland thought it was the right tone to strike — rational and pragmatic, sympathetic yet forceful — even if their chances of stopping the bill's passage were slim. Most lords in the House held little sympathy for the victims of automation.

Those lords recognized that industrial automation and factory-scale production offered a tremendous opportunity to corner new markets, and to further consolidate the nation's economic power. Those systems were well on the way to ensuring that Britain would be the uncontested world leader in textile production and exports. With the new machines, Britain could import cotton picked in America, process it, and sell the manufactured goods cheaply around the world. Mechanized factories meant big business and geopolitical might. Artisanal weaving shops may have been a cornerstone of a local economy, but they did not increase British power internationally, at a moment when it was hotly contested by Napoleon's army.

Britain's political system revolved around an oligarchy of landed aristocrats, where the lords had long held power, as the historian Eric Hobsbawm explained. But as soon as the factory owners "brought industry to their lands, it gave manufacturers political power. By 1750, manufacturers had political power over the merchants." At the same time, Britain was becoming the "workshop of the world." And it was with the backing of the state that the innovators and entrepreneurs moved to overcome the workers' opposition: "It was unreformed Parliament in its most ferociously conservative period, which introduced full laissez-faire into the relations between employer and worker."

This was precisely what was playing out with the artisans, weavers, and croppers now — over the first decade of the new century, the weavers' trade group had pushed Parliament to uphold regulations to restrict the use of machines in the cloth trade. It had refused. The weavers had politely petitioned for aid when two years of horrible harvests on top of the automated trend left them hungry and destitute. Parliament had refused that too.

Why? Members of Parliament were enjoying a period of unparalleled wealth and power. Most were so far removed from the citizens they governed that they could not plausibly relate to them. (Consider the US Congress in the twenty-first century, when it is unusual for a member *not* to be a millionaire. And consider their stance toward the tech sector since its boom in the 1990s. "National politicians used a remarkably light hand in regulating the data-gathering behaviors of an industry whose technologies they only vaguely understood, but whose hockey-stick growth boosted the domestic economy," historian Margaret O'Mara wrote. "When the government-built Internet finally opened up to commercial activity at the start of the 1990s, both Democratic and Republican politicians agreed that regulation should be minimal.")

"The British aristocracy and gentry were very little affected by industrialization, except for the better," Hobsbawm explained. "Their rents swelled with demand for farm produce, the expansion of cities (whose soil they owned) and mines, forges, and railways (which were situated on their estates).... Their social predominance remained untouched, their political power in the countryside complete." The first fifty years of mass industrialization were a "golden era for the landed and titled Briton." The trends rocking Huddersfield and Nottingham were benefiting the upper class tremendously.

Even if Byron could not empathize fully with the impoverished workers, he saw how unjust their circumstances were. Most ministers in London preferred to attribute the Luddites' actions to laziness, ignorance, or a fantastical thirst for chaos rather than to economic circumstance, a failure of policy, or technology-driven insecurity. Ever the provocateur, Byron resolved to make the frame-breakers' cause his own.

"P.S.," he wrote, puckishly appending his letter to Lord Holland, "I am a little apprehensive that your Lordship will think me too lenient toward these men, and half a *frame-breaker myself.*"

NED LUDD

Winter 1812

We lament to state, that the spirit of mischief and violence amongst the workmen continues yet unsubdued. Many letters, under the fictitious signature of "Edward Ludd," have been sent within the last week to different workmen, threatening them with death, at the expiration of a given time, if they do not desist from making certain descriptions of work, which the writers find obnoxious to the trade.

This morning a most daring outrage was committed by a party of frame-breakers, in West Street, Broad Lane Paddock, in this town: Between five and six o'clock, just after the watch had gone off, a party of men entered the dwelling house of Mr. Harvey...by means of a ladder, reared against the window of the second story, which they forced open, and drawing the ladder up after them, proceeded into the workshop, where there were seven frames, which they immediately began to destroy.

The Bill for inflicting the penalty of death on persons who may hereafter be found guilty of destroying stocking or lace frames, was read a second time in the House of Commons on Wednesday, and ordered for a third reading yesterday, when it would be sent up to the Lords. We sincerely hope that this act of the Legislature will have its due effect on the minds of the deluded men, who have so long set law and decency at defiance, by deterring them for the future from the commission of enormities, which, in their consequences, have amounted nearly to High Treason.

— Nottingham Journal, *Saturday, February 22, 1812*

Some newspapers were sympathetic to the Luddite cause. Conservative papers, like the *Nottingham Journal*, struck a tone of outrage that aligned with the

Crown. They tended to agree, however, that what had begun as an outbreak of attacks on the instruments of mass production had erupted into a full-blown upheaval. Luddites were rising in Nottingham, Leicester, Manchester, the West Riding, Derby. They were everywhere.

It may have been unclear what *exactly* the Luddites were to an ordinary newspaper reader in England, in a town unaffected by the "disturbances." A movement? A sect? Some kind of enraged trade guild? They were described by the authorities as "rioters" or "depredators," or even as terrorists. But if that were the case, why were so many of their actions accompanied by cheers from the crowds who looked on as a factory was raided or a machine was smashed? Why was no one — truly, it was incredible — *no one* giving the authorities information about these acts, which had by now accumulated thousands of witnesses? It helped, surely, that machinery, not people, were the Luddites' targets.

By exploiting the media technologies of the day — printing handbills and declarations, writing threatening letters, adopting a meme for a figurehead — the Luddites were able to make their uprising seem limitless. Think of modern decentralized movements animated more by an idea than a leader: Occupy Wall Street, Black Lives Matter. That the Luddites were faceless and monolithic helped excite people about their program; they really were a Robin Hood phenomenon. They also offered a mechanism for joining in, if you shared their grievances. After all, Luddism was based not on a specific platform or a list of demands (though there certainly *were* demands), or a charismatic leader. Wherever you were, General Ludd was there, an embodiment of the message: if the machinery of the rich and the powerful left you hungry and poor, you could choose to destroy it.

The openness of this framework allowed different regions with different grievances and politics to accommodate Ned Ludd as their crusading avatar. If you lived in Nottingham, and speculators were charging oppressively steep rents on the devices you relied on for work, forcing you to take your pay in truck, and using wide frames to ramp up production of cheap goods you had to match prices with, you could turn to General Ludd. In Huddersfield and the West Riding, if local factory owners and entrepreneurs were automating your job in the middle of an economic depression, King Ludd might help you dole out justice. In Manchester and Lancashire, where poor working conditions, power looms, and the imposition of large factories pressed on a hungry population, Ned Ludd could ride in with his hammer, threatening letters, and army of redressers. He was an able avatar of vengeance for each of the causes.

The message did not just appeal to weavers and cloth dressers and knitters, which perhaps worried the magistrates and the Home Office most. Artisans, hat-makers, shoemakers, bricklayers, small shop owners, and farmers joined the cause. So did coal miners and railroad workers, whose industries were on the *rise* due in part to the technology and automation — because the Luddite movement was not about technology; it was about workers' rights. Luddism started as a tactical strike against the technologies of control, but had exploded into a greater expression of the rage against a system where the privileged few with access to the right levers could lift themselves up at the expense of the many.

"It was a matter of surprise and distress to the authorities that such workers, not directly or personally affected by the new machinery, should have joined in the riots," one early historian of Luddism, F. O. Darvall, wrote. "Observers might well ask…why the colliers [coal miners] and the navvies [canal and railroad workers] should join the cotton weavers in breaking steam looms. Did not the looms burn coal, and consist of iron, which created a further demand for coal?" For these workers to join the uprising was, to the authorities, "astonishing and irrational."

In the case of the framework knitters, the obvious way to protest mounting inequality was to destroy the specific machine that was speeding the imbalance of wealth. The symbolism of that target would have been widely appealing, especially in a region where stealing from the rich to give to the poor was a cultural virtue, encoded in myth. To that end, the Luddites raised a subscription from their ranks to help support its poorest members.

Ned Ludd, Robin Hood — say them out loud, back to back — both were larger than life. As such, folk songs and ballads about Luddites began making the rounds. Written lyrics to "General Ludd's Triumph," sung to the tune of "Poor Jack," date to January 1812; it was likely one of the more popular Luddite songs of the day.

Chant no more your old rhymes about bold Robin Hood,
His feats I but little admire
I will sing the Achievements of General Ludd

Now the Hero of Nottinghamshire
Brave Ludd was to measures of violence unused
Till his sufferings became so severe
That at last to defend his own Interest he rous'd
And for the great work did prepare

The guilty may fear, but no vengeance he aims
At the honest man's life or Estate
His wrath is entirely confined to wide frames
And to those that old prices abate
These Engines of mischief were sentenced to die
By unanimous vote of the Trade
And Ludd who can all opposition defy
Was the grand Executioner made.

It's a raucous, spirited folk song, a battle hymn, even, that culminates in a call for a job guarantee and wage controls, and a day when machinery won't threaten anyone's livelihood. It's a call not for the destruction of all machines, but for a world where machines are not used by the "haughty" to oppress the "humble."

Not all writers who shared such a vision would remain anonymous. The man who was about to become the most recognized poet in the world was about to take up the Luddite cause, too.

LORD BYRON

February 27, 1812

"As a person in some degree connected with the suffering county," Byron began, delivering the lines in a booming, practiced tone, "I must claim some portion of your Lordships' indulgence, whilst I offer a few observations on a question in which I confess myself deeply interested."

The House of Lords was packed for the debut of the poet lord. Speaking in what Byron later described as "very violent sentences" with "a sort of modest impudence," he embraced his penchant for theatricality; his thunderous defense of the Luddites was impassioned and performative.

"To enter into any detail of these riots would be superfluous," he continued; they were by now well-known throughout England.

> The House is already aware that every outrage short of actual bloodshed has been perpetrated, and that the proprietors of the frames obnoxious to the rioters, and all persons supposed to be connected with them, have been liable to insult and violence. During the short time I recently passed in Notts, not twelve hours elapsed without some fresh act of violence; and, on the day I left the county, I was informed that forty frames had been broken the preceding evening as usual, without resistance and without detection.

Byron had been there, in the midst of the violence engulfing his hometown, which rendered it that much more dramatic when he pivoted to declare his sympathy for the men destroying the machinery.

> Whilst these outrages must be admitted to exist to an alarming extent, it cannot be denied that they have arisen from circumstances

of the most unparalleled distress [and that] nothing but absolute want could have driven a large and once honest and industrious body of the people into the commission of excesses so hazardous to themselves, their families, and the community.

"These men," Byron said, "never destroyed their looms till they were become useless, worse than useless; till they were become actual impediments to their exertions in obtaining their daily bread."

To the gathered lords, now either entertained or irritated, Byron took on an even less sympathetic tack. He mocked the massive military force that occupied Nottingham, drawing a scathing portrait of the Crown's soldiers as a hapless throng, always one step behind the Luddites.

Such marchings and countermarchings! from Nottingham to Bulnell — from Bulnell to Bareford — from Bareford to Mansfield! [Byron sneered,] and, when at length, the detachments arrived at their destination, in all "the pride, pomp, and circumstance of glorious war," they came just in time to witness the mischief which had been done, and ascertain the escape of the perpetrators; — to collect the *spolia opima*, in the fragments of broken frames, and return to their quarters amidst the derision of old women, and the hootings of children.

It was a long and blistering speech. Byron reminded the lords that even while they looked down on the Luddites as a mindless mob, "it is the mob that labor in your fields and serve in your houses, that man your navy, and recruit your army." That mob was the force "that have enabled you to defy all the world, and can also defy you when neglect and calamity have driven them to despair."

He warned that making the punishment more severe would only push the Luddites further into secrecy. The solution was not more violence, he argued, but addressing the root causes of the suffering. But Byron was clearly not as interested in diplomacy as he was in creating spectacle — he soon turned his scorn directly at the Lords proposing the law, and all those who might support more aggressively punitive measures.

"Is there not blood enough upon your penal code that more must be poured forth to ascend to heaven and testify against you?" Byron intoned, then scoffed.

> How will you carry this bill into effect? Can you commit a whole county to their own prisons? Will you erect a gibbet* in every field, and hang up men like scarecrows? Or will you proceed...by decimation; place the country under martial law; depopulate and lay waste all around you; and restore Sherwood Forest as an acceptable gift to the Crown in its former condition of a royal chase, and an asylum for outlaws? Are these the remedies for a starving and desperate populace? Will the famished wretch who has braved your bayonets be appalled by your gibbets? When death is a relief, and the only relief it appears that you will afford him, will he be dragooned into tranquility? Will that which could not be effected by your grenadiers, be accomplished by your executioners?

Can you threaten the man with no future by snuffing out his future? Byron took aim at this dark absurdity, of expanding the scope of capital criminality to ensnare the "wretched mechanic who is famished into guilt." Byron brought his speech to a crescendo, invoking ancient Greece before striking for the heart.

> The framers of such a bill must be content to inherit the honors of that Athenian lawgiver whose edicts were said to be written, not in ink, but in blood. But suppose it past, — suppose one of these men, as I have seen them meagre with famine, sullen with despair, careless of a life which your lordships are perhaps about to value at something less than the price of a stocking-frame; suppose this man surrounded by those children for whom he is unable to procure bread at the hazard of his existence, about to be torn forever from a family which he lately supported in peaceful industry, and which it is not his fault that he can no longer so support; suppose this man — and there are ten thousand such from whom you may select your

---

* A gibbet is any instrument of public execution, like a hanging gallows.

victims, — dragged into court to be tried for this new offense, by this new law, — still there are two things wanting to convict and condemn him, and these are, in my opinion, twelve butchers for a jury, and a Jeffries for a judge!

The last reference was to Baron George Jeffreys, who, as Lord Chancellor a hundred years before, had sentenced over seven hundred men of dubious degrees of guilt to the gallows for treason. Byron's message was clear: in seeking the death penalty for the Luddites, the state was growing paranoid, irrational, and bloodthirsty.

The speech was carried in England's newspapers — a signal, issued from the House of Lords to the Luddites and to the working poor, that someone with influence understood their crisis, and could take aim at the state's response. The John Booths and George Mellors the region over, across the Midlands and the West Riding, would have read those violent sentences aloud.

It would have been cause for excitement, granting the Luddites a feeling of solidarity that reached out beyond the cropping shops and cottages and factories, even if the audiences had never heard of Byron. Finally, someone with power understood that there was injustice being done here, and it was not being done *by* them.

But the state had dealt with organized dissent in precisely the manner Byron decried for centuries — with violence. And it was not about to change course on account of a florid speech or some splintered machines.

NED LUDD

February 27, 1812

William Hinchcliffe woke to the sound of a gunshot outside his home. It was the middle of the night, somewhere between 1 and 2 a.m., he would later tell the authorities, and the blast was so loud he almost fell out of bed. Now he was dazed and in shock. When Hinchcliffe peered out the window, he could see the dark shapes of a crowd of men. There were about fifty of them, he guessed. In the hazy moonlight, he saw that their faces were blackened, but could make out little else.

Hinchcliffe was a manufacturer, a master dresser and owner of a midsize cropping shop. Not unlike Ben's father, the master weaver William Bamforth, Hinchcliffe was torn between the customs of his trade—and the men he employed—and its future, which seemed to belong to the machines. Unlike Bamforth, he opted in favor of automation. He installed new shearing frames in his shop in Leymoor, just outside Huddersfield, and tonight he knew precisely what the Luddites had come for.

The banging began. A hammer to the door of his workshop—again, then again. He heard the door break and give way, and, in quick succession, the smashing of what he already knew was his shearing machinery.

Hinchcliffe ran out of his bedroom, rounded up his family, and ushered them into the parlor, where he closed the door as tightly as he could, trying to keep quiet.

The pounding started again. This time, it was a hammer to his front door. He heard that break, too, and the sound of the men spilling into his house.

"Where is he?" one voice called out. "Bring him out, let's kill him!"

The door handle to the parlor that Hinchcliffe and his family were currently hiding in started to twist.

"Let him alone," another shouted, and the pulling stopped. Hinchcliffe heard the footsteps of yet another man approaching the other side of the door, and then: "If you cause us to come again upon this subject, we'll take your life."

There was a voice from outside the house, urging the men to come out, followed by a command: "Fall in."

Huddled in the parlor, Hinchcliffe listened to the sound of a hundred boots stamping outside as they moved into formation.

After a long spell of silence, the shaken entrepreneur cracked open the door, was satisfied the Luddites were really gone, and set about surveying the destruction.

WILLIAM HORSFALL

February 27, 1812

Two days after the attack on Hinchcliffe's factory, a poster appeared on the walls of shops across the West Riding.

The handbill announcing the Committee for Suppressing the Outrages, a coalition of business owners working to combat the Luddites.

AT A GENERAL MEETING
OF THE
MAGISTRATES.
MERCHANTS, MANUFACTURERS,
AND INHABITANTS,
OF HUDDERSFIELD, AND ITS VICINITY,
THE FOLLOWING RESOLUTIONS WERE AGREED TO:

IT appears to this Meeting, that a violent and determined Spirit of Insubordination has gained much Ground amongst the Workmen employed in various Trades and Manufactures, and particularly amongst the Shearmen, and that the same is organized in a Manner, not only Alarming to Trade in general, but to the Peaceable Inhabitants of this Town and Neighborhood in particular.

THAT We cannot sufficiently deprecate all Attempts to limit the ingenuity of our Artificers, the employment of our Capital, and to prescribe the mode in which the different operations of our Trade shall be conducted, and particularly those now making in this Neighborhood, for the Destruction of the Machinery used in the finishing of Woollen Cloth.

This poster announced the formation of the Committee for Suppressing the Outrages in Huddersfield. The day of the Hinchcliffe attack, the incensed local magistrate, Joseph Radcliffe, called an emergency meeting at the George Inn. The manufacturers who'd been targeted by the Luddites, including Hinchcliffe, as well as William Horsfall, his brother Joseph, and the city's most prominent entrepreneurs and gentry, had gathered to debate how to respond to the campaign to destroy their machinery.

The owners were not about to sit idly by and watch their investments be dismantled. They framed the attacks as "attempts to limit the ingenuity of our Artificers," referring to themselves, infringing on their ability to "employ Capital" how they saw fit, and on the Workmen's attempts to "prescribe the mode in which the different operations of our Trade shall be conducted." They condemned, above all, the infringement on their freedom to innovate, to invest their money, and to conduct their business operations. They were outraged at the notion that they might be compelled, by force, to accommodate the requests of the working people whom their investments were affecting.

The men who formed the backbone of the committee had been active in lobbying Parliament to repeal regulations on apprentices and machinery; most of their names had been signed on a petition supporting an aggressively pro-business Tory candidate, Henry Lascelles, in an 1807 election that hinged in large part on the machinery question. (Lascelles lost.)

More recently, Joseph Radcliffe had written the Home Office a flurry of letters, one for every new Luddite attack. He'd beseeched Richard Ryder to send more troops to Huddersfield. Now that the Luddites had struck Hinchcliffe, a bigger business here, he wanted the Crown to administer a thunderous local response.

Born Joseph Pickford, Radcliffe had always been ambitious and striving, and when his unmarried uncle William Radcliffe had died, he saw an opportunity. The relative had left his considerable estate in the West Riding to Pickford, provided he changed his name and took up residence. Pickford eagerly obliged.

Now in middle age, Joseph Radcliffe was "short-legged and portly, with a Punch-like face and a long nose set between white muttonchop whiskers." He had happily made himself at home at the provincial estate, established himself in local politics, and managed to get himself appointed the local magistrate. He wasn't in the tech sector; his investments were in agriculture, as enclosure had dramatically increased his land holdings. (His lands were so large that they were known locally as the Radcliffe Plantation.) But he quickly allied himself with the tech founders and factory owners in town, understanding them to be the newly emergent locus of power, and even helped them secure labor. As magistrate, he personally located orphans like Robert Blincoe and sent them to work in the mills.

He had set his sights on the next rung of power: ennoblement. Radcliffe started thinking about how he might wrangle himself a baronetcy, figuring perhaps that such a dramatic, though achievable, elevation in class would require an act of great distinction. Proving instrumental in putting down a worker uprising that was loathed and feared by industry and the Crown would fit the bill.

So, now gathered into his committee were Horsfall and his brother, as well as

William Cartwright, the operator of the imposing Rawfolds Mill; Radcliffe; and men like Joseph Hirst and Francis Vickerman, manufacturers and merchants who all held an urgent self-interest in stamping out the wave of anti-entrepreneurial sabotage. Many had received threatening letters from General Ludd instructing them to dismantle their machinery. Cartwright had taken to sleeping in his factory every night, and most of the men had begun fortifying their operations with defenses.

Some of the committee men were more aggressive and outspoken than others — Horsfall made no bones about his animosity toward the Luddites, and the Reverend Hammond Roberson, vicar of Liversedge, wrote to Radcliffe that he had "good reason to think we should have had a visit from these Croppers if we had not been prepared — as we are. I almost wish they would make an attempt. I think We should give a good account of them."

Not only did it seem like there was an imminent threat to many of their operations, but there was a risk that the Luddites might win the war of popular opinion in their own ranks, too. Many of the smaller manufacturers were already succumbing to the workers' demands, and some openly voiced sympathy for the frame-breakers' cause. Recall that not all the entrepreneurs and shop owners were so sure that automation was moral, but recent events were forcing them to pick sides. At least one large manufacturer had closed his shop out of fear and empathy for the Luddites. William Horsfall called him a coward and refused to speak to him.

Joseph Horsfall was appointed to run the Committee to Suppress the Outrages, and Radcliffe would liaise with the Home Office in London and marshal the local security detail. They denounced the destructive acts that "evidently proceed from an illegal Combination of Desperate Men." Their document called on anyone with information about "who those Men might be" to come forward, and offered a reward of 100 guineas, roughly $10,000 today, to anyone who did. Finally, the committee started a subscription fund, each contributing to raise bounties, militiamen, and support for their operations. That subscription would exceed £2,400, or over $230,000 in 2023 dollars, for the industrialists to purchase defenses against workmen who could not earn enough money to feed their families.

GEORGE MELLOR AND NED LUDD

March 1812

One of the men wore a grotesque mask cut from a bloody calfskin. It stuck out, even among the elaborate camouflage of those assembled in the night's raiding party. Others were draped in black, with stockings pulled over their faces, or darkened with streaks of coal. All were members of the Luddites of the West Riding.

Will Thorpe was among them. So was Joshua Schofield, who kept watch, and George Lodge, who wore a sash and answered to Sergeant. Thomas Smith brought his hatchet. The red and white calfskin mask belonged to John Walker, "one of the most desperate of the Luddites," whose large family was on the brink of starvation. George Mellor may have cut the most imposing figure of all: a pistol in one hand and a scythe blade in the other.

The rain fell so heavily that one of the groups had to turn back, leaving a small band of sixteen men to push through the muddy, pulsing wet thick of night toward the target.

Sometime between one and two o'clock in the morning, the Luddites arrived at the house of master cloth dresser Samuel Swallow. They "knocked violently" and, when Swallow opened the door, demanded to be let in. One, in a blackened face, guarded Swallow "with a pistol in one hand and a candle in the other," while the others rushed in and quickly did the work: they smashed four pairs of shears, two shear-frames, and a brushing machine to pieces.

Next the squad moved to William Cotton's workshop, where they tied up the foreman, John Sykes, and his wife while they smashed ten shears and a brushing machine. For the "business," they used the great hammer Enoch.

They left Cotton with a threat: if he started using the machinery again, they'd return and blow the place up.

Word spread that the West Riding Luddites would often, peculiarly, extend courtesies when entering and exiting and let sympathetic machine owners off with a warning, if they promised to stop using the machines.

In a deposition later, Swallow recalled that one of the assailants wore a "most terrible mask," but that they politely bid him "good morning" as they left, and suggested he lock the door behind them.

LORD BYRON

March 1812

On March 2, a poem appeared in the *Morning Chronicle* and the *Nottingham Review*. No author was listed. The Frame Work Bill was on the verge of becoming law, which meant that breaking a machine was about to become a capital crime. Readers of the papers would have recognized Richard Ryder and his cosponsor, Lord Eldon, as the two parties most responsible for the now-imminent passage of the bill that many, including the city officials in Nottingham, had urged the state not to pass.

Its tone was pitch-black satire:

Oh! well done Lord El — n! and better R — er!
Britannia must prosper with Counsels like yours;
Hawkesbury, Harrowby, helps you to guide her,
Whose remedy only must kill ere it cures!

Those villains, the Weavers, are all grown refractory,
Asking some succor for charity's sake;
So hang them in clusters, round each Manufactory,
That will at once put an end to mistake.

The rascals, perhaps, may betake them to robbing,
The dogs to be sure have got nothing to eat —
So if we can hang them for breaking a bobbin,
'Twill save all the Government's money and meat.

Men are more easily made than Machinery,
Stockings will fetch higher than lives;

Gibbets on Sherwood will heighten the scenery,
Shewing how Commerce, how Liberty thrives.

Some folks for certain have thought it was shocking,
When famine appeals, and when poverty groans;
That life should be valued at less than a stocking,
And breaking of frames, lead to breaking of bones.

If it should prove so, I trust by this token,
(And who will refuse to partake in the hope,)
That the —— of the fools, may be first to be broken,
Who when ask'd for a remedy, sent down a rope.

It may not have been difficult to guess the identity of the author if a reader happened to be aware that a young, bombastic poet recently took the floor of the House of Lords to make many of these very points, in speech that only nearly rhymed.

Byron's "violent sentences" were not enough to slow the passage of the bill, as delighted as he initially was with his performance. "[I] abused every thing and every body, and put the Ld. Chancellor very much out of humor," he had bragged to a friend. He left the hall "glowing with success."

Lord Holland was less impressed. Byron's dramatic delivery left the impression that he was performing, not sincerely advocating for his position. "His speech was full of fancy, wit, and invective, but not exempt from affectation nor well reasoned, nor at all suited to our common notions of Parliamentary eloquence," Holland wrote to a colleague. Its floridity may have made it easier for Parliamentarians to dismiss him. His delivery, which came across in a "sing-song" style, drew further criticism. After his second speech, in defense of Catholic emancipation, was similarly dismissed, he considered his debut in politics something of a failure.

Byron really was angry. Publishing that anonymous poem taking aim at his peers in the House was not a calculated political maneuver. He was a Romantic, after all; his instinct was to throw in with the downtrodden, those fighting for their very right to exist. Byron would later say that he never had much interest in politics, and that Parliament and its squabbling repulsed him. He was bored by the proceedings: he ultimately attended fifteen sessions in total, and spoke at three of them. He glazed over at the speeches, and hated the naked favor-trading and deal-making. He later described Parliament as "the sickbed of the nation," remarking in a letter to his friend Leigh Hunt that "if you knew what a hopeless & lethargic den of dullness & drawling our hospital is during a debate & what a mass of corruption in its patients…you would wonder not that I very seldom speak but that I ever attempted it."

Still, he would not stew long. The day after his "Ode" mocked Ryder and the Frame Work Bill legislators, John Murray published the first part of *Childe Harold's Pilgrimage*, the epic poem Byron had been reworking since he'd returned to England.

To his surprise, *Childe Harold* was an immediate sensation. The first printings quickly sold out, and he was soon deluged with letters from poets and readers commending him on his achievement. "I awoke one morning to find myself famous," Byron remarked.

Ryder and the Tories, meanwhile, were undeterred by Byron's sneers and protestations. On March 5, two days after *Childe Harold* was published, the Frame Work Bill passed both houses of Parliament.

Anyone convicted of dismantling a machine could now be put to death.

GEORGE MELLOR AND NED LUDD

March 1812

George Mellor's raiding party answered the new statute with Enoch's hammer.

After Swallow and Cotton, the Luddites turned to the shop of George Roberts of Nether Moor, where they destroyed two frames and threatened to blow out his brains if he set up his machines again. They visited John Garner in Honley, and smashed two shearing frames and seven shears, then Clement Dyson of Lockwood, where they shattered "two frames, one brushing machine, seven shears, tubs and other utensils."

On March 9, a prominent entrepreneur in the West Riding received a blunt, detailed, and threatening letter from General Ludd. "Information has just been given in that you are a holder of those detestable Shearing Frames," it read, "and I was desired by my Men to write to you and give you fair Warning to pull them down." If the machines weren't taken out of use by the end of next week, this Ludd wrote, "I will detach one of my Lieutenants with at least 300 Men to destroy them." He threatened to burn the whole factory "down to Ashes," adding that if anyone fired on his men, "they have orders to murder you, & burn all your Housing." Clearly incensed, this letter-writing Ludd went to greater lengths to explain himself than most. The views of "me and my Men have been much misrepresented," he wrote, and our grievances "are not to be made sport of."

> The immediate Cause of us beginning when we did was that Rascally letter of the Prince Regents to Lords Grey & Grenville, which left us no hopes of any Change for the better, by his falling in with that Damn'd set of Rogues, Percival & Co to whom we attribute all the Miseries of our Country. But we hope for assistance from the French Emperor in shaking off the Yoke of the Rottenest, Wickedest and most Tyranious Government that ever existed . . . and we will be governed by a just Republic.

There were, he insisted, 2,782 "Sworn Heroes" in Huddersfield, and twice as many in Leeds, who planned "either to redress their Grievances or gloriously perish in the attempt." Workers from Manchester to Sheffield and from "all the Cotton Country" were prepared to join them, he said, and so were the weavers in Scotland and Ireland.

> We will never lay down our Arms till The House of Commons passes an Act to put down all Machinery hurtful to Commonality, and repeal that to hang Frame Breakers. But We, We petition no more. That won't do — fighting must.

> Signed by the General of the Army of
> Redressers
> Ned Ludd Clerk

The West Riding Luddites had build a formidable reputation taking on that "machinery hurtful to commonality," but they'd done so by mostly targeting smaller manufacturers, relatively safe targets. This would change soon. They had been preparing for a larger attack ever since they'd learned that Horsfall, Radcliffe, Cartwright, and the factory bosses had formed the Committee to Suppress the Outrages and put a bounty on their heads.

A few days before the ides of March, another letter was thrown onto the premises of a member of that committee: the factory owner Francis Vickerman. "We give you Notice," it read, threatening to destroy his automated machinery "if it were not taken down."

Vickerman was a larger manufacturer, and, as part of the committee, he'd secured two troops of armed dragoons to guard his factory. They arrived promptly at 9 p.m. and stayed until morning. When the Luddites came for him, it would have to be with a plan, and one executed with "military precision."

George Mellor asked three young men, a cousin of Will Thorpe's and his friends, to join him at the pub. By the time the night was through, George had convinced one of them to stand watch at a nearby bell tower, and to give it a ring when the Crown's military was nearing Vickerman's shop. The other two were to be posted near where the dragoons gathered; their job was to fire a pistol shot when they began to march.

At eight o'clock, fifty Luddites gathered in a field near the factory, and by eight thirty they were at Vickerman's doorstep. The rebels fired a single gunshot and burst into the home. Vickerman came out to meet them.

"Ned Ludd of Nottingham has ordered me to break this clock," one of the Luddites said, and smashed the ornament with the muzzle of his gun. Vickerman turned and ran upstairs.

The Luddites quickly moved to the doors to his factory and broke them open with Enoch's hammer. Once inside, they destroyed just about everything in sight. George Mellor himself took up his scythe and sword and smashed every pane of glass in the shop.

The church bells began to ring.

"Out! Out!" someone yelled.

Someone else threw a piece of cloth on the stove in a slipshod attempt to start a fire. Others fired guns into the factory. A shot rang out in the distance. The dragoons were close. The Luddites formed ranks and filed off. They were gone by the time the first militiaman made it to Vickerman's.

In less than half an hour, they had reduced the place to ruin.

This kind of destruction was still the exception, not the rule. In previous encounters, the Luddites were surgical and more restrained. The violence visited on Vickerman's factory may have been an effort to punish him for joining the Committee to Suppress the Outrages. It may have been because he was particulalry disliked by his workers — or it could have been an escalation. A message that not even the more powerful entrepreneurs, who most vocally opposed the Luddites, were safe from their campaign.

It may even have been George letting his anger get the better of him. He'd proven a restrained, methodical, and forceful leader so far, and his dedication, zeal, and charisma were infectious. George was a natural. But he still had a temper, and his conviction that the Luddites were enacting moral justice was deep.

In *Ben o' Bill's*, Ben, who had joined the Luddites too, credits his cousin George as "always best and brightest in action.... He was obeyed, without question, by many a hundred men; all bound together by a solemn oath, who had implicit trust in him."

That same trust propelled actions like the decisive attack on Vickerman, helping the Luddites rack up victories and grow in confidence. "The military and the special constables were only our sport," Ben said. "They were never any serious hindrance, at first, to anything we took in hand."

What's more, the campaign was proving successful in Huddersfield, as it had in Nottingham. "The mill owners were in fear for their machines, and would rather any night pay than fight," Ben recalled. "And for the great mass of the people, those who had to work for their living, they believed in General Ludd. In some way they could not fathom nor explain, the Luddites were to bring back the good times, to mend the trade, to stock the cupboard, to brighten the grate, to put warm clothes on the poor shivering little children."

So as the government moved to deter and punish the Luddites, the movement itself picked up momentum. A series of noteworthy successes had emboldened frame-breakers across the Midlands. The meetings were larger and more spirited, the aims more ambitious, the organizations expanding, becoming emboldened. All of which brought back some hope to the working people involved. A much-needed reason to celebrate. Something to drink to.

Spirits were high as the young Luddites gathered at a pub one night. "A brother had left the room, and now appeared with an immense jug of ale," Ben Bamforth recalls in *Ben o' Bill's*. "Pipes were produced and coarse tobacco. Who paid the shot I do not know. But I have heard tell that some masters who were threatened paid quit money, and others even gave money that their neighbors' mills might be visited."

For now, filled with moral purpose, with agency, they were flush of spirit, and even flush with cash. For the first time in who knows how long, they felt in control. They had beaten back their disrupters, for the moment.

"This I know," Ben said: "there was no lack of ale."

TWO CENTURIES OF DISRUPTION

Technological development is fluid: humans are constantly improving their tools, experimenting with devices, writing new programs. Gravener Henson was infatuated with this process; he kept detailed notes of the innumerable improvements made to cloth-producing technology over the decades by the workers who used it.

Yet technology is often described as undergoing "explosive" periods of "rapid acceleration," like the Industrial Revolution, or the so-called Second Machine Age of artificial intelligence and computerized automation said to be taking shape in the twenty-first century. The explosion comes not when technology itself advances to any particular point, but when it's *used* to disrupt the way we work and live.

Technological disruption is not an accidental or inevitable phenomenon, either, but an intentional one. Two hundred years ago, like today, aspiring entrepreneurs and nascent tech titans saw an opportunity to deploy technology to do work, more cheaply, more efficiently, and at greater scale than it had been previously done by skilled workers. They saw an opportunity for disruption, and disruption was the point. They knew that their machines would upend communities and traditions, but also make them money. Sometimes they knew they'd be trampling regulations, but reasoned that such laws were old and outmoded, and that they could justify it later. Motivated by competition with their peers and the promise of profits, often with privileged access to resources, they imposed their technologies and standards, top-down, on communities with their own long-standing relationships to technology.

This is a remarkably undemocratic way to decide what kinds of technology a society might want to live with, and one that's apt to elicit anger of the explosive variety — why wouldn't it?

The Luddites did not randomly pick up hammers in a rage one day and decide to smash any technology they saw. Their campaign was, if anything, a *logical* response. It was the product of years of accumulated grievances: years spent watching entrepreneurs disrupt their livelihoods through methods that were in many cases illegal, pressing their government to uphold laws *it already had on the books,* and seeing their political leaders do nothing.

The cloth workers of England at the outset of the Industrial Revolution had every reason to be angry; they were not "unthinking" in their opposition to machinery. They even proposed plans to help cushion the introduction of automation in a way that would be more stable for workers and employers alike. And they were shut out of the process altogether, often ignored or derided, and ultimately left to starve.

Yet they tried in vain to see that technology was rolled out fairly, even democratically. And their saga has deep parallels to the modern economy and the tech industry of the twenty-first century, when workers are time and again ignored by regulators and governments in favor of entrepreneurs and their technologies of disruption.

For centuries, Britain's cloth industry had been governed by a raft of norms, regulations, and statutes that determined, say, how many workers could be employed in the trade and the standards to which the cloth should be produced. One law banned the use of gig mill technology outright; entrepreneurs simply ignored it. Some of these rules were indeed old, and had not been appealed to for years, while others had been amended relatively recently. In the regions where they held sway, they were well-known; they governed framework knitting in Nottingham, cotton weaving in Manchester, and cropping in the West Riding.

"However obsolete the statute of Edward VI prohibiting gig mills may have been, it is important that the croppers were aware of it and held that protection against displacement by machinery was not only their 'right' but also their *constitutional* right," E. P. Thompson wrote. "They also knew of the clause in the Elizabethan Statute of Artificers enforcing a seven years' apprenticeship, and of a Statute of Philip and Mary limiting the number of looms which might be employed by one master. Not only did they know of these laws: they attempted to put them in force."

This is why the hymn of "General Ludd's Triumph" could declare that the frames were sentenced to die by "unanimous vote of the trade." The workers believed they were illegal, because technically they were, even if those statutes had long been spottily enforced at best, and that dismantling the machines was therefore legally sanctioned. But long before General Ludd made any such pronouncements, and despite the laws against forming unions, the workers raised enough money to hire lawyers, train representatives, and to take their case to Parliament beginning in the early 1800s.

By the time the Luddites rose up in 1811, cloth workers had made one impassioned effort after another to press Parliament to uphold those rules and to enact bare-bones protections like minimum wage laws. They believed that by working in good faith and by frankly presenting the ample evidence of their economic plight, they could persuade their government to prevent its skilled and industrious workers from falling into poverty.

So, starting in 1802, representatives from the cloth trade began to petition Parliament. Thousands of signatures were collected, a well-written petition was submitted, and representatives were sent to London. But year after year, they were denied hearings. In both 1803 and 1806, cloth workers made big, full-hearted appeals to Parliament. They presented data about their productivity, economic importance, and falling wages. Frustrations soon began to boil over. Their earnings were declining, and they knew that the factory system and machinery were making matters worse. Parliament remained unmoved.

"The debate over the old laws was not simply a question of removing obsolete impediments to economic growth," the historian Adrian Randall wrote. "It revealed a powerful ideological struggle between an old political economy based on order, stability, regulation, and control supplemented by custom and the new political economy based on a faith in free-market forces and the power of capital." He continues: "Here we can see the real issues which the Industrial Revolution raised. It was not just a question of more and more machines. It involved a complete reorientation of perceptions of economic and social relationships."

The cloth workers were pressing the case that they would lose a lot more than wages if the nation surrendered its economy to factories and machines, and subordinated themselves to the bosses who ran them. Many hoped that they could put forward an alternative "to that of the image of progress offered by the innovators." So, they "sought not only to combat machinery itself but also to refute the whole ethic of laissez-faire industrial capitalism which it represented."

And it was indeed the case that there was sympathy for their cause, just not among the very highest rungs of power. Many lords and aristocrats, even the land-owning elite, were made deeply uncomfortable by the encroaching of upstart, ambitious entrepreneurs and their embrace of automated machinery.

And, of course, lots of merchants and smaller shop owners were, too. This was why, as Thompson points out, "the croppers seem to have cherished some hope of a general negotiation within the trade, and were chiefly indignant at the attitude of a few masters, motivated by 'Revenge and Avarice.' "

But prime minister Spencer Perceval, most of the influential Tories, and a number of Whigs opposed any regulations that would impede the free market, and they easily won the day. In the 1800s and 1810s, as in Silicon Valley today, entrepreneurs could generally count on the national government — which above all saw the tech-driven industry as a powerful boon to its economic interests — to side with their cause.

Parliament finally took up the cloth workers' petition in 1806 and formed a special commission. After having the cloth workers investigated for trade union activities, it appointed a cadre of deregulation advocates to study the issue.

The cloth workers were not only proactive, legally minded, and dogged in seeking their fair shake. They were creative, too. They recognized technology was improving — cloth workers themselves were often the ones that improved it — and were on the lookout for ideas as to how machines might be more harmoniously introduced into workplaces to benefit them all.

Take, for instance, this idea for blunting the pain of automation by taxing technology: "Proposals were in the air for gradual introduction of the machinery, with alternative employment found for displaced men, or for a tax of 6d. [sixpence] per yard upon cloth dressed by machinery, to be used as fund for the unemployed seeking work." They suggested placement and retraining programs. They also proposed phase-in periods, or waiting for economic conditions to improve, so that automated machinery could be introduced less disruptively.

In fact, these 1800s cloth workers put forward just about every idea that's gained prominence in the twenty-first century to blunt the pain of automation.

Retraining programs, for instance, are a popular corporate remedy proposed by companies that are automating tasks typically performed by their workers. A 2018 McKinsey report found that a majority of executives polled were prioritizing "upskilling" precarious workers. In 2019, Amazon began investing $700 million in retraining the employees it would automate out of work. Both then and now, such initiatives have a discouragingly low success rate. But some retraining and job placement has been effective, if, for example, it's guaranteed in a union contract.

The Culinary Union in Las Vegas represents sixty thousand casino and restaurant workers; it's the most powerful union in Nevada. When it was fighting for a new contract in 2018, one of its chief objectives was to win protections against automation. The major casino companies like MGM were beginning to roll out automated serving kiosks that threatened thousands of servers and bartenders.

"We were researching a lot about how technology comes in other countries," Geoconda Argüello-Kline, the secretary-treasurer for the union, explained.

> We look at countries that are very advanced, with technologies like Japan, where you have robotic servers bringing food to the table that costs $900 — and we see how it's going to be implemented. So many restaurants will have that technology — so if you're implementing the bartender machine, how will the workers get affected? This is going to happen here. It may be faster than we think. We see it moving little by little now, but . . . what we want is to protect the jobs.

Their solution was a raft of mandatory checks on automation, stipulating that the union must be notified 180 days before a new technology will be adopted, and informed of "who's going to be affected," Argüello-Kline says. Companies must provide a retraining option for employees who risk redundancy from those technologies, as well as six months' severance if they choose to let the robot take their job. Those automated out of work get priority in rehiring.

The negotiations in 2018 were intense — the union voted to approve a Las Vegas–wide strike of service workers before the casinos acquiesced, and the automation-resilient contract was ultimately ratified. "We know nobody's going to stop automation," Argüello-Kline says, "but how can this be an opportunity for the members, so they can make choices — maybe I'm close to retiring, and I want severance and health care. And the retraining part is so important — if you having a salad-making robot, someone has to prepare the salad, maintain the machine." And that's a job a worker could be trained to do; one that would minimally disrupt their way of life. "We are not opposed to technology," Argüello-Kline says, but "the machines can replace our jobs if we don't work to protect them."

Now consider that proposal from the 1803 cloth workers, for a small tax on each bit of cloth made by a machine, to fund a safety net for those put out of a job by automation. This is a lot like proposals put forward for a so-called robot tax, or what former presidential and New York City mayoral candidate Andrew Yang describes as his tech-focused value-added tax (VAT), which would take a small fraction of every transaction made by a corporation. "A VAT makes it impossible for them to benefit from the American people, automation, and infrastructure without paying their fair share," Yang's campaign website states. "By implementing a VAT, the American people will get a tiny sliver from the transactions of the big winners from the twenty-first-century economy, the trillion-dollar tech companies." The income generated by the VAT would pay for a universal basic income program designed to insulate working people from adverse impacts of automation.

The more direct robot taxes, those specifically targeting automation or robotics, have been put forward by some unlikely bedfellows — New York City mayors, tech monopolists, and South Korea. Bill Gates came out in favor of taxing industrial robots at the same rate that the humans they replace would have to pay. "The human worker who does, say, $50,000 worth of work in a factory, that income is taxed and you get income tax, Social Security tax, all those things," he said in an interview with *Quartz*. "If a robot comes in to do the same thing, you'd think that we'd tax the robot at a similar level."

Meanwhile, when Bill de Blasio was running for president in 2019, the former New York City mayor advocated for making any company that institutes automation and eliminates a job pay five years' worth of payroll taxes, and to offer retraining for those laid off. De Blasio also proposed creating an authority, the Federal Automation and Worker Protection Agency, that would gather data on automation and employment and help protect workers. He argued that the bureau should cover work displacement, not just from automation, but from app-based companies as well.

"We've had some initial fights over autonomous vehicles already," de Blasio said;

> companies that want to come onto the streets of New York that we would not allow for safety reasons alone, that might have a huge negative impact, employment-wise. But on just the for-hire vehicle

sector, you know, Uber and Lyft — it's not automation in some ways in the fullest sense, obviously, but simply a new technology that automated some functions — and had a massive dislocating impact on our transportation sector.

South Korea is so far the only country to institute something akin to a robot tax. The nation eliminated incentives previously offered to companies investing in technology, but only for automation tech. It was an effort to curb the appeal of further investing in robotics, and seen as a signal that the government predicted a decline in taxable income from those automated out of work.

That the cloth workers proposed some version of each of these ideas, especially what is essentially a value-added tax for automated textile technology, lands them in the annals of history as some of the earliest policy futurists.

George Mellor spent seven years apprenticing at the cropping trade because that was the law, and that was tradition. He was playing by the rules; many entrepreneurs purchasing machinery and adopting the factory system in the first decades of the nineteenth century were not. Then as now, the most ambitious deployed their technologies and expanded their factories regardless of whether they might be running afoul of some dusty old laws, and *then* lobbied the government to overturn them outright when the workers complained.

They argued that the regulations on the books, which had for generations clumsily if importantly protected workers and helped stabilize labor markets, did not apply to factories and machine-made textiles.

"The old legislative controls preventing the establishment of loomshops and those regulating apprenticeship were attacked as outmoded," Randall wrote. The entrepreneurs argued that "the fears that repeal would be followed by the universal adoption of factories were absurd. Most innovators, many disingenuously, denied any intention of establishing them. The West of England clothiers, it was said, preferred domestic weaving, provided they could prevent embezzlement!" It was just that those old apprenticeship laws "were unnecessary for trades easily learned, as Adam Smith had shown, and unfairly restrictive of liberties."

So: the regulations on the books are out of date, should not apply to new technologies, and are stifling innovation; they're really not that important, and no one plans on violating them anyway; the new tech simply enables more secure and transparent processes, and to constrain it is an affront to the freedom to do business. Sound familiar?

These arguments parallel the ones advanced by Uber and other major gig economy companies like Instacart and Doordash throughout the 2010s and 2020s. Uber's chief innovation is not that its app summons a car to your location with a smartphone and a GPS signal. It is that it used this moderately novel configuration of technology to argue that the old rules did not apply whenever it brought its taxi business to a market that already had a regulated taxi code.

Uber claims that it is not a taxi company but a software company. It says its workers are not, in fact, taxi drivers, or even *workers,* but independent contractors, who are incidental to its proprietary software. Uber argued exactly this as it moved into city after city after city, setting up shop, recruiting drivers, running ad campaigns, and launching the service whether or not it was sanctioned by local laws. Led by its brash and regulation-despising founder, Travis Kalanick, Uber deployed a strategy to undercut the local taxi drivers, establish its service as the cheaper, sleeker, modern alternative, and gain a foothold if not a dominant share — steamrolling any laws already in place to protect workers in the industry and dealing with local officials after the fact. *New York Times* reporter Mike Isaac documented how Uber's operation became aggressive and sophisticated, even in cities like Portland that put up a resistance. Uber ran a surveillance operation in new markets that could, among other things, identify and blacklist accounts of investigators trying to hail a then-illegal ride from the company, to avoid picking them up.

As a result, livery drivers and cab drivers were hit hard; some drivers' entire livelihoods were wiped out or degraded irrevocably. As Uber was introduced around the world, cab drivers protested and begged city officials to regulate the company as a taxi service, which it is, even if users hail its cars through a smartphone. Time and again Uber battered down the doors.

In both the 1800s and the 2010s, the strategy worked. With Uber, users in city after city grew accustomed to the convenience and cheaper prices, and the service became popular; soon the proposition of regulating it grew untenable for a lot of policymakers. Never mind that Uber has never been a profitable business. It could afford to undercut the taxi companies in so many cities because it was loaded with a huge war chest of venture capital that gave it the resources to operate at a sustained loss.

In 1800s London, the well-funded technological system entrepreneurs hoped to use to control workers was different — automated factories, not algorithmic gig apps — but the circumstances, not so much. When the findings of the 1806 special commission came back, it was the workers who were targeted with the deepest scorn. They were accused of engaging in trade union activities and publicly derided, in a hint of what was to come, for opposing progress. One MP said it was not even worth stating the case for machines as their benefits were so obvious: why should we bother to "trouble them on the general utility of machinery in a nation which has profited so much by it"? Machinery was "the principal means by which we have excluded other nations from the foreign trade. To restrict its use was madness." Another simply said workers' arguments against machinery "were hardly worth noticing."

"Thus the repealers' case demanded that the woollen trade should be freed from restriction," Adrian Randall wrote, and that "Parliament should 'encourage competition.'"

So the factory owners kept churning out product, displacing and degrading skilled workers, and lobbying their friends in power for favorable treatment. The machines did indeed do a shoddier job of making cloth products than the skilled weavers, they just made a lot more of them. Owners could make up for the lower value goods by moving more inventory. Human weavers and small operations had no such luxury, and were left competing on price against the cheap machine-made stuff. In 1809, Parliament repealed the regulations altogether, leaving suffering workers with no protection against the gale forces of automation, factorization, and industrial capitalism.

"The bitterly resisted repeal of this legislation in 1809," Randall noted, "commenced the process of the destruction and deconstruction of old concepts of custom and ultimately of community."

PART III

BREAKING FRAMES, BREAKING BONES

THE PRINCE REGENT

Spring 1812

Much of Prince George's country was occupied territory. From Sherwood Forest to the Pennine hills of Yorkshire, England was patrolled by thousands of Crown troops—yet there's little indication that the man ruling England was invested much in their campaign, if he knew the details of the occupation at all. With his soirees, vacations, and fixation on the arts, he seemed hardly to notice the cataclysm engulfing his land.

His subjects noticed. "It seemed to many people in the South like the whole of the North was going up in flames," according to the historian Jenny Uglow. "It was a small industrial civil war."

Yet the Prince Regent was in Brighton, with his own military regiment, the 10th Hussars, known as the Prince of Wales's Own, entertaining the officer corps with parties, hunting, and horse racing. Between the travel, the carousing, and the flow of news about Wellington's progress in the war against Napoleon, the Prince Regent did not seem to have fully grasped the fact that north of London, insurrection was rampant.

The home secretary, Richard Ryder, was, of course, well aware how little progress the troops were making. With the paranoid, badgering letters from magistrates like Radcliffe, and the relaying of wild reports from spies like B, he had ample evidence that many considered the Luddite rebellion a threat to the state itself.

Ryder had now moved thousands of troops north. "Both Leeds and Huddersfield were suddenly beginning to look more like garrison towns bracing themselves for war than peaceful, profitable centers of the woollen industry," the historian Robert Reid wrote. Ryder helped push through the Watch and Ward Act, which let local magistrates compel any citizen over seventeen to "watch by night and ward by day" on behalf of the state, against the Luddites. They would be obligated by law to serve as anti-machine-breaker police.

Huddersfield alone, with a population of some ten thousand people, was occupied by a thousand soldiers. There were no military bases or official encampments; by decree of the state, the soldiers were put up in the inns and pubs of the city, at the owners' expense.

In Leeds, one military leader, Sir George Armytage, assembled a militia and marched them straight through town to their training grounds. The townspeople lined up and shouted insults as the regiment passed. A full-scale riot nearly broke out, and the mayor of Leeds wrote to Ryder pleading that future troops be routed around the town center to avoid further disturbance.

Still, Ryder couldn't get all the generals to listen to his demands; for every Armytage willing to make a public display of power, there was a William Dyott, who commanded several thousand troops in the Midlands but couldn't be bothered to be roused from his country home. (Men like these, from the local districts, may have been sympathetic to the Luddites, or at least unconcerned for the factory owners.)

So the military was as ineffective as ever at slowing the Luddites down. The machines kept breaking, and the countermarching soldiers kept showing up too late to stop them. For the most part, the soldiers' presence in town only heightened tensions and added to the chaos. Hundreds of soldiers with little oversight, posted up at pubs all day and night, was a recipe for disaster.

Those soldiers were stationed all around Huddersfield, where the dangers of the occupation were on full display.

On Ben Bamforth's eighteenth birthday, his family threw him a party at their home. It was a cheerful and festive day, according to *Ben o'Bill's*, with George, his friend Soldier Jack, his cousin Mary, and Ben Walker all playing parlor games and singing songs at the small piano. At one point, Mary stepped out to get fixings for dinner with Ben Walker. When she returned, her dress was torn, her face flush, and she was wailing at Walker, who was pleading behind her. Everyone immediately thought the worst, but Walker was guilty only of being a coward.

As they were walking, the two passed a soldier stationed in town. The soldiers had flirted with Mary before when she would walk by, but harmlessly, she had figured, and in broad daylight. She'd thought little of it. Tonight, one soldier approached her, probably drunk, backed up by his comrades, who were egging him on, and grabbed her. While she protested, he stuck his tongue down her throat. So she bit it and ran. Walker, her escort, did nothing. Back home, the celebration ground to a halt. Mary berated Walker, as cousin Ben stood behind her, and the family consoled her and spoke words of outrage. George stood up and took the chance to rail against the injustice of the occupation, and "made us a grand speech against the army and officers and men."

Meanwhile, in Brighton, the Prince Regent took his officers, the Hussars, on a hare-hunting trip. When they couldn't find any, he suggested they do some racing instead, jumping over the equipment they left out.

Legend has it that this was the birth of horse hurdle jumping. The Prince Regent had helped to mint a new luxury pastime.

WILLIAM GODWIN AND PERCY
BYSSHE SHELLEY

March 1812

In the early days of 1812, William Godwin received a surprise letter from a young fan. In exuberant, florid prose, the writer announced himself as a stout disciple of *An Enquiry Concerning Political Justice*, and told how his adherence to its principles had nearly gotten him kicked out of university. He was a writer, a poet, and freethinker, and had clashed over all of the above with his father, who had disowned him because of his political views.

This was only partly true, but Godwin read the letter with intense interest, as it became clear that this was no idle fan, but a man who was a gifted if overzealous writer, genuine in his admiration, and the son of a wealthy lord in Sussex. The writer painted himself as an outcast, misunderstood and wronged by a society he had no choice but to reject — along with the lucrative baronetcy his father was hoping to foist on him.

The writer's name was Percy Shelley, then nineteen years old. His letters detailed his dramatic fortunes, misfortunes, political stands — just a teenager, he was already a scabrous critic of the Prince Regent and the Tories — and selfless crusades. Shelley quickly had the entire Godwin household rapt; "to Mary, Jane, and Fanny, their father's new correspondent sounded like the hero of a romance," one historian wrote. They were "enthralled."

Shelley had in fact been run out of college at Oxford for publishing a pamphlet called *The Necessity of Atheism*. And he was estranged from his father, whose patience was wearing thin, between Percy's expulsion, his threatening to reject the baronetcy, and his recent elopement with an heiress, Harriet Westbrook.

The letters were long and full of drama. Shelley, who had only recently consumed all of Godwin's books, at a friend's urging, asked the older radical to impart his wisdom, to teach him, and to let him serve as Godwin's patron. Godwin had received a lot of letters from a lot of young men hoping to be his protégé, but fewer who wanted to support him financially. And Godwin was still very much in debt.

It's easy to see why a young idealist was excited by Godwin's best-known work of political theory, which advocated not for revolution but anarchy. "Political complexity is one of the errors that take strongest hold on the understanding," Godwin wrote in the introduction; the French Revolution had allowed him to see "the desirableness of a government of the simplest construction," free from onerous and excessive restrictions and repressions.

Mary, meanwhile, was suffering from eczema and was losing interest in her studies. Doctors recommended a seaside retreat, so Godwin discussed sending her up to Scotland to stay with some friends who lived simply on the coast; they were Glassites, communitarians who rejected the accumulation of wealth.

As the dark winter progressed, and Godwin and Shelley exchanged more letters, the family was drawn further into the intrigue. "You cannot imagine how much all the females of my family, Mrs. G and three daughters, are interested in your letters and your history," Godwin wrote on March 4; in another letter, he shared how they thrilled to the sight of "the well-known hand" and were "on the tiptoe" for news of Shelley's latest exploits.

Godwin wrote to Shelley in March, describing how proud he was of what impressive women his daughters were turning out to be. "Seeds of intellect and knowledge, seeds of moral judgement and conduct, I have sown, but the soil for a long time seemed 'ungrateful to the tiller's care,'" he wrote. "It was not so. The happiest operations were going on quietly and unobserved, and at the moment when it was of the utmost importance, they unfolded themselves to the delight of every beholder."

Finally, Godwin agreed to Shelley's request, making what he viewed as an agreement to accept Shelley's patronage until his death. The two would be bound for the rest of their lives, more inextricably than either could possibly anticipate. Mary, meanwhile, prepared for an ocean voyage to Scotland, to live among the egalitarians.

GEORGE MELLOR

A ticket used to gain entry to a secret Luddite meeting, recovered by a spy and delivered to the Home Office.

The code was simple, and taken dead seriously. Enter the swinging inner doors to the Saint Crispin Inn. Make eye contact with the sentry posted across the hall. Gesture the hand signal. Do not arouse attention.*

When the guard, who will appear bored, accompanied only by a drink and a pipe, motions in return, head not to the reliably crowded main rooms to the left or the right, but straight ahead. Walk up the narrow staircase to a smaller meeting chamber on the second story. There, utter the code words (reportedly, "win, work") to the second sentinel. Succeed, and be admitted to the secret meeting place of the Luddites.

Rituals like these were being carried out across the country, in pubs, homes,

---

* The account of the secret code and meeting in this section is according to Frank Peel's oral history, *The Risings of the Luddites* (1880). The dialogue is from Peel and from Sykes's account in *Ben o' Bill's*.

and open fields, and the secrecy was more important than ever. Despite being a cell-based movement lacking heavy central organization, the Luddite movement was remarkably airtight. Despite the state employing an army of spies, very few had seen any success. Most of the Luddites had worked together, lived in town together, and grown up together, forging natural bonds that made any outsider suspect. For the rest, there was the practice of oath-taking and secret organizing that had developed for years, thanks to the Combination Acts, and had driven that organizing underground. A ticket for one such meeting was stamped with the motto, "Enter, no General but Ludd means the poor any good."

The Crispin was a pub in Halifax named after the patron saint of shoemakers. Tonight, a few dozen Luddites had given the signals and code words. Along with a system of codes, oaths, and rituals, a more developed movement was emerging. Delegates from different sources of uprising, especially from Nottingham, still the locus of the movement, were meeting with upstart and expanding Luddite chapters.

Tonight, in that private upstairs room, gathered around a long table used in the past for settling trade disputes, the Luddites — George Mellor, Will Thorpe, weavers, blacksmiths, and other oath-bound men — and local reformers were anxiously awaiting news from the expected delegate from Nottingham. The dank air was filled with excitement.

The proceedings were being led by John Baines, a hatmaker, who, with two of his sons, headed up the local republican club of Tom Painers, progressive democrats who aspired to reform Britain's monarchical political system. They were sympathetic to the Luddite cause, naturally, even if Baines was twice as old as the oldest among their number tonight, as he would point out himself. Nearing seventy, with thinning hair and a "deeply wrinkled" face, Baines had lost none of his vigor for the democratic project.

"I should like to have the privilege of saying a few words to the delegates assembled before our friend Weightman from Nottingham gives his report," he said, calling the men to attention. "I am glad to see so many young men present, and hope their enthusiasm may carry them right to the end of this movement —"

The man from Nottingham cheered.

"And that end, what is it?" Baines continued. "Is it the destruction of the cursed machines that are robbing your children of their bread? Well, that is one great object, but, my friends, is that the *end*?"

"Down with the bloody aristocrats!" someone yelled.

"Amen!" Baines agreed, and launched into a speech castigating elites. "The vampires have fattened too long on our heart's blood." He condemned the landed aristocrats — "all the offices in the land are held by them and their friends" — and lamented the taxation and exploitation the working man suffered at the hands of this "Old Corruption."

"Let the people rise in their majesty!" Baines shouted. "Old as I am, I shall yet see the glorious triumph of democracy."

A number of the Luddites stood, applauding rapturously. Some were so moved by the older man's monologue that they reached out and pressed his hands where they gripped the table. George leapt to his feet. He was not feeling quite so idealistic. It's not that he disagreed with the politics, but he was committed to more decisive modes of action.

Young as he was, he would have known the history of the last ten years of the weaving industry's efforts to persuade Parliament to uphold trade regulations, and how futile that effort proved. It was only in 1809, just over two years earlier, after all, that Parliament had torn those regulations up and sided with the machine owners. As he'd impressed onto John Booth in the cropping shop, democratic reform was a nebulous and far-flung aim. Meanwhile, their neighbors were hungry, now. He'd made this argument before — and won — and he was sticking to it.

"We'll reckon with the aristocrats in London in due time," George said, "but, friends, is there not some work nearer home to be done first? I know of no aristocrats who are bigger tyrants than our own masters, and I'm for squaring with them the first."

"Our friend is right," Weightman, the Nottingham delegate, retorted, "and he is also wrong. Right in his longing to strike down the tyrant who would rob him of his daily bread and turn him out to starve in order that the gains may be poured with redoubled speed into his bursting coffers; but wrong in thinking he is his greatest oppressor." How, Weightman asked aloud, did they think this conflict would really end? Between "decaying commerce" and "our starving workpeople," conditions were so dire, he suggested, that the time was ripe to move beyond breaking frames and sabotage.

The question of what role conventional political organizing played in the Luddite movement has been a source of debate among scholars. Few internal documents exist, and the writings of Luddites are primarily intended as threatening, blunt negotiating tools. They never produced a clear, united political manifesto. Champions of democratic reform, like Baines, were natural allies. There's evidence that groups like the Luddites and republican reform causes overlapped, and a revolutionary spirit was indeed present in many of the letters and proceedings, especially as the Luddite movement progressed.

As the movement and its audacity grew, so did the awareness among many of its limitations: that its popular aim, to smash the owners' machines that were paving the way for the factory system, was at best a rallying cry and a temporary salve. Calls, and not just from reformers like Baines, were growing to expand the movement's purview.

A hymn by an anonymous author, for instance, began making the rounds in the spring of 1812. It began,

> *Well done, Ned Lud, your cause is good*
> *Make Perceval your aim,*

and proclaimed,

> *We're ready now your cause to join*
> *Whenever you may call,*
> *To make foul blood run fair and fine,*
> *Of tyrants great and small.*

The fight should be taken to London, to the prime minister's doorstep.

"Now that we know our strength and have proved ourselves able to crush our local tyrants, let us go down the root," Weightman said in the Halifax pub.

Spencer Perceval was well-known as a staunch opponent of workers' rights. He had killed the weavers' petition for industry protections in Parliament, and later admitted that the public hearing he'd granted them was only for show. He knew that if he offered them a hearing, he could at least give the appearance of being concerned. "It was better that the cotton weavers should be disappointed after a discussion of the merits of their application," he explained, "than by a refusal...to submit it for consideration." The state had already made up its mind about how it would rule.

So, the delegate from Nottingham believed, it was time to raise the stakes and do what the British state most feared the Luddites would do — start something more like a revolution. "Our movement is everywhere powerful; we have thousands of weapons collected and stout arms to wield them," Weightman continued. "Our council is in daily communication with the societies in all centers of disaffection, and we urge a general rising in May."

It may have seemed tantalizing, if daunting, to the Huddersfield Luddites like George Mellor and Will Thorpe, who'd just begun organizing raids in the past month. "Since John Westley was shot at Arnold, the feelings against the masters and their frames has redoubled," Weightman said. "The people everywhere sympathize with us."

While Horsfall and Cartwright openly expanded their operations, their neighbors, even those the entrepreneurs were friendly with, like the Bamforths, were pushed toward poverty. Yorkshire was becoming a hotbed of visceral inequality, and the most vocal factory owners were indifferent or worse. Public sympathy naturally lay with those willing to push back, which may have further emboldened the Luddites. "Many in our ranks advocate the policy of shooting such of the masters as are engaged in hunting and harrying us, but some hardly like the idea of murdering them in cold blood," Weightman continued.

"Let them do what's right then," George cut in. "If we're to stick at such squeamish nonsense, there will be naught done. If they shoot at us, why shouldn't we shoot at them?" George was speaking metaphorically, for now — it was time to set their sights on a major target that would send an unambiguous signal to the factory owners. "There's two in this neighborhood that will have to be taken underhand, and at once. I mean Cartwright, of Rawfolds, and Horsfall, of Marsden. Most of you know how these two brag, day after day, at Huddersfield market, and threaten what they'll do to Luddites if they come near their places."

"What's to hinder us from attacking Rawfolds Mill next?" George asked. "We've done little in the area, and we're middling strong there."

There was at least one thing hindering them from an attack of that scale, they agreed. They'd need more guns.

Cartwright was turning his massive mill into a fortress; he hired mercenary soldiers to stand guard, and was building up the site's defenses. Will Thorpe suggested they attack Horsfall instead, as he was more aggressive in his threats toward Luddites. "I am just about sickened with the reports of his brag and threats brought into our shop," he said. "Cartwright isn't half as bad as him."

Some others concurred, and an argument broke out. According to Peel's oral history, the matter was settled with a coin toss. Heads, it'd be Cartwright, tails, Horsfall.

It was heads.

There was more than just vengeance at work here. If the Huddersfield Luddites could move successfully against a major operation in the region, it would imbue them with a new sense of power — a considerable strategic asset.

"As the man you speak of is setting you at defiance, sooner you deal with him the better," Weightman said. He then offered a window into how collective bargaining by riot works in practice:

> We have inspired such a wholesome dread of us at Nottingham, amongst all classes, that the threat of General Ludd is almost invariably sufficient. Our motions have been so rapid and our information respecting the possession of arms so accurate that few now dare to speak of us defiantly. As usual rumour magnifies our deeds, and we are credited with much that we never do; but the authorities, I can assure you, are very much dissatisfied with the result of their efforts to put us down. Many who do sympathize with us taunt them with the little progress they make in crushing us.

Reclaiming power was a central factor in the Luddites' success. They had turned the tables, but to continue to do so, Weightman argued, they needed to show that they possessed the power to register a threat to even the richest and most established entrepreneurs. They had to take the fight to the tech titans.

George, Thorpe, and the local Luddites turned to the plan at hand. They had a fortress-sized factory to infiltrate and empty of automated machinery. Destroying that operation alone could restore hundreds of jobs to their peers and, more importantly, bring them a major step toward winning the bargaining power they would need to turn the tide with the rest of the industrialists.

They would canvas the shops and houses throughout the area for arms and munitions. They would rally the neighboring regiments of Ludd's Army to the cause, and plan for a large-scale assault.

It would be the Luddites' boldest effort yet.

A date was marked. On Saturday, April 11, they would gather force and hundreds would march by night for Cartwright's massive factory at Rawfolds.

B AND THE SPIES

April 1812

The workers of Nottingham, the West Riding, and Manchester may have been ready to join a revolution, B said, but not those in London. Not yet. There were so far only 7,400 Luddites twisted in there (or so B wrote to Colonel Fletcher in April), numbers that paled in comparison to the tallies mounting in the north.

The leaders, however, had begun strategically dividing up turf there, he had heard, to better organize and lay the groundwork to make inroads into the nation's capital. The month before, B had written about a meeting with men from Stockport, outside Manchester. He described them as "dangerous" and "daring fellows," and said that no less than four of them had been "in the Rebellion in Ireland and no doubt but the[y] wish to be at the same game again."

Most pressing was word of a secret meeting held at the King's Arms, in Salford, attended by another one of Fletcher's spies, John Stone. Its aim was to coordinate attacks on the major factories in the region. The meeting was attended by Luddite delegates from Bolton, Stockport, Manchester, and the smaller surrounding towns. Support was rampant.

One delegate, from Eccles, announced that they had only been recruiting men for two weeks, and had already twisted in 180 men, including a militia leader who would give them access to a stockpile of arms. The tone of the meeting was tactical and more aggressive than past meetings — the time for fact-finding and for hatching strategies to raise wages was through. All those present agreed that the aim should be to "put the great men down that had trampled them under for so long."

This "firing of factories" was organized to be a simultaneous action, taking place at once in each of the cities represented at the meeting, on Monday, April 13, at midnight. The talk was that a factory in Westhoughton, where a steam engine ran 200 power looms, should burn.

According to the spies, someone suggested at this meeting that the firing of factories include an outright assault on the military officers and magistrates, who should be executed during the attack. But when the plan was put before the Manchester committee at large, it was deemed too aggressive, according to B. The Luddite leadership rejected such an action.

In a curious footnote, Stone relayed that a delegate had brought along the pieces of some kind of model. While everyone wondered what they were, the man took the parts on the table, and then pieced them together into a pike. It slowly dawned on the men what he was doing — the Luddite was demonstrating how a harmless-looking pile of parts could become a deadly weapon.

WILLIAM HORSFALL

Spring 1812

The attack on Vickerman had strengthened the resolve of the largest manufacturers in Huddersfield. The owners of smaller operations may have been uneasy, but with Horsfall taking the lead, the Committee to Suppress the Outrages was pushing the local entrepreneurs to defend their right to do business whatever the cost. The organization had increased their reward for anyone who came forward with information about Luddite operations, too.

A dozen shops had now been broken into across Huddersfield, their machinery destroyed. The Luddites also intercepted a large shipment of machines bound for Cartwright's mill and smashed them all. William Horsfall was furious, and made it a point to broadcast his anger in public. He had by now repeated his threat to spill "Luddite blood" so often that it had become a catchphrase around town, used to mock him behind his back and occasionally to his face. The children continued to taunt him as he made his weekly ride to the market from his factory. But that mockery did not make his threat an empty one.

He'd gone to considerable expense to turn his factory into a fortress. On a hill facing the road leading to Ottiwells, he had constructed a vast stone wall that extended the entire length of the premises. It was castellated along the top to accommodate gun loops, so his men could take cover while firing on any advancing forces. He had armed those men, too — his will to violence and dedication to the industrial future had endeared him to a certain kind of man, soldiers and militia members, willing to watch over the mill day and night.

Additionally, Horsfall had summoned the troops of dragoons that had recently arrived at the direction of Richard Ryder. Armed cavalry would flank any mob or army that engaged the factory from the front.

Horsfall's aggressive militarization of his private enterprise set him apart from even his most industrious peers. Perhaps the state's move to render frame-breaking a crime punishable by death emboldened him to prepare to do the same on a vigilante basis, or perhaps he was simply a pioneer in his efforts to protect productive machinery at any cost. There was a zeal to Horsfall's belief that it was a man's right to use technology to increase profits, whatever the consequences, that efficiency was paramount, that bordered on religious fervor — on prophesy.

Cartwright, meanwhile, was more subdued. Taciturn and reserved by nature, he did not inspire men to take up arms on his behalf; but he'd hired mercenaries to guard his own establishment and was dedicated to continuing to import shipments of automated machinery. He displayed few outward signs of making war against the Luddites.

Horsfall, meanwhile, had cut gaps into the wall large enough to make room for his deadliest implement — a wheeled cannon he had bought precisely for the purpose.

LORD BYRON

Spring 1812

To the surprise of no one who knew him, Byron very much enjoyed his new place at the center of attention among London's upper crust. He struck up friendships with literary celebrities like Sir Walter Scott, appeared at balls and galas, and began one of his most gossiped-about trysts.

In late March, he was invited to Melbourne House, a manor that was "physically as well as sexually perilous, with its elevated public spaces and 'dark and winding passages and staircases' behind them." Byron's first visit was at the invitation of Caroline Lamb, a well-known aristocrat and writer who had developed an interest in the poet. As he was walking down a curved staircase along with the poet Thomas Moore, he slipped, joking that this was "a bad omen."

Byron and Lady Caroline began an intense affair, which soon became the subject of public interest. To Byron's consternation, she was not concerned with being discreet, and not as invested in cultivating his reputation as he was. In one of his earliest letters to her, Byron addressed "your heart, my poor Caro (what a little volcano!), that pours lava through your veins; and yet I cannot wish it a bit colder."

The best-selling epic poem, the scandalous liaisons with high-society women, the defense of the downtrodden Luddites — it all contributed to the beginning of what would later be called Byronmania. Thanks to his groundbreaking and sensational *Childe Harold*, his self-mythologizing Romantic heroism, his carefully managed public image, his pale, striking good looks, even his singular limp, Byron became a figure without peer. In fact, he was on his way to becoming the most famous man in Europe this side of Napoleon Bonaparte.

Byron became notorious on a level that some scholars say is all but incomprehensible to us today. (Imagine if there were just *one* pop-culture icon who was a household name.) Byron was in a league of his own. His works sold out fast, and artisans stamped his face on merchandise, a practice virtually unheard of for a figure other than a king.

So, regardless of whether his speeches were hissed at by his peers in the House of Lords, Byron had become a political force to be reckoned with. Regency politics were overwhelmingly conservative, he was anything but, and the Tories were worried about his growing influence. There was true indignation at Byron — if not over Caroline Lamb, or the alleged affair with his half-sister, or the speculation that he took gay lovers — then because he was a threat to Tory power.

Byron is "mad, bad, and dangerous to know," Caroline said shortly after their first meeting, and the phrase would endure. The line succinctly conveyed the truth about his capacity for harm toward women — he could be reckless, cruel, predatory, even violent — while the conservative Parliament saw him as a threat for different reasons.

That reckless intensity would imbue his myth — and capacity for mythmaking — with more power than either he or Caroline recognized. His interest in the Luddites may have waned as his celebrity was ascendant, not that his sympathy with their cause wasn't genuine. But that celebrity was a crucial tool in spreading Luddite and working-class solidarity through the vessel of Romantic prose and popular thought.

THE BATTLE OF RAWFOLDS MILL

Ned Ludd and George Mellor

April 1812

It was time. The West Riding Luddites began to gather at eleven o'clock, well after dark had settled, by a stone obelisk known as the "dumb steeple." There were dozens, then scores, then over a hundred of them.

George Mellor was in command, his sturdy frame and blazing confidence a beacon to the men arranged in the gloom. Thorpe and Thomas were there, along with Ben Walker, John Booth, and even George's cousin Ben, who'd been reluctant to join many Luddite raids. So were around a hundred and fifty others, with more on the way.

The men organized in the pitch dark, half a mile from the factory outside Huddersfield. They had come with hatchets, pikes, guns. Pistols, blunderbusses, muskets. They'd assumed military formation, more natural now, after months of practice. The minutes ticked by, yet there was no sign of the Leeds contingent, which had agreed to join. The decision was made to press on regardless. Perhaps they'd gone ahead. If not, they'd catch up.

As the Luddites marched through the field, the formidable Rawfolds complex came into view, obscured by a guard wall and moat. Their recon told them there were at least two sentries out front, a guard dog on the premises, and armed guards posted inside. The only way in was through the thick wooden front door.

First, two teams of the Luddites' best combatants were stealthily sent in to take care of the sentries. Simultaneously, with barely a sound, they snuck in and disarmed them, and removed them from their posts. Another team started working on the barred outer gate, prying the hinges apart until the massive door fell backward "with a fearful crash." Another door, also barred, lay ahead, level with the ground-floor windows of this six-story behemoth.

"Hatchetmen, advance," George yelled. "Now men, clear the road!"

The men ran forward, pikes and hatchets raised. Somewhere inside, a dog started barking. The Luddites with guns fired a volley inside, shattering the nearby windows.

The hammer men made for the main doors. Ben Bamforth, as large as any man there, was one of them, per the account in *Ben o' Bill's,* where he describes rushing Rawfolds with the other men:

The glass was falling from the frames with crash upon crash, sticks and stones were flying above our heads as we streamed forward. The volleys of musketry made their din, and now from loop holes and from windows came answering shots. We could see the streak of fire from the barrels and hear the sharp ping of the bullets as they whizzed about our heads. Our men roared and roared again and yelled with frenzied cries.

The Luddites began pounding the door with hammer and hatchet.

Like Horsfall, Cartwright had fortified his mill for such a defense. He and ten or so men slid their guns onto flagstones attached to pulleys he had rigged to be lifted and dropped to provide cover or an opening to fire through. It would be almost impossible for the Luddite riflemen to hit them from outside.

George Mellor summoned his men to try an attack on the rear.

"To the back!" he yelled.

Then, a voice rung out from inside:

"Come round, we'll meet you," a voice shouted. It was Cartwright himself.

A contingent of Luddites ran around back, but the river and the mill dam kept them from getting close enough. A Luddite named Thomas Brook slipped and fell into the river, losing his hat, and had to be rescued by his brothers in arms.

They'd have to attack the front, after all.

"Way for Enoch!" the Luddites yelled. "Way for Enoch!" Ben and a comrade again reached the door, and swung the great blacksmith's hammer.

A bell rang on the factory parapet — a signal to alert the dragoons stationed outside town.

"Damn the bell," George yelled, "shoot the bell!" The Luddites opened fire. For a moment, it fell silent. They cheered, but it resumed a moment later — they'd hit the rope, but the bell could still be rung.

Ben swung the hammer — over and over again — with little success. The front door was fortified with great bolt heads studding the panels.

"There was room only for us two," Ben relates in *Ben o' Bill's*, referring to himself and Enoch, the hammer,

> and above the roar of the mob, above the yells and curses and cries, above the thud of stones and the crash of falling lime and glass, above the clanging of the mill bell, above the din of gun and pistol, rang out the mighty sound of Enoch's echoing thunder. With every blow that fell quivering shocks ran up my arm as the hammer dithered in my grasp, and still I pounded at the door, and still the stout timbers yielded not a jot.

Several Luddites crouched outside the factory windows, which they'd broken with pikes and gunfire. George Mellor urged them to invade: "In with you lads!" he yelled.

Inside Rawfolds Mill, Cartwright was managing the defense. He sent two men up to the roof to keep that bell ringing. He noticed one of the men on the defense was not firing, and asked if his weapon was jammed. The man said it was not.

"Then why don't you fire?" asked Cartwright.

"Because I might hit some of my brothers," the man replied.

Before long, the Luddites managed to smash a hole in the front entrance. "The door is open!" someone yelled.

As the men reached their hands inside and tried to maneuver to open the door, one of Cartwright's soldiers fired a round through the gaping hole. John Booth fell backwards into the dirt, screaming and bloodied, his leg shattered by a musket ball. Another shot threw the hammer from the blacksmith Jonathan Dean's hand as he tried to widen the hole in the door. Ben Bamforth took a shot to the upper arm and collapsed on the spot.

Samuel Hartley, a twenty-four-year-old handloom weaver who'd been laid off from this very factory just six months earlier, was shot through the chest and went down, vomiting blood from his mouth.

Bells rang, dogs were unleashed, more bullets flew. The assault was raging for nearly half an hour, and the Luddites had not yet breached the factory. They had no idea of the size of the armed force inside Rawfolds, and their men were falling.

It risked turning into a massacre. One of the Luddites remembers George acting "like one off his head" — he leapt to the sill of a lower window and grasped its frame, jagged glass all around it, and shook and gnashed his teeth. It took the news of Ben's being shot to snap him out of his mania.

"Cease fire!" George screamed, despairing at the carnage around him. He ordered a retreat.

By now, the Crown's soldiers were almost certainly en route. The Luddites dragged their bloodied comrades out of range and gently lay Booth and Hartley on the dusty ground, where they writhed. Both were bleeding profusely, and neither could move on his own — their fate seemed all but certain.

Decisions had to be made quickly. They were miles from safety, there was no way they could be carried. Ben was wounded too, but not nearly as badly — George would be able to help him.*

With tears in his eyes, George returned to tell the others that they had no choice but to leave the injured men behind — nothing could be done. He reminded them of the oath, vowed that they would be avenged, wept for what he feared would come of John Booth, fired his pistol into the air, and joined the other men in making an escape.

It was the Luddites' first major defeat, and more devastating than any could have imagined.

Shortly after George Mellor, Will Thorpe, and the other survivors fled, Cartwright emerged from his factory fortress to examine the bloody scene on his doorstep. Neighbors, townspeople, and the militias, drawn by the noise, gathered around as the mill owner demanded the wounded men tell him who organized the assault.

The crowd begged Cartwright to call for a doctor, or to let one through, but the entrepreneur said he'd refuse until the wounded named names. Booth and Hartley, moaning and bloody, twitching in pain, would not speak, and eventually Cartwright sensed so strong a disgust percolating around him that he sent them away for medical help, not to the closest doctor in town, but to one in the next city, down a winding, lengthy path.

Booth and Hartley were taken to the Star Inn at Roberttown, where they were

---

* Ben's injury is the focus of the account in *Ben o'Bills,* and gives us insight into the turmoil that confronted the Luddites — many of whom were shot — as they faced their first defeat.

treated by doctors and the local clergy. A crowd gathered around the inn as word spread; by one account, there were soon hundreds of people outside the doors, kept in check by a local garrison.

The Reverend Roberson, the Tory preacher who had been working with Radcliffe to organize the local militias against the Luddites, pressed the dying Luddites to confess and give up the names of their comrades. Hartley refused to answer at all. All that Booth would say was that he had joined the Luddites "in a moment of weakness," and that he regretted it, but would not elaborate.

As the night wore on, the doctors decided to amputate Booth's leg, but he'd lost so much blood already that he went into spasms during the operation.

Roberson continued to press Booth for a confession until the end. According to those present, Booth finally gestured for the preacher to lean closer.

"Can you keep a secret?" Booth said.

"I can," Roberson said.

"So can I," Booth said.

Shortly after, at 6 a.m. on Sunday morning, the young idealist passed away.

Hartley died in the early hours of Monday. In the end, neither man broke his oath. A jury was convened to legally determine their cause of death. They delivered their verdict within the day: "Justifiable homicide."

Jonathan Dean and at least one other Luddite were believed to have died as well; both were believed to be buried outside town in Hartshead, in an unmarked grave, in a ceremony quietly officiated by Pastor Patrick Brontë, whose daughter Charlotte later recounted the incident in her novel *Shirley*.

Hartley's body was transported to Halifax, and a throng of supporters gathered to follow along with it, in a parade of condolence, a show of compassion and empathy for the fallen Luddite among the townspeople. Jonathan Saville, the preacher who gave the sermon after the funeral, described the scene in his memoirs:

I preached one Sunday to the largest congregation that ever assembled in Halifax Chapel. A young man had been shot while assisting to make an entry into a Mill near Cleckheaton, for the purpose of breaking machinery; and he was brought to Halifax Chapel to be buried. On the Sunday after the funeral, the people came from far and wide to show their sorrow for the deceased [or rather, to make a political demonstration]. They filled the Chapel to overflowing; hundreds stood on the outside, unable to get in, and constables walked before the doors to keep the peace.

Such a turnout doubtless unnerved the authorities. Remembering that the funeral procession for John Westley, the Luddite killed all those months ago, had kindled so much solidarity, the magistrate Radcliffe and the authorities of Huddersfield decided to bury John Booth — who was not only popular, idealistic, and young, but the son of a luminous preacher — under the cover of night.

Headlines trumpeted news of the attack, the bloodshed, the triumphant William Cartwright, and the Luddites' first major loss.

Charlotte Brontë, who grew up hearing dramatic stories of the momentous battle from her father, the townspeople, and perhaps even from Cartwright himself, described the fallout, from the point of view of the entrepreneurial elite, in *Shirley*:

> The rioters had never been so met before. At other mills they had attacked, they had found no resistance; an organized, resolute defence was what they never dreamed of encountering. When their leaders saw the steady fire kept up from the mill, witnessed the composure and determination of its owner, heard themselves coolly defied and invited on to death, and beheld their men falling wounded round them, they felt that nothing was to be done here.... They dispersed wide over the fields, leaving silence and ruin behind them.

Yet even those who sided with the victor saw this as "no cheering spectacle," Brontë wrote.

> These premises were now a mere blot of desolation on the fresh front of the summer-dawn. All the copse up the Hollow was shady and dewy, the hill at its head was green; but just here, in the centre of the sweet glen, Discord, broken loose in the night from control, had beaten the ground with his stamping hoofs, and left it waste and pulverized. The mill yawned all ruinous with unglazed frames; the yard was thickly bestrewn with stones and brickbats; and close under the mill, with the glittering fragments of the shattered windows; muskets and other weapons lay here

and there; more than one deep crimson stain was visible on the gravel: a human body lay quiet on its face near the gates.

The blow was as dramatic and ruinous as Brontë's depiction suggests, which helped cement the battle's enduring legend. The Luddites had been received by their peers as heroes, mythical and nearly unstoppable in their machine-shattering crusade for the working class. But though concessions had been won, the failed assault made the greater trend lines painfully clear. No matter how forebodingly poverty loomed, the state would defend the machines' owners, not the men who worked them. And the Luddites clearly did not have arms or strategy yet to mount a meaningful large-scale resistance against a militarized and organized opponent.

Many of the Luddites were wounded, and the escaping men had fanned out across the fields, leaving the marks of their defeat along the way: lost articles of clothing, trampled vegetation, and dark splashes of red.

Few men outside the fallen would feel the loss at Rawfolds more than George Mellor, who had staked everything on leading this movement. Blood could be found on the trail back to Huddersfield, four miles away.

WILLIAM CARTWRIGHT

April 1812

When Cartwright assessed the damage to Rawfolds Mill, he found that fifty panes of glass around the factory had been shattered, and eight windows and their frames had been destroyed. The front door was beyond repair, so thoroughly had it been battered by the hammers and axes of the Luddites. They had almost broken through.

Among his peers, Cartwright was hailed as a hero. His victory over the Luddites was stunning and decisive: he had triumphed not just in the physical battle, repudiating their tactics for the first time, but legally as well. Since the deaths of Sam Hartley and John Booth were ruled "justifiable homicides," neither Cartwright nor his men faced legal repercussions. Quite the opposite.

With frame-breaking already a capital crime, the aftermath of Rawfolds sent a signal to factory owners that the way to defeat the Luddites was to crush them outright, with as much force as possible. The Luddites may have been ferocious in their attacks on machinery, but they had not drawn any human blood. The trick was to treat them as if those offenses were one and the same. The state would send troops, yes, but just as importantly, it would publicly and legally sanction the violence, and imbue the entrepreneurs with the right to protect their property with gunfire.

If he was celebrated by the region's elite, Cartwright was despised more than ever by the town's working people. It was widely known that he'd refused to give medical attention to the young men who were bleeding to death while he pushed for a confession. Rumors circulated that he had sent the wounded Luddites needlessly far away for medical help; and worse, that the authorities had tortured the men in their waning hours. All this earned the factory boss a new epithet — "the bloodhound."

At least two men, but probably four, were gone, and many more were injured.

And once again it seemed perfectly clear who was the kind of person who wound up dead.

Even so, Cartwright had little interest in cultivating sympathy in the community. He reported the militiaman who'd refused to fire on the Luddites to the authorities at the first opportunity, and rode into town the next Saturday to give evidence in his court martial. Cartwright knew how much he was hated — he took every precaution when he set foot in public; he refused to eat out for fear that he'd be poisoned.

George Mellor, meanwhile, was furious that the authorities had buried his friend John Booth in secret, so his funeral couldn't be properly attended. George's friends had to drag him back to the cropping shop so he didn't bring himself or anyone else to harm — or give them away. There, at John Wood's, George, Thorpe, Thomas Smith, and the rest of those present discussed the news that Cartwright would be riding to the Plough Inn the next weekend to give evidence about the attack on Rawfolds Mill. They knew exactly the path he would take. They drew straws to decide who would lie in wait.

The trial lasted less than two hours. Cartwright gave his account; the soldier did not deny the charges. A judgment was handed down to the man who had said he did not want to fire on his brothers — 300 lashes, to be administered in public, for acting as a traitor. The soldier gasped when the sentence was read; it was a death sentence. Even Cartwright asked for leniency. The presiding officer agreed to none.

Returning to Rawfolds, at the point where the town began to thin out into open spaces, Cartwright thought it best to hurry his pace a bit. He drove his spurs into his horse and they went into a gallop at about the very moment that a pistol ball went flying over the back of his horse. He ducked low, close to his horse, and rode on at full speed. A second shot rang out from the other side of the path. Cartwright escaped unscathed, but it was now clear he was a marked man.

Brazen as he was, Cartwright knew what bad optics it would be if the punishment for the disobedient soldier was carried out on the grounds near his mill. He tried, without success, to get it moved. The public flogging was clearly meant as a deterrent to others who might consider wavering in the Crown's mission to stamp out the machine breakers. But the event would also sear into everyone's brains a link between Cartwright and wanton violence against those sympathetic to the workingmen. His concern was well-founded.

Hundreds of locals came to witness the flogging, joining the dozens of soldiers who encircled the site — a very public spectacle.

The convicted soldier's shirt was stripped off and his hands were bound. As the official who was to carry out the punishment stepped forward with the whip, "the onlookers watched the preparations with evident concern," according to Frank Peel's oral history. The official raised the whip:

> It whistles swiftly through the air and descends on the white back of the soldier on which a broad red line appears, and beneath the muscles quiver visibly.

The torture began in earnest.

> Again and again the whip is raised and descends, and bye and bye the onlookers are shocked to observe that the skin is broken, the blood begins to trickle slowly down, and the sight soon becomes sickening. The women in the crowd, for there are many present, turn their eyes from the sight, and even stout-hearted men cannot forbear to express their pity for the poor wretch, who, with pale face and firmly compressed lips, suffers the dreadful torture.

He had disobeyed orders not from a military superior, but from an entrepreneur. Cartwright knew that this connection could easily set in, not only among the townspeople, but the soldiers, who were workingmen, too, and might begin to question the service their arms were being put to.

There was a commotion at the outskirts of the crowd, and Cartwright, tightly guarded, approached the official. "Many make way sullenly, gazing upon the stern man with dark and threatening brows and muttering fiercely." Most assumed the bloodhound had come to gloat, and more than a few were surprised when he asked the officer to stop the lashings.

The official listened for a minute, then signaled for the whipping to continue.

> Again the whip descends, and at every stroke the skin seems to be striped from the shoulders to the loins. Only twenty lashes have been given as yet; two hundred and eighty more are required to complete

the sentence. It is plain that the man will never live to receive them. The cries of the women wax louder; the ominous muttering of the men grows fiercer. The doctor examines the sufferer and feels his pulse. The surging crowd gather nearer to hear his report, but the stolid functionary simply steps back to his place saying nothing, and the signal to proceed is given. Five strokes more are inflicted and the crowd surges wildly, angry exclamations filling the air.

The scene was so brutal, the man's suffering so palpable, that Cartwright again stepped forward to call for an end to the lashings. If the soldiers were to kill this man, in front of his mill, with Cartwright present, who knows — he might meet with even more enraged attackers, with no one willing to defend him. The state relents at last. The sympathizing soldier is "dazed and almost unconscious," his back a bloody mess, when they finally unbind him.

Shortly afterward, Cartwright received word that thirteen pairs of shears that he had shipped to the blacksmith in Huddersfield for sharpening were smashed to pieces.

GRAVENER HENSON

April 1812

The purpose of the bill that Gravener Henson and the Framework Knitters Committee were preparing to put before Parliament was simple: rein in the excesses of unscrupulous machine owners and stabilize the industry for workers everywhere. Simple — and, at a moment when support was widespread for the Luddite insurgency — popular, too.

While the machine-breaking reached explosive heights in Yorkshire, the attacks had subsided in Nottingham. An uneasy calm had settled over the medieval city as the framework knitters organized for legislative reform. Henson had spent the last few months meeting with representatives across the Midlands, drafting the language, and readying a strategy to bring the bill to London. The effort was being supported by stockingers, lacemakers, weavers, and artisans across the region, and Henson was working tirelessly to gin up support and signatories.

The result was the bill "For Preventing frauds and abuses in the frame-work-knitting manufacture, and in the payment of persons employed therein." True to its catchy title, the bill would ban cheap, low-quality goods like the cut-ups that were, in their eyes, ruining the reputation of tradesmen, driving prices down, and swindling consumers. It would set minimum prices for an array of cloth goods to ensure that a fair wage was paid to workers and merchants, and it prohibited the practice of the bosses paying in truck, or goods that workers would have to resell or barter.

On April 19, Henson arrived in Leicester, where he began collecting samples of the industry's output to show to politicians in London, to demonstrate the finery of their trade, and meeting with the organization of knitters there, soliciting input and support. He wrote back to the committee HQ at the Sir Isaac Newton pub in Nottingham with his progress. Gravener was obsessively marshaling every scrap of good clothing and bit of data about how it was produced that he could get his hands on. He was gathering thousands of signatures in support of the bill. His meticulous efforts were reflected in the letter he sent that day:

> When you send the Red Book, send the Minutes of the Evidence which Blackner* has got; Endeavor to obtain every Species of Warp Nett, Raw and Got up; Cotton Stockings made of Single Cotton, Light Silk Hose, Stiff Silk Hose, Mr Alliotts pair of Pantaloons, from Mr Tutin; The Average Statement of Double Lap which you took. Count the Courses of the few German Rib Frames you have, Cause the Cotton and Silk Hose to be further counted. Take the Length of time that Frames have gone and their original Price if it can be obtained. Count the Jacks and Courses of the Silk Ribbs, at Woods in York Street, send to Mr. Columbell to do the same send us the Statements.

Henson and the United Framework Knitters weren't the only cloth workers desperately seeking peaceful solutions. On April 24, a petition signed by eighty thousand cotton weavers in Lancashire was sent to the Prince Regent, pleading for assistance.

"We feel it an indispensable duty to solicit the attention of your Royal Highness to the unparalleled and increasing sufferings of the poor manufacturers of this once-flourishing district," the weavers wrote. "From our knowledge of the benevolent, liberal, and constitutional principles which your Royal Highness has always professed, we are encouraged to hope, and you will not turn a deaf ear to the complaints of any class of his Majesty subjects." They pointed out that "immense numbers" of them were working for "less than one-fourth of what they earned previous to the commencement of the war with France, whilst the necessaries of life are, since that period, nearly doubled in price." Pressing for direct relief, or a policy intervention, from the state, they described "their pale and ghastly countenances — their squalid and ragged clothing — their houses emptied of furniture — their half-starved and half-clad children crying for bread, or begging with piteous moan from door to door for the dole of charity."

---

* John Blackner was a writer, editor, historian, and accused Luddite sympathizer who went on to run the *Nottingham Review*.

Henson's approach was more proactive. With his full court press of figures and examples, he would demonstrate to the Prince Regent and his ministers just how singular England's knitting trade was, the dire shape it was in, and how, exactly, it could be healed. Oh, and how many thousands of people supported such an effort; Henson's tally of signatories to the petition now sat above ten thousand, and it inlcuded a great many shop owners and hosiers, too. With the bill's contours sketched out, and a plan to lobby Parliament and get the bill considered by the House of Commons in place, Henson instructed the committee in Nottingham to circulate the text and raise funds for the campaign:

> Send some Subscription Sheets for our Bill to the Leicester Committee, Cause the Town to be Collected, Direct the Circulars to the Gentlemen of the Counties of Derby and Nottingham, send Copies from the Committee Book of the Gentlemen I have wrote to, send a considerable Quantity of the Circulars to Southwell, Derby, and Mansfield and almost every Village that they may distribute them among the Gentry.

Tomorrow, Henson would make for London.

LADY LUDD

Spring and Summer 1812

The battle of Rawfolds Mill was a harrowing defeat, but the Luddites raged on. The great loss may, in fact, have further *fueled* the anger of thousands of precarious workers across England. The next week saw a hive of activity set loose; attacks on machinery, factory owners and their factories, even food riots.

Just three days after the attack on Rawfolds, there was another, altogether different episode of uprising, in the cotton-weaving city of Stockport, Lancashire. Factory owners there had "made a sport" of falsely promising to increase wages, and the weavers' rage finally boiled over. Thousands of people took to the streets, taking aim at the entrepreneurs who had lied to them. They flocked to the mansion of the man who owned the biggest factory in the city and smashed every one of its windows. Then they moved on to the next one. Some of the largest bands were led by men in drag who called themselves "General Ludd's wives."

This was a radical gender inversion for the time, but not uncommon among Luddite parties. In fact, the practice led to some of the most enduring imagery of the uprisings. The most famous surviving illustration of the Luddites from the time may be the picture of a giant cross-dressing man waving his followers to battle while a factory burns in the distance.

The cloth work that was historically done by women — spinning and carding cotton and wool into yarn — was automated first, decades before the rise of the Luddites. This automation had been resisted, too, if less visibly — the inventor of the spinning jenny, James Hargreaves, had seen his home attacked as early as the 1760s, and Richard Arkwright himself had faced so much opposition that he moved a planned factory from Manchester to the East Midlands to avoid constant assault. Spinning was an important contributor to household income, and its automation dealt a serious blow to family wealth and stability. And it wasn't forgotten.

"The Leader of the Luddites." A rare surviving contemporaneous illustration of the Luddites, first published in May 1812.

In their crusades against mechanization, Luddite leaders donned women's clothes in what has been interpreted as solidarity with the spinners who lost their livelihoods. While also serving as a disguise, female costumes reflected the sense of hope in these protests, that a show of popular power was capable of challenging and defying patriarchal norms. While weaving, framework knitting, and cropping were trades dominated by men, a number of Luddites, and many of their sympathizers, were women. It's unlikely that women joined the nightly raids, but the movement had begun to expand, and women led protests, organized food riots, and advanced the Luddite cause.

On April 30, the owner of a Manchester cotton factory received a letter signed by an Eliza Ludd. It's one of the most eloquent of the Luddite missives, comparing the ongoing uprising to the American Revolutionary War. "Doubtless you are well acquainted with the Political History of America," the letter begins, and "if so you must confess that, it was ministerial tyranny that gave rise to that glorious spirit in which the British Colonies obtain'd their independance by force of arms." The writer bemoans the loss of "torrents of blood, yes British blood!" and implores the entrepreneur to stop his machinery. "Let me persuade you to quit your present post, lay by your sword, and become a friend to the oppress'd — for curs'd is the man that even lifts a straw against the sacred cause of Liberty."

There are few if any known instances of women participating in frame-breaking outright, but "during the troubles of 1812 women continued to be very prominent," wrote the historians Malcolm Thomis and Jennifer Grimmett.

They gave leadership to market disturbances in Manchester and Maccles-field, and the "poor misguided creaters" who took part in the Sheffield food riots of August were said to have been principally led on by women. [During later Nottingham food riots,] one of the assailing divisions bore a woman in a chair, who gave the word of command, and was dignified with the title of Lady Ludd, and a similar thing had happened in Leeds… There a party of women and boys had also been led by a woman with the name of Lady Ludd and had paraded the streets, attacking meal shops and millers' premises.

In Lancashire, General Ludd's wives "led a crowd of men, women, and boys on a rampage through the market area of Stockport, destroying food shops, and resetting prices for bread and potatoes." The mob was fierce and relentless — one factory owner came out to defend his looms and was promptly beaten and chased out of town.

Another target was the major tech founder John Goodair, whose factory, full of the automated steam-powered looms, had been fired into weeks ago. When his wife saw the mob forming, she packed up and fled. Goodair, who had perhaps sensed the way the winds were blowing, was already in London. General Ludd's wives, and the men who followed them into its doors, burned the factory, full of power looms, to the ground.

INSURRECTION

Ned and B

April 1812

"*'Vengeance for the blood of the Innocent'* is written on every door," a colonel wrote in a report to his superior in the days following the deaths at Rawfolds. The metropolis of discontent had not taken the death of the local Luddites well. On top of the threatening scrawl found on doors across Huddersfield, handbills were printed with an ominous message: "Innocent blood crys for vengeance. Now or never."

"The country is in a most perturbed state," the colonel concluded. It was an understatement.

The Luddites broke into more factories and destroyed yet more machines, looms, and shears.

Hundreds of armed men and women broke into a warehouse and stole the meat and flour stored inside (the military intervened and arrested forty people, "mostly girls"). A barn containing a threshing machine was torched. A spinning mill was set on fire in Bolton. More machines were crushed.

Secret societies of workingmen and illegal trade groups met during the nights. B, the spy, was at many of the Lancashire meetings, sending word to Fletcher (who in turn passed it on to Ryder) that "a general rising" was planned for May 1, and a half million men were ready to begin a revolution.

Food riots swept through Manchester, the hub of industrialization, and the surrounding areas. Weavers, Luddites, groups of protesters, furious artisans, and the desperately poor took to the streets and instigated what came to be known as auto-reductions; instead of robbing shop owners and food suppliers outright, women stormed the markets and demanded that the goods be sold at a fair price. Elsewhere, crowds broke into boarded-up shops and distributed the food there among their peers. One older man, helping to distribute the food, had written "General Ludd" on his hat.

Just over a week after the battle of Rawfolds, that rage coalesced around one target in Middleton, just five miles outside Manchester: a factory owned by Daniel Burton. He and his sons knew that discontent was growing — their factory was full of coal-powered looms — and so the Burtons, inspired, it was said, by Cartwright's initiative, had armed some fifty of their men and trained them to shoot. The Burtons, too, were prepared for war.

Over the next two days, the fury and anguish of the poor and the armed and now emboldened factory owners met.

Thousands of food rioters joined with agitating weavers and ordinary townspeople and rising Luddites outside Burton's Mill. Most did not carry guns, just "sticks and bludgeons," though someone fired a pistol from the throng. When the crowd surged forward to try to enter the mill, Burton's men fired blanks in an attempts to scare them off. The crowd realized they weren't real munitions, and they again charged the mill.

"If you dare to attack this factory I will resist with force of arms," Daniel Burton yelled out above the fray. According to Samuel Bamford, a weaver, reformer, and writer who lived in Lancashire and was there that day, the crowd "continued to throw stones and to use clubs."

Burton and his servants opened fire.

Five people were killed. One was a sixteen-year-old boy. Eighteen more were injured. The Luddites fled, "vowing that they would burn down Burton's dwelling house as a punishment for what they considered as a cold-blooded murder."

The next day, the Luddites regrouped and amassed an even larger force. They returned to Middleton, and this time, they were armed with guns, scythes, swords, and pitchforks.

Instead of heading directly to the factory, some of the factions split off. One made for the Burtons' mansion, which the marchers knew would be deserted. The Luddites ransacked the place. Bamford describes the younger rioters "licking out preserve jars" and eating clumps of plain sugar. They broke every piece of furniture inside. Then, Clem and Nan, two daughters of a local elder weaver, took over.

"Let's put a finish to this job," one said to the other. They grabbed some cloth that lay on the floor and took it to the fireplace, which had been left smoldering.

They threw the fiery cloth onto the sofa, which burst into flames. After half an hour, "not a beam nor a board remained unconsumed in the whole building."

The Luddites did the same to three other prominent supporters of the Burtons, piling the furniture in the street and lighting it ablaze.

General Ludd himself made an appearance, as reported in the *Leeds Mercury*:

A body of men, consisting of from one to two hundred, some of them armed with muskets with fixed bayonets, and others with colliers' picks, marched into the village in procession, and joined the rioters. At the head of the armed *banditti* a man of straw was carried, representing the renowned General Ludd whose standard bearer waved a sort of red flag.

As forces were gathering at the factory to make another attempt at destroying the power looms, the Crown's soldiers, the Scots Royal Greys, arrived, joining the hired hands defending the mill, and moved to disperse the crowd.

It was a bloodbath. Some of the armed workmen retreated to higher ground and opened fire on the military guarding the factory. The military returned the fire, quickly overpowering the workers. Again the bodies started falling. "A man named John Nield, from Oldham, was shot through the body by one of the Greys whilst attempting to escape," Bamford wrote. "Another man was shot by one of the Greys, and left for dead, near Tonge Lane; a woman, also, who was looking through her own window, was fired at by another of the same party, and a bullet went through her arm."

For all the carnage, one killing enraged Bamford more than the rest. "A Serjeant of the Militia earned deathless execration by shooting an old man," he wrote. The man, whose name was Johnson, hadn't been anywhere near the action during the day. He'd waited out the day's violence with his family at home, drinking beer in the kitchen. When dusk fell and it'd grown quiet, he assumed the riots were over and went for a walk. He'd stopped at a graveyard outside a local church, and was reading the inscription on a headstone, his hands in his pockets. A sergeant in the private militia spotted him there. "He went down on one knee, levelled, fired, and killed the old man dead, the ball passing through his neck."

The death toll from the two days of violence is unclear. One newspaper reported that twelve civilians had been killed, and another said the number dead could have been as high as thirty. It's been described by historians as "one of the most bloody and prolonged examples of civil insurrection in the history of the UK."

In the wake of the destruction and massacre, the owners of the coal-powered factory that lay at the center of it all issued a brief statement: "D. Burton and Sons have determined not to work their looms anymore."

They laid off their four hundred workers and closed the shop.

All of this would have nicely fit the portrait of impending revolution painted by spies like B: growing numbers and boldness, violence swelling to a crescendo, and acts of open defiance of military authority. Yet the targets were still factories stocked with the automating machinery, the homes built with profits made from running them, or places that stored or distributed food. Even at its most violent, the Luddite uprising had kept its focus fixed on the implements of inequality, or on a means of evening the scales.

On April 24, fifty Luddites, armed with pikes, scythes, and large sticks, made their way to Westhoughton, just outside Manchester, where a factory that housed two hundred power looms had made the jobs of nearly a thousand workers redundant. It was the same militia guards, fearing such an action, had fled. The owner had left, too, to try to rouse the military; on his way, he saw the procession headed toward his factory.

The Luddites smashed the windows to gain entry, brought in a torch, and pressed it to the cloth still on the looms. Within minutes, the entire building was ablaze.

Like Burton and sons, the mill owners quit the town for good.

As easy as it is to forget the technologies that have been rejected — whether automated cloth-weaving devices, nuclear power plants, or contemporary facial-recognition tech — "No" is, and has always been, an option — whether by policy or by force.

Power looms did not return to West Houghton for thirty years.

GEORGE MELLOR

April 1812

If you are George Mellor — young, strong-willed, expectant of just circumstances, and looked to as a leader in a fight to protect the only future that seems available to you and your peers — what now?

You watched as lines were drawn around your land and claimed by the rich. You watched as men moved machines into factories that foretold the end of your trade. You watched as they set them to run at a moment when suffering and starvation afflicted the people you pass on the streets. You organized a powerful response that gave hope to the working poor. You won the respect of friends and colleagues who followed you into battle. And now you have watched them die in the dirt.

For a time, George Mellor's Luddite army had seemed unassailable. A young man in his early twenties, especially a hotheaded and righteous one, does not always grasp the full weight of consequence before he acts. That weight now came crushing down. If he had been enraged before, well. Mellor's muscular six-foot frame was spotted around Huddersfield, wandering. He carried with him on his wrist the cropper's hoof, beaten into his body by a skill now soon to be performed by machinery; he carried the stigmata of obsolescence. The last hopeful path, as faintly outlined as it was, had been violently wiped off the map.

After the failure at Rawfolds Mill, George spiraled into a frenzied state of loss and fury. According to Frank Peel's history, George's feelings had become "monopolized by his dead friend Booth," whom he had personally convinced to become a Luddite. After Booth's death, in the raid he led, George "seemed altogether to lose his balance. He grew perceptibly day by day more bitter against the masters who had adopted the new machinery, and his subtle brain was always planning and scheming for their injury or destruction." He looked deeply unwell.

"Booth had exercised a restraining and beneficent influence over him, and now he was dead," Peel wrote.

> Mellor's whole thoughts appeared to be engrossed with what he deemed the wrongs of his class, and the whole subject of his conversation with his fellow workmen was how to avenge them. His outward appearance was a true index of the fierce and tumultuous passions that reigned within his soul. He had grown careless and slovenly of late, his cheeks had become pale, his brow careworn, and his lurid, bloodshot eyes were now habitually fixed upon the ground.*

The factory world loomed, and it could not be stopped by force. A life governed by the traditions, standards, and culture he'd grown up with had vanished like a chuff of smoke. In its place, the prospect of toiling at the whims of someone else's machinery; technology that has the power to disappear your work, your earning power, and your identity. The prospect of a life spent guessing how they'll change the algorithm next — and how you will have to serve it.

George was out one day walking alone, near Horsfall's factory at Marsden, when he encountered the wife of a cropper who had been unable to find work. He did not know her, but she was sobbing and hyperventilating, and George stopped to see what was the matter. According to the story relayed in *Ben o' Bill's,* her husband had been out of work so long he had been unable to find a way to feed them. And she had just given birth to a baby girl.†

The mother was severely malnourished and could not produce enough breast milk to feed the child. The young family tried giving the baby cow's milk, but it was in vain. The infant starved to death.

And there, according to Luddite legend, was the starved newborn, wrapped up in a blanket, still held at the mother's breast, while she keened in sorrow. As

---

* Frank Peel, *The Risings of the Luddites.*
† Given that this remarkable story's provenance is *Ben o' Bill's,* it must be treated skeptically, but it has become part of the Luddite legend.

George was trying to comfort her, Horsfall was making his way back home to his factory on horseback. He would later say he did not know what came over him, but George took the dead baby in his hands and stepped into the middle of the road.

Horsfall shouted at him to get out of the road. George stood firm.

"Look at this work, William Horsfall; look at this work, and be glad," he yelled, raising the tiny corpse to where Horsefall could not avoid seeing it. The horse reared back, and George saw a "start" in Horsfall's eyes.

As the stallion reared, Horsfall raised his horse whip, and lashed it across George's face.

"Out of my way," Horsfall said as he did so, and drove the horse on, shouting as he left: "I marked you, George Mellor. I marked you, and know you for what you are."

NED LUDD

April 1812

The hunt for Luddites was on. Huddersfield city officials, led by Joseph Radcliffe, had launched an aggressive search for anyone who had joined the attack on Cartwright's factory. Survivors of Rawfolds had already begun to slip out of town to avoid detection. Others, like Ben Bamforth, were laying low as they nursed their wounds. Some Luddites swore off the effort, deeming it hopeless, too risky, or both. Others still were furious and desperate as ever.

Radcliffe sought out all of them. One of his first promising leads came when he interviewed a man named James Haigh. The magistrate got word that Haigh had been to a doctor about a serious and unexplained wound on his shoulder. Haigh told Radcliffe that he'd fallen on a rock, but the magistrate didn't buy it, and held and interrogated him. Ultimately, he had no proof, and had to let him go.

Paranoia and fury ruled, and some Luddites turned as brutish as their oppressors when threatened. A woman named Betty Armstrong, a local believed to have informed on the Luddites, was beaten so badly after she left Joseph Radcliffe's offices that her skull was fractured.

Radcliffe himself had received death threats and, like B, was convinced things would worsen unless the military acted more decisively. (In London, Richard Ryder was continuing to respond the only way he seemed to know how — by ratcheting up the force and working to expand the litany of crimes now warranting the death penalty to include "illegal oath-taking" along with property destruction and burglary, too.) Radcliffe received one particularly colorful letter that doubled as a prediction of the ultimate outcome of automation, and a promise of retribution:

"If this Machinery is suffer'd to go on it will probable terminate with a Civil War, which I could wish to be avoided," it read. The writer warned that unless certain measures were immediately adopted, like the abolition of the Watch and Ward Act, "great Destruction" was imminent, beginning with their chief persecutors. Mr. Atkinson, a large manufacturer in Huddersfield, and Mr. William Horsfall, the letter threatened, "will soon be number'd with the dead, and summoned before the awfull Tribunal, and that God who will Judge every man according to the Deeds done in the Body. And Jesus knew their thoughts and said unto them, every Kingdom divide against itself is brought to desolation; and every city or house divided against itself, shall not stand."

More than a threat, it invoked the coming of holy war: Horsfall and his machinery, automating inequality, were dividing the Kingdom against itself and hurtling both sides toward conflict. These sentiments were not uncommon: that cataclysmic "general rising" that the spies kept warning about was supposed to unfurl in a matter of days.

In Birmingham, a city whose rapid industrialization would come to rival Manchester's, a series of handbills reflected the general attitude. The most recent singled out Prince George himself:

> How long ye Wretches will ye
> Grind The Faces of the poor —
> But the day is at hand, I will help
> to the last expiring breath in assist-
> -ing you — nothing but victims
> will do why; will you drag
> out such miserable lives
> dash away, Fear Not
> overturn the Prince
> and His Rubish —
> Fly, Fly, your
> Aid is Strong

Overturn the prince and his rubbish. There was a genuine sense that anything could happen, to the machinery transforming the land—or to its masters.

One of the military officers who witnessed the uprising against Burton's factory in Middleton put it bluntly: "If more military be not sent into the country, they will not be called upon to prevent it, but will be required to reconquer it."

WILLIAM HORSFALL

April 28, 1812

On Tuesday, April 28, 1812, the day his factory was finishing its first order for black cloth, Horsfall undertook his weekly pilgrimage to the market. He emerged, done with his business, in the early evening. The street would have been lined with underemployed weavers and beggars. He may have been in high spirits still.

After the Luddites' defeat at Rawfolds, at the next meeting of the Committee for Suppressing the Outrages, Horsfall had enthusiastically congratulated Cartwright on his "pluck and determination" in defending his factory.* He then "loudly expressed with great heat his own determination to defend to the utmost his machines, if it should prove necessary."

His ire extended not only to Luddites, but to many of his peers in the manufacturing trade. Jonathan Brook of Longroyd, for instance, ran a local blacksmith shop that built shearing frames. He received a letter from a certain Captain Blunderbuss warning that "if he made any more such machines, Ned Ludd would fire his premises and lay his body in ashes." Brook quit building the machines and posted a handbill announcing his decision. He even ran into a cropper on the street (maybe George Mellor) who told him that, if his intention was true, then "you need go no further."

When Horsfall heard of the deal, he was livid. He made his way to the foundry where the master blacksmith worked, and when Brook put out his hand, Horsfall reportedly said "no," and stepped back. "I understand you dare not make any more cloth-dressing machines. I won't shake hands with a coward."

Outside the market that Tuesday night, Horsfall stopped for his usual drink at the Warren House, a two-story stone building outside of town, at a quarter to six. After sharing a round with a fellow mill boss named Eastwood, he mounted

---

* Frank Peel, *The Risings of the Luddites,* 69.

his horse and made for home on the Manchester road, a main passage lined with a stone wall.

He'd ridden barely a quarter of a mile when just ahead of him, two men, hidden inside the bordering plantation, slid the barrels of their pistols into apertures that had been cut in advance into the wall, then signaled to two others positioned twenty yards away. Horsfall saw two muzzles swing into sight and take aim, but before he could so much as tug on his horse's reins, they opened fire. As many as five shots struck him, cutting through both thighs, his groin, and lower abdomen.

"Murder!" Horsfall yelled in agony. It was still light out. A witness who'd been walking on the road, and two startled boys who'd been collecting dung alongside it, ran toward the commotion. Eastwood, who'd been watering his horse, ran to his side. "Good man, I've been shot," Horsfall said to Eastwood, then lost consciousness and fell from the neck of his horse. "The blood flowed from the wounds in torrents," and his thigh began to swell monstrously.

Horsfall was carried back to the Warren Inn, where the final hours of his life passed in conscious torment.

The next day, everything changed. Horsfall was about to become the industrial state's first martyr to the Industrial Revolution.

Here's what was known of the murderers. After carrying out the "sanguinary deed" in front of multiple witnesses, they walked — *walked* — many yards across the plantation, before finally breaking into a jog toward a forested area known as Dungeon Wood. No one followed them, or searched for them, until the Crown's soldiers arrived nearly an hour later.

There was a sense that their identity was known and being protected. The men, most knew, were Luddites. And Horsfall had promised "ride up to his saddle girths in Luddite blood." Just as he had recognized the barrels of the muskets too late, witnesses thought that he recognized the men who found him in their sights.

"What?" Horsfall had shouted at his assassins, with what would prove to be one of his final exhalations. "Art thou not contented yet?"

When the next edition of the *Leeds Mercury* ran, the headline splashed on the front page consisted of just two words: ATROCIOUS MURDER.

The paper described one of the assassins as "about six feet high, another as a low portly man, and the two others as about five feet six or seven inches high, and rather slender; they all wore dark coarse woollen coats, and appeared to be working men."

In an early sign of how the news would affect public opinion, the *Leeds Mercury* editorialized that

> the machine destroyers, who knowing his premises were too well defended to justify an attack on his property, committed a crime against his person, that will embitter every future day of their existence, and, that will, in all probability through the retributive justice of that Being, from who no secrets are hid, bring the blood-stained perpetrators of this worst of crimes, to an ignominious end.

The paper announced a reward of £2,000 to anyone who might contribute information.

Not mentioned in the report was that, as Horsfall lay in the dirt by his mare, bleeding out, a number of the townspeople took it upon themselves to scold the dying man for how he had treated the poor.

An etching of the assassination of William Horsfall, by the famed Victorian artist "Phiz," aka Hablot Knight Browne.

AN INVOLUNTARY MACHINE

Charles Ball

1800s

Charles Ball was out hunting for snapping turtles in the deep reaches of the swamp when he heard a haunting chime ringing through the thick. There shouldn't have been anyone else out this far. Alarmed, he crouched down by a pond and listened. The "mysterious bells" sounded like they were coming right toward him. He was struck with horror; he was at least three or four miles from his housing on the plantation, too far to run, and too far to explain his presence away without inducing a flogging if he were caught.

The bells chimed nearer, and the figure that came into view only terrified him further:

> I saw come from behind a large tree, the form of a brawny, famished-looking black man, entirely naked, with his hair matted and shaggy, his eyes wild and rolling, and bearing over his head something in the form of an arch, elevated three feet above his hair, beneath the top of which were suspended the bells, three in number.

It was a black iron collar, fastened around the man's neck, padlock still attached. He carried an iron spear, also black. Charles was unsure if what he saw was real or a ghost, an emissary from purgatory, the reanimated dead sent back to earth. As he watched, "the black apparition moved past me, went to the water and kneeled down."

This was the end for him, he thought. "I now gave myself up for lost, and began to pray aloud to heaven to protect me." When the figure heard him praying, though, *he* jumped up, startled. The man begged for mercy, and beseeched

Charles not to return him to his master. Charles was flooded with relief, then mystified at himself for being so afraid in the first place. "As to carrying him back to his master, I was more ready to ask help to deliver me from my own, than to give aid to any one in forcing him back to his."

Charles had been enslaved for all of his twenty-some years, and he'd been conditioned to treat every unknown figure in his environment as a threat, because much of the time, it was.

"The suddenness with which we pass from the extreme of one passion," he would later write, "to the utmost bounds of another, is inconceivable, and must be assigned to the catalogue of unknown causes and effects, unless we suppose the human frame to be an involuntary machine, operated upon by surrounding objects which give it different and contrary impulses." The antidote — the path to debugging this involuntary machine — was, perhaps, recognizing one's fellow humanity: "I had no sooner heard a human voice than all my fears fled, as a spark that ascends from a heap of burning charcoal, and vanishes to nothing."

Ball was four years old when he was sold and taken from his mother. He grew up on a plantation in Maryland, under a master he deemed tolerable, until he was twelve, when he was sold to a brutal one. He was loaned to the navy, was sold again, moved again. Amidst these hellish shifting sands, he fell in love with Judah, an enslaved woman held by a master who lived in the neighborhood, and eventually they were married and started a small family. He was able to visit her and their children every week; it was their refuge, a sliver of good in a violently unjust world.

That sliver was soon removed. One morning, Charles's master told him to come into the house to have his breakfast. As he ate, he overheard his master talking in a low voice with a stranger at the door. Then, suddenly, "this man came up to me, and, seizing me by the collar, shook me violently, saying I was his property, and must go with him to Georgia. At the sound of these words, the thoughts of my wife and children rushed across my mind, and my heart died away within me." He was not given a chance to say goodbye before he was bound and chained to a gang of other enslaved men and women who were to set march, immediately, to the south.

This abrupt, traumatic departure was facilitated by men seeking to profit from a commodity — the same one that inspired countless events just like this family separation across the nation. That commodity was cotton, and the industry that produced it was booming, and transnational.

Although the Luddite uprising unfolded in England, British cloth workers were far from the only ones suffering under the rise of the factory system, automated technologies, and unrestrained industrial capitalism. Those systems, deeply interlinked, were fast spreading across the globe. At multiple junctures of the emerging industrial supply chains, early automation technology offered unscrupulous innovators and masters the opportunity to gain advantage and dominion over workers. It led to great wealth, hideous exploitation, and atrocious unintended consequences.

So it was that a terrible confluence of historical events, guided by racism and the emergent forces of global capital, by plantation owners, traders, and the cotton barons of Manchester, drove the repeated enslavement of Charles, the blood in its machine. For workers like him, it led to punishing extremes and suffering. For the elites, it solidified their power.

Nowhere was all of this more evident than in the American South, where slavery was joined with automation to produce the raw material for factories in Manchester to spin into cloth.

Recall that for centuries, Britain's major industry was wool production — an economic infrastructure comprising thousands of weavers, artisans, merchants, mechanics, and exporters. But from around the middle of the eighteenth century, a new industry emerged: cotton cloth production. Cotton had a key advantage — machines could process cotton fibers better than wool. Production was easier to automate. And cotton could be cheaply grown in a number of countries under British colonial rule.

Cotton was a small industry in Britain in the early 18th century. By the end, it was a force gathering nearly unstoppable steam. Thanks to the spinning jenny, the water frame, and the power loom, England gained the ability to process cotton into cloth products at an unparalleled pace and scale. Wool had laid the groundwork, but it was cotton — imported from the United States, where it was picked almost entirely by enslaved workers—that unleashed the Industrial Revolution.

It would have been impossible to profitably pick cotton at the scale required — the machines needed huge volumes of raw material — had it not been for a young American named Eli Whitney, who, in 1793, just a few years before Charles was kidnapped, chained, and sent south, invented a machine called the cotton gin. Whitney was looking to make a fortune and pay off his student loans by solving a problem that plagued business in the south. Tobacco plantations were in trouble, and the greatest promise lay in cotton. But cotton plants that grew inland had sticky green seeds that were a pain to pick out of the fluffy white fibers that were used to make clothes. The industry was bottlenecked.

Whitney's machine separated those seeds from the white cotton balls with a hand crank or a horse pull. "One man and a horse will do more than fifty men with the old machines," Whitney wrote to his father. "'Tis generally said by those who know anything about it, that I shall make a Fortune by it." The plan was to patent the device, build units himself, and then charge plant owners a portion of the profits they made by using the gin. Whitney even suggested that his device could help end slavery, since laborers would no longer have to do the unpleasant work of picking the seeds out by hand.

That is not what happened. Instead, the cotton gin is one of the original sins of automated technology, and the most disastrous case of unintended consequences unleashed upon the world this side of the nuclear bomb. Whitney's machine was widely pirated, modded, and adopted by plantation owners, who saw little need to compensate the inventor. The cotton gin worked so well that it wildly increased the demand for workers to do every other part of the cotton production process, especially the hoeing and the picking. Slavery, an institution whose future was at the time in question — Northerners wanted it abolished, and were drawing close to legislating restrictions — received a lifeline, then an economic raison d'être. The export of cotton became the biggest industry in the United States, so economically powerful, generating so much wealth for plantation owners, that it helped sustain the institution of slavery for another seventy years.

And that is what Charles saw taking shape across the fields and valleys of the southern United States as he was forced to march in bondage at the turn of the century. "The tobacco disappeared from the fields, and the cotton plant took its place, as an article of general culture," he noted, as they left Virginia. They took a ferry across the Yadkin river, and his new master told them they had arrived. "We staid this night in a small town called Lancaster; and I shall never forget the sensations which I experienced this evening, on finding myself in chains, in the state of South Carolina." (The town was named after the House of Lancaster, which also gave Lancashire its name, and where the cotton production industry was centered in England.)

Along the way, Charles heard time and again how valuable cotton, and those who were able to pick it, would be. The value of cotton, one landlord told their master, "had not been higher for many years, and...the boys and girls, under twenty, would bring almost any price at present." This was because even automated mass production still needed many human hands. (The same is true today.) "Slavery provided the raw material for industrial change and growth," according to a BBC survey of period historians. "The growth of the Atlantic economy was an integral part of the growth of exports — for example manufactured cotton cloth was exported to Africa. The Atlantic economy can be seen as the spark for the biggest change in modern economic history."

"British cotton imports rose from £11 million in 1784 to £283 million in 1832," the historian and sociologist Eric Williams wrote. "The New World, thanks to Eli Whitney, had come, not for the last time, to rescue of the old." And the growth was exponential. The automated cotton gin and the labor of enslaved workers on one side of the Atlantic met the booming demand of the automated factories churning out the woven cloth on the other. The South Carolina plantation owners and the Manchester factory chiefs got richer, while the slaves and the children, women, and unskilled workers who ran their machines increasingly suffered. Days grew longer, breaks fewer, to keep up with the pace of production. As Williams wrote,

In 1785, the exports of British cotton manufactures exceeded £1 million in value; they were £31 million in 1830. The cloth printed in Great Britain increased from 20 million yards in 1796 to 347 million in 1830. The population employed by the industry rose from three hundred and fifty thousand in 1788 to eight hundred thousand in 1806. There were sixty-six cotton mills in Manchester and Salford in 1820, ninety-six in 1832. Cotton was "raising men like mushrooms."

Meanwhile, it was causing an explosion in slavery. "The first federal census of 1790 counted 697,897 slaves; by 1810, there were 1.2 million slaves, a 70 percent increase."

The profits cross-pollinated; so did the modes of work. The factory system, pioneered in England and fine-tuned in factories employing unskilled and child workers, was emulated first in the West Indies sugar plantations, then in the American cotton plantations. It used the brutal "gang system" of slave labor, in which enslaved people were regimented into groups, as if forming an assembly line, and overseen by taskmasters. "Whilst on the cotton estates, I have seen four or five hundred, working together in the same vast field," Ball recalled. The system was even more ruthless than the already horrific, less regimented "task system" that at least permitted the enslaved some freedom of motion to complete their work. It was so brutal that men routinely and desperately tried to escape, such as the chained and bell-laden man whom Charles encountered in the swamp.

That man introduced himself as Paul. He said he hailed from the Congo, where he'd been ripped by slavers from his wife, widowed mother, and children. He had picked up some English, but spoke haltingly, and his back bore layers of scars so thick and lacerated that they had hardened into a knotted mass. Paul said that his master had taken an irrational disliking to him almost immediately, and he was whipped and beaten for no reason at all most days, and he couldn't take it anymore. He'd escaped to these swamps before, and was recaptured and saddled with the hangman's bells.

Charles said he'd do what he could to help. If Paul could just stay hidden here, Charles would return to the spot next Sunday, he said, with an instrument to break the manacles. Charles gathered the terrapins and eggs he'd collected that day, then sparked a fire from the iron around Paul's neck, and used it to cook them dinner. Charles told Paul to hold tight, and found his way through the dark, back to the sprawling South Carolina plantation complex.

The words *South Carolina* had long been a bugaboo for Charles and his friends and family back in Maryland. It was shorthand for a special kind of hell, even in the hellish context of slavery in general. When he arrived in the South for the first time, he immediately felt those fears being borne out. He recalled surveying his new station: "I was now the slave of one of the most wealthy planters in Carolina, who planted cotton, rice, indigo, corn, and potatoes; and was the master of two hundred and sixty slaves." As to the plantation, "the description of one great cotton plantation will give a correct idea of all others."

There were thirty-eight cabins for the enslaved population. Ten to twelve enslaved workers toiled in the garden and the house. "At a distance of about one hundred yards from the lines of cabins stood the house of the overseer; a small two-story log building, with a yard and garden attached to it.... This small house was the abode of a despot, more absolute, and more cruel than were any of those we read of in the Bible." Beyond the garden was a towering building, "constituting the principal feature in the landscape of every great cotton plantation." That, of course, was the cotton gin, and the adjoining sheds to contain the cotton.

Most worked the field, like Charles. The overseer would lead a "wretched-looking troop" off to the rows of cotton plants. "There was not an entire garment amongst us. More than half of the gang were entirely naked," men, women, and pubescent boys and girls alike. The first day of work, no one but Charles had headwear. He still had the hat his wife made for him back in Maryland, but he got rid of it to avoid drawing attention to himself.

It was a regime predicated on control, mechanistic work expectations, and constant surveillance: the embryonic factory system taken to a sudden and ferociously barbaric extreme. In some ways, the hellish vision of factory organization that the weavers, Luddites, and workers of England feared would devour their way of life had already been realized, in its most punishing and sadistic form, in the United States. It would not be correct to say that the driver of both circumstances ultimately rose from the same root source—the legacy of racism and colonialism did not bear down upon English workers in at all the same way—but the unprecedented, automatizing engines of profit spurred demand for both the expansion of the African slave trade and the destruction of the British weaver's way of life.

Many in both groups were active in staging resistance, too. They understood deeply the nature of the inequalities they were being made to suffer under. However, the sufferings were not at all equal. The Luddites may have been starving and overworked in oppressive conditions; but they were not experiencing a total racist domination and subordination as was enacted on enslaved Black people, who suffered constant physical abuse and were afforded no freedoms at all.

Yet on both sides of the Atlantic, they dreamed similar dreams of equality—and of vengeance, of rising up.

"The idea of a revolution in the conditions of the whites and the blacks, is the corner-stone of the religion of the latter," Charles said,

> and indeed, it seems to me, at least, to be quite natural, if not in strict accordance with the precepts of the Bible; for in that book, I find it every where laid down, that those who have possessed an inordinate portion of the good things of this world, and have lived in ease and luxury, at the

expense of their fellow men will surely have to render an account of their stewardship, and be punished, for having withheld from others the participation of those blessings, which they themselves enjoyed.

The passage rhymes with E. P. Thompson's chronicling of the distressed workers in England, too. Thompson notes that "faith in a life to come served not only as a consolation to the poor but also as some emotional compensation for present sufferings and grievances: it was possible not only to imagine the 'reward' of the humble but also to enjoy some revenge upon their oppressors, by imagining their torments to come."

"The slave sees his master residing in a spacious mansion, riding in a fine carriage, and dressed in costly clothes, and attributes the possession of all these enjoyments to his own labour," Charles notes, "whilst he who is the cause of so much gratification and pleasure to another, is himself deprived of even the necessary accommodations of human life."

As with the Luddites, these ideas and dreams were not only the province of the imagination; enslaved workers resisted their oppression by breaking machines and challenging the owners. America's "enclosures were conquest of Indian lands and its Luddites were insurrectionary slaves," the historian Peter Linebaugh wrote. "The destruction of farm implements by those working them on American plantations belongs to the story of Luddism, not just because they too were tool-breakers, but they were part of the Atlantic recomposition of textile labor-power. They grew the cotton that was spun and woven in Lancashire."

And those who rose up were just as canny and aggressive in breaking the machines as the Luddites. "Slaves engaged in a remarkable variety of acts to demonstrate their discontent," the historians John Hope Franklin and Loren Schweninger note. "Many openly defied the system." They regularly engaged in "day to day" resistance, most of which

involved "crimes" against property. Slaves pulled down fences, sabotaged farm equipment, broke implements, damaged boats, vandalized wagons, ruined clothing, and committed various other destructive acts. They set fires to outbuildings, barns, and stables; mistreated horses, mules, cattle, and other livestock. They stole with impunity: sheep, hogs, cattle, poultry, money, watches, produce, liquor, tobacco, flour, cotton, indigo, corn, nearly anything that was not under lock and key — and they occasionally found the key.

When conditions became intolerable, many refused to work or demanded concessions, threatened overseers and their masters, and physically fought back. "Verbal and physical confrontations occurred regularly, without regard to time and place."

In January of 1811, just before the first outbreaks of Luddite action unfolded in Nottingham, one of the largest rebellions of enslaved people in US history exploded in what is now Louisiana. Hundreds of enslaved laborers who worked the sugar plantations along the Mississippi River marched toward New Orleans over the course of two days, armed mostly with planting tools and axes. They gathered recruits as they marched, set fire to five plantations, and began a full-blown insurgency. They killed two white men who worked on the plantations in the process, and by the end ninety-five slaves were wounded, killed, or executed.

These events occurred in the same year as the organized machine-breaking that spurred the largest domestic occupation in England. It was a raw, barbarous flashpoint for industrial capitalism. Perhaps it should be little surprise that the violent subordination of human workers into mass-production systems and the first industrial supply chains was met everywhere with a great paroxysm of resistance — and in neither case did punitive measures deter that resistance.

After the week passed, Charles made his way back to the clearing where they'd agreed to meet. No one was there.

There were footprints, and the bones of frogs and turtles eaten around the remnants of that fire, but they were smoothed over by the rain that had fallen a few days ago. As he set off looking for Paul, worrying that he'd been recaptured, a putrid smell overran him. A calf, perhaps, that Paul had knocked over and killed for food, was rotting now in the swamp. That would explain the "cloud of carrion crows" circling overhead. Charles walked toward it, regardless.

He recoiled when he saw its source. Paul's lifeless body, "mangled and torn," hung from a cord made of hickory bark tied around his neck and bound to a tree branch. The stench was overpowering, but Charles lingered long enough to see that the body's "identity was beyond question, for the iron collar, and the bells with the arch that bore them, were still in their place." It appeared that Paul had stripped the bark, climbed the tree, tied the noose, and jumped to his fate.

The iron bells, in fact, had prevented his corpse from being devoured and thus returned to nature.

"I observed a crow descend upon it," Charles would later say, "and make a stroke at the face with its beak, but the motion that this gave to the bells caused them to rattle, and the bird took to flight."

PART IV

MORE VALUE THAN WORK OR GOLD

THE PRIME MINISTER

May 1812

On May 9, the lyrics to a hymn that had been circulating that spring were posted around the streets of Nottingham.

Welcome Ned Ludd, your case is good,
Make Perceval your aim;
For by this Bill, 'tis understood
It's death to break a Frame —
With dexterous skill, the Hosier's kill
For they are quite as bad;
And die you must, by the late Bill —
Go on my bonny lad! —
You might as well be hung for death
As breaking a machine —
So now my Lad, your sword unsheath
And make it sharp and keen —
We are ready now your cause to join
Whenever you may call;
So make foul blood, run clear & fine
Of Tyrants great and small!
P.S. Deface this who dare,
Shall have tyrants fare,
For Ned's Every where,
To both see and hear
An enemie to Tyrants

That the prime minister, Spencer Perceval, would be made a target of populist battle hymns was hardly surprising. He was despised among the working class; his administration was blamed for the rampant, unchecked poverty, and he was seen as callous and out of touch. He was the disciple of his predecessor, the free-market apostle William Pitt, and favored no interventions or assistance to impoverished working people. He had vigorously continued waging the war against Napoleon, and he taxed Britons to pay for it; he'd handed the Home Office to the punitive Richard Ryder, and he gave off the air of being uninterested in the well-being of the common citizen.

"He has looked at human nature from the top of Hampstead Hill" — the richest neighborhood in London — "and has not a thought beyond the little sphere of his own vision," the cleric Sydney Smith once wrote of Perceval.

Back in February, that missive signed "Ned Ludd Clerk" had emphasized that "the immediate Cause of us beginning when we did" was news of the Prince Regent's letter informing his former Whig allies of his decision to keep Perceval in power. That move left the Luddites with no hope of change, as it was Perceval "to whom we attribute all the Miseries of our Country."

On May 11, an unremarkable-looking man named John Bellingham sat on a bench in the lobby of the House of Commons, watching the statesmen enter and exit the building. When he finally spotted the prime minister, he stood up, walked over to him, and shot him through the heart at point-blank range.

"I am murdered!" Perceval shouted.

Bellingham returned calmly to the bench, behind the fallen minister, and placed the literal smoking gun on his lap. He was apprehended without a struggle and taken into custody. Perceval died within minutes. He was the first and, to this day, only prime minister of England to be assassinated on the job.

Bellingham later said he thought that Britain would cheer the murder as an act of justice, and he was not entirely wrong — by the time he was led away from the crime scene, a crowd had gathered and many called out their support. They swarmed the carriage that he'd been loaded into to take him to prison, and threw open its doors, encouraging him to escape.

Sir Samuel Romilly, a member of Parliament, was there during the event. "Among the multitude...whom the news of so strange and sudden a catastrophe had soon collected in the street, and about the avenues of the House, the most savage expressions of joy and exultation were heard: accompanied with regret that others, and particularly the attorney-general, had not shared the same fate."

Speculation arose immediately that Perceval's assassination was the work of the Luddites — it had been only two weeks since the murder of William Horsfall — or of a radical Jacobin conspiracy. Those apocalyptic predictions of a "general rising" in early May were suddenly a whole lot more credible. A letter to Lord Granville from a correspondent in Wolverhampton, a city in the Midlands between Birmingham and Manchester, captures the mood after the event:

> Every serious well-disposed person is struck with horror; but I am sorry to say that numbers of a quite different description have been shewing marks of rejoicing, by firing Guns till near midnight, & the greater part of this Day! Boys in the street are taught to exclaim — *now the great Man in the Parliament House is dead, we shall have a big Loaf!* My ears are assail'd as I pass along the streets with declarations of distress, & almost threats; and I have too much reason to dread that it cannot be long before some serious Event must take place, as the lower classes seem quite ripe for it.

And the historian Frank Peel noted,

> up to the time of the shooting of Mr. Perceval no clue had been obtained to lead to the discovery of the actors in either the Rawfolds mill fight, or the assassination on the Crossland Moor, it was but natural to conclude, at first, that the shooting in the lobby of the House of Commons was part and parcel of the tactics of the Luddite conspirators; and almost universal among the elites was the alarm and dread as to what would happen next, or whose turn it might be to fall by the hands of the assassins.

Elites may have been horrified, but vast swaths of the country *celebrated* the prime minister's death. In Nottingham, for example, "a crowd assembled with a band of music and paraded the streets for fully two hours, leading public rejoicing."

The assassination was no revolutionary act, however. Bellingham soon announced that he had acted out of a personal grievance over the government's unwillingness to help compensate him for wrongful imprisonment during a business trip. It's impossible to say whether the general mood of discontent — the circumstances that inspired Luddites' repeated calls for Perceval's removal, the public threats on his life, and so on — helped coax the idea into reality.

In fact, a number of magistrates and officials — among them Colonel Ralph Fletcher, the employer of B the spy — simply refused to believe that Bellingham's actions were *not* in some way related to a coming uprising.

While "the mob" was celebrating the news there, Fletcher wrote, "The Loyalists here cannot accede to what is stated in the Public Prints — viz that Bellingham had no political motive for committing the foul deed. We here, from the general language of the disaffected and from secret Information, of a Revolution…expected to have taken place early in May, cannot refrain from entertaining an opinion that Bellingham's motives were revolutionary."

Bellingham was rushed to trial, where he offered a bizarre but lucid account of why he'd been moved to kill the prime minister. He explained that he was doing business in Russia, when another businessman falsely accused him of holding a vast sum of debt. Bellingham was thrown into a Russian prison, where he languished for five years, and the British ambassador did nothing to help. The event left him bankrupt and furious. His grievance, in other words, appeared to be justified, even if his murder of a head of state was clearly not.

Bellingham declared himself insane in his trial, showed no remorse, and calmly recounted all the steps he had taken to obtain remuneration, including multiple letters to Richard Ryder at the Home Office, all of which were denied and ignored.

Remarkably, he cast his case as a lesson in the importance of a responsive government, proclaiming "that justice refused to me which is the duty of government to give, not as a matter of favour, but of right,…and I trust this fatal catastrophe will be warning to other ministers." It was not exactly a winning legal strategy, but Bellingham persisted. "If they had listened to my case," he continued, "this court would not have been engaged in this case, but Mr. Perceval obstinately refusing to sanction my claim in Parliament I was driven to despair, and under these agonizing feelings I was impelled to that desperate alternative which I unfortunately adopted."

There is more than a whiff of the desperation of the times in his act, though his circumstances are radically removed — feeling that he had had nowhere else to turn, "bereft of all hopes of redress" as he put it, and so took extreme action.

It also bears noting that one of Perceval's final official acts may have been to help appoint judges in another case — that of the Luddite uprisings in Stockport and Manchester, in which working people felt (more credibly) they were bereft of all hopes of redress — to personally help ensure the defendants were punished.

"The trials for offences in Lancashire took place before a Special Commission at the end of May and beginning of June 1812," the historians John Hammond and Barbara Hammond write. "The Judges were Baron Thomson and Mr. Justice Le Blanc. It seems possible that their selection was one of Perceval's last acts before his assassination." Perceval wrote to Ryder saying that he would choose the judges if an inquiry determined he should do so; the hope was clearly that said judges would not be sympathetic or lenient toward the Luddites.

Despite Bellingham's numbing testimony, and his less-than-revolutionary aims, he became a folk hero to many. The public raised a subscription for his family that was reportedly ten times what they would have been able to obtain otherwise.

With the prime minister dead, it was, some have argued, among the ripest moments for a full-scale popular revolution that England has known. "If there was ever going to be a revolution in Britain," the neo-Luddite historian Kirkpatrick Sale wrote, "it would have been at that moment."

And the letters kept coming. Richard Ryder received one as soon as the next day, informing him that "every frame Breaking act you Make an amendment to" — he had recently sought to append making secret oaths to the list of crimes punishable by death — "only serves to shorten your Days." He should prepare, the letter said, to go to the devil, to be "Secraterry for Mr Perceval theire." Great troops are amassing, it insisted, to destroy all the obnoxious forces in both Houses, as they had been at "a great Deal of pains" to destroy the country. As such, "It is now your turn to fall. The Remedy for you is Shor Destruction Without Detection — prepaire for they Departure and Recomend the same to thy friends." It was signed, "Luddites."

Bellingham was not allowed to enter any evidence of his insanity; the jury deliberated for just fifteen minutes before convicting him of murder. He was sentenced to hang the next week.

LORD BYRON

May 1812

Lord Byron rented a window seat to the execution. Intrigued by the momentous event, the poet booked a flat by Newgate Prison, where Bellingham's capital punishment would be carried out before an enthralled public.

Byron had traveled to London with two of his old friends from school, John Madocks and Mr. Bailey, but when they arrived at the residence that they'd rented at 3 a.m., they found it locked. While Madocks went to find a way in, Byron and Bailey sauntered up the street, arm in arm. Along the way, they came across a homeless woman lying on a stoop.

Byron stepped over to offer her a few shillings, as his friend would later relay in a letter; but "instead of accepting them, she violently pushed away his hand, and, starting up with a yell of laughter, began to mimic the lameness of his gait. He did not utter a word; but 'I could feel,' said Mr. Bailey, 'his arm trembling within mine, as we left her.'"

Byron, whose fame was growing by the day, and who considered himself an advocate of the poor, was shaken by being the target of her scorn.

The next day, Bellingham was hanged before a packed public square. The Crown's troops were there in force, stationed to guard against a "Rescue Bellingham" movement rumored to have been organized by his sympathizers. Just before 8 a.m., Bellingham was led to the scaffold, where he was blindfolded and the noose placed around his neck. The trapdoor dropped as the clock struck eight. Hours later, after his body was taken from the gallows, the surgeons who dissected his body found that his heart was still faintly beating.

Byron watched the execution from his room. "On Monday, after sitting up all night, I saw Bellingham launched into eternity, and at three the same day I saw turmoil launched into the country."

While Byron was in London, he went to a ball also attended by the Prince

Regent, who asked to meet the young celebrity poet lord. (Prince George almost certainly did not know that just a few weeks earlier, Byron had mocked him in the anonymous poem, "To a Lady Weeping," that had been published in a major London paper.)

Byron described the meeting with the prince in a letter to Sir Walter Scott, who was, if only for a little while longer, the nation's most eminent poet.

"He ordered me to be presented to him at a ball; and after some sayings peculiarly pleasing from royal lips, as to my own attempts, he talked to me of you and your immortalities: he preferred you to every bard past and present," Byron wrote. The prince spoke "with a tone and taste which gave me a very high idea of his abilities and accomplishments, which I had hitherto considered as confined to manners, certainly superior to those of any living gentleman."

Prince George invited Byron to attend a levee, a formal meeting where royalty mingles with the public, but Byron never did. "My curiosity was sufficiently allayed," he wrote, "and my politics being as perverse as my rhymes, I had, in fact, 'no business there.'" There were similarities between the prince and the poet, to be sure. Both were plagued with scandals brought on by predatory womanizing and infidelity, and both enjoyed basking in their literary celebrity. Byron seems to have been legitimately charmed by the Prince Regent during their meeting. The ease with which he moved through elite social circles, the pleasure he took at hearing the prince's praise; it serves as a reminder of the distance between Byron and true solidarity with the subjects of his defense.

Byron's politics may have seemed "perverse" — sympathizing with romantic underdogs, and not with the state, that is — but they comported neatly enough with the social stratum he traveled in to grant him comfortable access. Part of his aim was to raise eyebrows, to burnish his own legend. He was a celebrity happy to speak up to power in a public venue, but unable to do so in private, when power was staring him in the face. It speaks to the remove caused by power, the often performative nature of liberal politics, and why it is so difficult to bridge that remove; even the Luddites' most famous, popular, and strident public defender privately cherished the praise of the very prince who had authorized their destruction.

To Byron's credit, this would not be his last return to publishing broadsides against the Prince Regent and his policies. And next time, he'd do it openly.

CAPTAIN FRANCIS RAYNES

May 1812

The thirty-five-year-old Captain Francis Raynes was on leave from military service when he got word that the British army was mobilizing to put down the "disturbances," the Luddite uprisings in the North. The home secretary, Richard Ryder, had gone all-in on a strategy of military occupation and suppression. In his notorious speech to Parliament to press for the Frame-Breaking Act, Ryder noted how the force sent to Nottingham and the Midlands was the largest domestic occupation ever on British soil. That was nothing compared to the force sent north now. General Ludd had held sway long enough.

As Raynes would later write:

> In the earlier stages of this insurrection, the civil authorities, aided by the local militia and yeomanry cavalry, were deemed sufficient to check its progress: but as the outrages increased, the legislature, after passing acts for establishing a police, according to the ancient principle of watch and ward, and for making the destruction of machinery a capital offence, punishable by death, ordered a strong military force, consisting of cavalry, infantry, and artillery into the disturbed districts, under the command of Lieutenant-General the Right Honourable T. Maitland [and] his second in command, Major-General Acland.

Maitland, a well-respected and eccentric military strategist, took 7,000 soldiers, 1,400 of them mounted, under his direct command. He would also control the 4,000 men in the Midland District troops, and the 1,800 men stationed in the West Riding. As the historian Robert Reid points out:

Certainly many generals had fought major campaigns on foreign soil with less and that number is far more than the force Wellington sailed with to the [Iberian] Peninsula in 1808. To appreciate the significance of this number in terms of the newly discovered deep anxieties of the British government and of its newly adopted policies, it has to be seen in the context of Richard Ryder's speech to the House of Commons three months earlier. Then the Home Secretary had described the regular force of 3,000 acting the Midlands as being larger than ever before in history to be used in a local disturbance. Maitland now had under his command more than four times that number. But even these figures do not fully represent the immense force of arms at his disposal. At meetings shortly to be convened by him in Yorkshire, he was to assess that he had access to 12,000 soldiers of local militia regiments; a similar number was also available to him in Lancashire and Cheshire.

In total, Maitland controlled some thirty-five thousand soldiers.

The general's first move was to mobilize thousands of troops to Manchester, where rioting had been intense, in a show of force. He marched the armed soldiers, rocket detachment, and mounted artillery directly into the hub of industrialization as if it were a hostile foreign capital. He had enough troops to repeat such a maneuver at every other hub of the uprising.

Raynes had been on leave on account of a fever that had rendered him too sick to fight abroad, against Napoleon. Now, like other soldiers and officers who found themselves in England as the campaign against the Luddites was taking root, he eagerly signed on. Raynes was excitable and impetuous, and he left his wife and family to catch up with Maitland's forces. He was promptly given charge of two companies. In his memoir, he described what the soldiers and militia were up against, and how little the Luddites were cowed by the Crown's occupation:

The Luddites had attained a military system of organization, and held their meetings upon commons and moors, for the purpose of drilling, &c. and posted their sentinels to give the alarm, in case of discovery. Signals were made by the firing of a gun, and not unfrequently, by rockets, and blue-lights. Their musterrolls were regularly called over; not by names, but by numbers; each man answering to his own number. In the neighbourhood of Ashton-under-Line, on a Sunday, during divine service, they were seen drawn up as a regular battalion: in short, a most extraordinary degree of concert and organization existed amongst them, which only strong measures could counteract or subdue.

The Luddites may have been well-organized, fearless, and tactical, but the force that the Crown deployed against them now was enormous. At the time, there were around thirty-five thousand people living in all of Nottingham. A force that size, clearly marked by the soldiers' imperial red uniforms, and carrying the most sophisticated weaponry of the day, marched into civilian districts, sent quite a message to those who lived there.

As Reid put it: "The purpose to be achieved by this grand total of upwards of thirty-five thousand men, as Maitland saw it, was quite clear: the subjugation of the north of England."

B

More delegations from Ireland. More promises that thousands were ready to rise. More alarming news from Huddersfield. And in Manchester, disarray.

B wrote to Colonel Fletcher to tell him that he had met a delegate called Taylor from Staffordshire, just west of Nottingham. Seven thousand men were sworn in there, and they had over 2,300 guns and swords. Pikes, too. Delegates from Ireland were passing through as well, bringing word that hundreds of thousands were ready to rise. The same story, essentially, that B had relayed to the magistrate — and the magistrate had relayed to the Home Office — for months.

There was a different refrain: the latest report also seemed almost awed by the power of the Yorkshire faction of Luddites. B had met another man from Huddersfield, who said that all the local croppers were twisted in to swear an oath along similar lines they already knew about, but with the following difference: "They are to help aid & assist to the best of their power & ability in destroying all Mechenery who May be determental to them or deemed so and Bound to keep the secerit at the forfeit of their life."

Not only this, B wrote of the Huddersfield croppers, but "it is impossable to stop them at present."

It is, as always, hard to know whether B was embellishing his stories, and if so how *much* he was embellishing, or whether the delegates that passed through these meetings were overstating their case — or whether there were tens of thousands of men awaiting a unifying thread to bind them together, and that a nationwide insurrection was imminent. It was certainly *possible*.

The military now occupied the spaces where the front lines of the Industrial Revolution were drawn. Across those lines, every factory that pushed ahead with automated production, or exploited existing workforces, or threatened to produce a future where workers had no say over the conditions in which they worked, was a site of potential and explosive conflict.

And the Luddites had lit a spark. Their campaign against technologized oppression had dominoed into food riots, protests, and mass unrest. And if ever it was clear that the Luddite cause was not based on some misplaced fear, it was now. For a glimpse of the future that the state endorsed, working people had to look no further than the military encampments laid out alongside the automated factories staffed by unskilled laborers and children.

It was *because* the working men and women so intuitively understood the cause of their exploitation, the direction the country was headed in — toward more subservience of factory managers and less freedom, less say over their daily lives — that King Ludd directed his troops so vigorously against it.

It may have been lucrative to be a spy in such cataclysmic times. It was also dangerous. Word spread that on May 9, an informer was discovered by the Manchester Luddites — and that, after he'd refused to be twisted in, he was drowned.

B was doubtless alarmed, and not merely for his immediate safety. The men and women he moved among were desperate, maybe, but they were also opening up a space, amid the chaos, where change, perhaps even radical change, felt possible. They were filled with anger, anguish, and exhilaration, and all of it was targeting a relatively small elite. And B happened to be a member of that elite.

His name was John Bent, and he was a merchant of cloth waste in Manchester.

And everywhere he went, Bent still saw upheaval coming. At the next meeting, he counted thirty-seven delegates hailing from all across the region. The weavers were calling for a general strike, to begin in June. There were reports that Luddites were successfully twisting in local soldiers, that the local Watch and Ward Committee had to warn the troops to refrain from drinking with locals. There were plots to gather and stockpile arms.

Though not a craftsman himself, Bent was trusted enough by the men he did business with to be allowed into their inner circles, secret committees, and Luddite meetings. There was no guarantee, of course, that the trust would last.

GEORGE MELLOR

May 1812

"Justice Radcliffe never rested," Ben Bamforth proclaimed in *Ben o' Bill's*. "The least rumor that reached his ear was sufficient to justify an arrest, and no one knew when it would be his turn to be summoned to Milnsbridge House and have an ugly half-hour in the sweating room."

Radcliffe had moved his operations into overdrive. Anyone he suspected of being a Luddite he held in a chamber in his mansion that he'd converted into a holding cell, sometimes for days on end, in conditions that were worsening rapidly. One such suspect was James Haigh, who Radcliffe was certain had been wounded by a musket shot at Rawfolds and not, as Haigh claimed, by a fall. Radcliffe had already shipped Haigh off to York prison to await trial.

The magistrate had leaned into the Watch and Ward Act with abandon, directing his conscripts to survey and round up any suspected Luddites in town. "A vigorous body of police was formed," as Frank Peel explains. "So closely were [suspected Luddites] watched indeed that they found it dangerous to meet as before."

The reward for information leading to the discovery of Horsfall's murderer was roughly forty times what a cropper made in a year, and still no one came forward. Radcliffe had to haul in the suspects and witnesses and passersby and persons of interest himself; in they came, to his private prison at Milnsbridge, sometimes for hours or longer. "I do not know what warrant Justice Radcliffe had for such examinations," Ben continued, "probably none. But, then, how were ignorant folk, half frightened out of their wits, to know this; or if they knew it, how was their knowledge to serve them?"

Radcliffe's investigation was aggressive and probably illegal. Probable cause or evidence were not necessary to justify hauling suspects before the magistrate, but they were necessary to detain them there. When that Luddite letter-writer pointed out the oppression inherent in these tactics, this is what they meant, and this is what they feared. Their home had become an absurdist police state, with, thanks to the Watch and Ward Act, working men and women, many of whom were certainly Luddites, being conscripted into militias to try to round up Luddites.

So Milnsbridge became a local interrogation center and jail; men were brought in, released, brought in again. Yet weeks after the Rawfolds battle and the assassination of Horsfall, Radcliffe had made precious little progress.

The factory owners and manufacturers lived as though they were at risk of assassination themselves, and some of them probably were. The blacksmiths Enoch and James Taylor, the brothers who built the shearing frames for the local factories, as well as the hammers the Luddites used to destroy them, "were looked upon as marked men." They built up their own factory grounds into an armed base; "it was fortified as if for a siege; soldiers sleeping in the mill at night."

Horsfall's heirs, for their part, had removed the automated shearing frames from Ottiwells Mill, knowing full well the entrepreneur would have turned in his grave.

Meanwhile, life on the weavers' cottages and small farms was more depressed than ever. The subscriptions the Luddites raised to support the poorest among them was often simply not enough.

Ben Bamforth had not seen his cousin George Mellor for weeks after Horsfall's murder, as he recounted in *Ben o' Bill's*. After Ben was shot, George personally helped him home to safety, and he had to stay hidden, pretending to be sick to avoid arousing suspicion from Radcliffe and the investigators. Ben was home, still recovering from his wound, when his cousin finally came.

George looked as if he hadn't slept for any of those weeks. His hands were stained with dirt and dried blood, as if they'd been cut raw with thorns. At that moment, Ben knew in his gut who had arranged to pull the trigger. George slunk two steps into the kitchen and limply lifted his hand. When Ben did not lift his in response, George collapsed into a chair, head in hands, and sobbed.

After the loss at Rawfolds, George had snapped. Despairing, he swore revenge. According to Peel's account, on April 28 George was "exceedingly violent in his language in talking to his fellow workmen," and continued "his invectives in conversation," especially with his friend Thomas Smith as they worked in the shop that day.

There at John Wood's, George worked himself into a frenzy. He started pacing the room, making loud threats and curses. His colleagues had mentioned Horsfall's latest volley of taunts against the Luddites, and Mellor couldn't take it. "All his pent up evil passions are let loose, and he positively roars with fury," Peel wrote. "By and by he becomes quieter and gradually sinks into one of his moody fits."

They'd lost Hartley and Dean, and they'd lost Booth, their idealistic friend, driven to their cause out of empathy and a belief that a better world was possible — and driven there by George. They received word of violence against Luddites in Middleton, Manchester, and Stockport. Militias were opening fire on crowds of protesters; the military was marching through the biggest cities north of London.

"We must give up this frame breaking — it's no use," George said, sounding like the Nottingham delegate he'd shot down at the secret meeting just weeks before. "Now…there's only one way. Smith and I have settled it. We're going to shoot Horsfall today, and we want thee, Thorpe, and thee, Walker, to help us. Two could not hit Cartwright the other day, let's see if four can down Horsfall."

It may not have taken anyone in the room by surprise, but it was still a shocking act for young men in their early twenties to grapple with.

"I'll make one, George," Thorpe said. Thomas Smith was in, too. They convinced Ben Walker to join them shortly after that.

Another Luddite gave them the arms, and George loaded his own gun, reportedly a Russian pistol he'd gotten while traveling abroad during his stint in the military. The four of them set off, and the rest was known, in varying degrees, to everyone in Huddersfield. After Horsfall's murder, they split up, hid the weapons, swore everyone to secrecy who had been present at the cropping shop while they were making the plans, and worked to stake out alibis across town. Now, given Radcliffe's dogged investigations, the expansion of the military presence in the region, and the death and defeat they'd already suffered, the young Luddites were wracked by grief, anger, guilt, and despair.

George looked like a ghost as he sat there in that chair in Ben Bamforth's kitchen. His face was pale, his eyes sunken in a "hunted, shifting look — and when they looked at you, which by rare times they did, they seemed as though they asked a question and feared the answer."* He finally spoke.

---

* This dialogue between Ben and George is from *Ben O' Bill's*.

"Tha knows all, Ben?"

"All I fear, George," Ben said.

"And tha flings me off?"

"I fling thee off."

George stood up in a brief flash of rage, color returning, for the moment, to his face. He inveighed against "the oppressors of the poor," yelling that he'd "struck the blow that others feared to strike," but he seemed to Ben to lose faith with every uttered word. George then turned on his cousin for abandoning the Luddites, issuing threats for his oath-breaking. But even he seemed to know that it was empty bluster. George was blown apart.

He heaved a feverish, defeated sigh.

"All this night I have wandered the fields and in the lanes," George said. "A hundred times I have set my face over the hills to leave this cursed country."

But go where? He was bound to this land, this community, this trade; by generations, by tradition and by blood. This was home, even if the land was closed off, his neighbors riven by poverty, his work undercut by disrupting entrepreneurs.

George left his cousin's cottage and did not come back.

That same month, in May of 1812, at eleven o'clock in the morning, a massive explosion rocked the earth in North East England. It could be felt for miles around the coal mine in Felling, where 120 men and boys worked underground. As the demand grew for more coal to run the steam engines that powered the machine-filled factories, miners had carved deeper and deeper into the earth, bringing them into contact with highly flammable gases. Those miners carried candlelit lamps to see in the cavernous dark.

Reverend John Hodgson's parish encompassed the mine, and he recalled reeling from the explosion, and its aftermath. The dust that erupted from the mine was so thick, he wrote, that it "caused a darkness like that of early twilight, and covered the roads so thickly, that the footsteps of passengers were strongly imprinted in it." The wives, children, and siblings of the workers came running through the debris: "Wildness and terror were pictured in every [face]. The crowd from all sides soon collected to the number of several hundreds, some crying out for a husband, others for a parent or a son, and all deeply affected with [a mixture] of horror, anxiety, and grief." More than ninety miners, some as young as eight, perished in the explosion.

"It seemed at times as if the whole proud industrial world itself was murderous," one historian remarked of the accidents and upheaval of the period, "its casualties as terrible as a burning battleship."

NED LUDD

"The Luddites continue actively stealing fire-arms in this neighborhood," the *Leeds Mercury* reported, and "not a night passes but we hear of them prowling in great numbers." It seemed a change was underway in Luddite strategy. The ferocity and frequency of machine-breaking attacks had slowed. In Nottingham, they had ceased altogether. A campaign to gather arms elsewhere, meanwhile, had taken up the slack.

June 16th, 1812. Ashton —

On Sunday morning last, about eleven o'clock, between 2 and 300 Luddites assembled suddenly, like crows coming over fields, hedges, and ditches, upon Hough Hill, in the vicinity of Newt and Staley-Bridge, which commands a view of the country to Manchester, and several miles round.

They formed themselves into ranks, two or three deep, and appeared at a distance to be performing some military evolutions. No stranger was permitted to approach them....After continuing together about three quarters of an hour, they dispensed by the signal of a gun or a piston being fired, the report of which was heard through the neighborhood.

If they merely intended to alarm the Magistrates, and bring some dragoons from Manchester, their object is accomplished; for messengers were sent off immediately to that place, and early on Monday morning, a party of the Scot's Greys arrived at Ashton.

During the last week, scarcely a night has elapsed without fire-arms being collected in this neighborhood by the Luddites. They suddenly appear at the door of the house, but do not break in, without previously demanding entrance; if resistance be made, instant death is declared to be the consequence; but if the arms be peaceably delivered up, they use no violence, and in some instances have promised to restore them in a very short time.

— Statesman (London), *June 23, 1812*

The Luddite campaign to stockpile arms was reportedly such a "great success" that magistrates across Manchester were authorized to seize arms themselves, to prevent the Luddites from getting them first. The report noted that two thousand Luddites were seen splitting up near Stockport, and that one Tuesday night a particularly large number of arms were stolen. Arms seizures were continuing apace in Huddersfield and Wakefield in the West Riding, too, where hundreds of men were spotted drilling at night.

"The practice of forcibly obtaining arms still continues to prevail to an alarming degree in the neighborhood of Leeds, Huddersfield, and Wakefield," the *Bristol Mirror* reported. "Luddites carry off every article of lead, such pumps, water spouts, &c. for the purpose of casting musket balls."

The Luddites were melting down the metal products of the Industrial Revolution and casting them into bullets.

GRAVENER HENSON

June 1812

After the second reading of his bill in Parliament, Gravener Henson was feeling optimistic.

"There seemed from the Cheering a great Majority in the House in favor of our Bill," he wrote back to his peers in Nottingham. Bills were typically read aloud and debated three times before they were voted on, and the reaction in the House of Commons seemed overwhelmingly positive. So far, so good. Two key ministers voiced the "opinion that Regulations should take place in our Trade," he wrote.

Henson's legislation built on a series of meetings in which he'd received strong encouragement from some of the most powerful politicians in London. Henson had visited the new secretary of the Home Office, Henry Addington, the Viscount Sidmouth, who had replaced Richard Ryder, and now made a show of pronouncing his and the Prince Regent's support for the knitters and their plight. Gravener excitedly wrote that Addington "assured us that it was the Inclination of the Prince Regent to give our Manufacture every encouragement in his Power, that he had no Doubt that the Prince Regent, would wear our Stockings, Ornamented and give encouragement to that Fashion." The viscount himself put in an order for six pairs of silk hose for his daughters, and bought a double-press cotton shawl, though the knitters insisted they'd gift it to him. "He said repeatedly Success to your 'Manufacture'," Henson wrote elatedly. "Therefore my Lads there's no opposition in the Lords."

The next day, they were off to try see the Prince Regent, to present to him their finest wares — stockings, veils, black presses, and handkerchiefs — to demonstrate what was at stake if machinery and corner-cutting hosiers were allowed to rule the trade.

It had been a rough start, however. Henson and his colleagues had arrived in

London back in April, delivered their bill to Parliament, and waited while the body formed a committee to look into the matter. Henson had been aggravated at the lack of urgency, while violence was consuming the regions back home. "I have just seen the Statesman and dont find a Single Syllable respecting our Business," he'd written in May. He'd also been aggravated at the lack of money the United Framework Knitters were willing to send him.

Parliament did form a committee, however, and in May, Gravener gave his testimony. He skillfully built up a narrative, appealing to Parliament's nationalist sensibilities, talking up the superiority of British knitters over the French. He detailed the falling rates of pay for the stockingers behind that superiority, explained how spinning machinery had lessened the quality of thread, and discussed the practice of investors buying up frames and charging poor workers rent just to use them. Fellow working knitters spoke out to back up his case, underlining the fact that entrepreneurs were selling the goods produced by the wide frames for the same prices as their quality work, and they were suffering for it. The testimony stretched over two days, and Henson was nervous; he was frustrated that more examples of lace and knitted goods hadn't been sent down, and wrote Nottingham urging them to send more. "If any Man in the Trade refuses to do his Duty in the making of Articles for the Recovery of his Trade Knock his Teeth down his Throat instantly," he wrote.

As Parliament drafted its report in response to the hearings, Henson and the United Knitters continued to bolster support everywhere they could. This time, he traveled to Ireland and convened a meeting with the framework knitters there, though not without some difficulty. "I had considerable difficulty during the afternoon in finding the Persons I wanted, owing to the Difference between the Pronouncing the Street I wanted and the Spelling of it, it being wrote Malpas, and pronounced Maypas," he grumbled, apparently annoyed enough to open an account of his meeting the Irish weavers with this episode of semantic-inspired inconvenience.

His annoyance only grew. The Irish were convinced that they had matters under control; *their* trade was tightly regulated, and no one was suffering there (yet). "We do not practice any of those evils which you so loudly and Justly complain of," the Irish emissary wrote back to Henson. "No Sir, we have no cut up work or fraudulent work made of any description the evil Originated with your Selves We have no Coults nor Women working with us, each Man must Serve his

Regular Seven years before he will be Allowed to get Journey work." That, Henson surely tried to impress on them, was exactly how it was supposed to work in England, too, before the entrepreneurs used a time of bad trade to press their advantage and disrupt those standards. Owners using machinery to automate and degrade work was the future coming for them all.

But now, in the first week of June, having spent the better part of the year traversing England and appearing before Parliament and cajoling his colleagues to send money, signatures, and intel, it felt like the effort was finally paying off. Addington's effusive support clearly meant most of all, given his influence and power.

The most vociferous opposition so far had come not from the conservative Tories, who had pushed the draconian bill casting frame-breaking as a capital offense, but from the advocates of Adam Smith and laissez-faire, men like Joseph Hume, and even Lord Holland, Byron's mentor.

"We have only Dr. A Smiths Disciples to contend with, whose principles are execrated all over the Kingdom," Henson wrote. He was right that, particularly now, the idea that an invisible hand should be left to guide the economy was immensely unpopular, even "execrated" among most Britons, with poverty rates and inequality sky high and desperate pleas for assistance pouring into Parliament from across the nation. The ringing cheers in the House, the assurances from Lord Sidmouth, and the proxy backing of the Prince Regent eased Henson's mind. It should prove enough, he thought, to overcome the disciples of Dr. Smith.

"I have every reason to think that the Ministry will support the Bill," he wrote. "Lord Castlereagh has signified his approbation of the Bill, and so has Lord Sidmouth."

A peaceful, democratic solution, a bridge between aggrieved and suffering workers and the machine-owning class, seemed within reach.

MARY GODWIN

Summer 1812

The five months that Mary Godwin spent in Dundee, on the "blank and dreary" east coast of Scotland, where she had been sent by her beleaguered anarchist father to live with her father's friends, the Baxters, was, she would later write, when her creative powers first began to take shape.

The northern shores of the Tay inlet were "the eyry of freedom and the pleasant region where unheeded I could commune with the creatures of my fancy."* "It was beneath the trees of the grounds belonging to our house, or on the bleak sides of the woodless mountains near, that my true compositions, the airy flights of my imagination, were born and fostered."

For decades, Dundee had been a hotbed of radical and revolutionary thought; her father had hoped that the region's politics would rub off on her. The Baxters were still "stout Jacobins" — a radical affiliation that was rarer, at least in the open, in London after the concerted repression of political speech that Pitt, Perceval, and the Tories had pursued since the 1790s. In Dundee, apparently even the youngsters were radical: "Isabella, the daughter who rapidly became Mary's favorite in the family of four girls," a biographer wrote, "knew the events of the French Revolution so well that she almost seemed to inhabit the past."

Mary very much liked living with her hosts. She found them to be kind and inquisitive, and she enjoyed her role as an adoptee into their unusual and egalitarian religious sect. In turn, she was embraced by the family. "They were all fond of their new companion," another biographer, Florence Marshall wrote; the Baxters described her as "agreeable, vivacious, and sparkling; very pretty, with fair hair and complexion, and clear, bright white skin."

The Baxters prized education and culture, were well read, and passed time

---

* An eyry, or aerie, is a lofty nest, or a residence at an elevated vantage point.

painting and drawing. Mary joined the teens in their studies, and in attending mass. Given that they were Glassites, a strict Christian sect that dissented from the Church of England, there was no shortage of either. Mary had to practice a vegetarian diet, observe their laws against vices like gambling, and deliver a holy kiss to the elders. Still, Mary relished those months; and Dundee's industrial coastal world — the view from her house included deforested mountainsides and empty, foggy ocean skies — would stick with her for the rest of her life, even though she would never physically return.

"A great deal of time was spent in touring about, in long walks and drives through the moors and mountains of Forfarshire," according to Marshall. "They took pains to make Mary acquainted with all the country round."

The desolate environment that served as a canvas for idle thoughts, the freedom from financial stress, the din of London and her sharp-mannered father, the company of young, bright, radical minds, and the portal into a more just existence. All of these helped set the stage for Mary's most famous and history-bending creation: *Frankenstein.*

EDMUND CARTWRIGHT

June 1812

For one short day the world forego,
Its noise and cares and follies flee —
That short unclouded day bestow
On friendship, solitude, and me.

This poem, written by Edmund Cartwright and sent to a friend in a letter dated June 4, 1812, conveys the mood of the inventor of the power loom a few weeks after that invention had been the target of the bloodiest day of the Luddite uprisings. Cartwright, now in his sixties, had retired, comfortably if not extravagantly, to the countryside.

Despite inventing a machine just as disruptive as Richard Arkwright's water frame, Cartwright never saw the same level of financial success. He'd opened a factory in Doncaster, but it was repossessed by creditors after a few years. In 1792, a factory owner named Grimshaw installed dozens of Cartwright's power looms in Manchester. He had planned to install two hundred, but the mill had burned down under suspicious circumstances — it was assumed, but never proven, that the blaze was started by agitated weavers. Regardless, Grimshaw didn't find the operation promising enough to try it again, and power looms were abandoned until 1803, when another inventor made the necessary improvements on Cartwright's model to make them profitable.

After that, Cartwright watched as factory owner after factory owner adopted the technology and the power loom rose to prominence in industrial Britain. The loom's initial failure left Cartwright with, by his own telling, a "barren reputation, accompanied by ruined fortunes." He blamed the machine breakers.

At the turn of the century, Cartwright petitioned Parliament for compensation due him as "the author of various mechanical inventions, of great utility to the manufactures of his country." He argued that had the machine breakers not torched that power loom factory in Manchester, other entrepreneurs would have purchased his device.

His petition, filed in 1801, stated that "Mr. Cartwright dates the origin of his misfortunes from the burning of a mill, which…there was strong reason to believe had been the premeditated act of evil-minded persons." This act of sabotage, he claimed, "had the effect of deterring other manufacturers from attempting to adopt the use of the patent looms." With his those looms, he claimed, "one child of ten years old could execute as much as two experienced men, and in a style superior to any hand-weaving." He concluded by boasting that his power loom gave users "the same number of threads to every yard of work, with mathematical accuracy." (Cartwright provided no actual evidence that the quality of the work his tech produced was better than that of the handloom weavers, only that it was theoretically more mathematically consistent.)

But after that initial setback, he never had a chance to make the looms profitable, and a parade of subsequent entrepreneurs infringed on his patent or improved it without crediting him, and now those looms belonged to the public, and men across England were growing rich from his invention — to the tune of what would soon amount to "one million and a half sterling per annum."

Cartwright and his lawyer, Mr. Taylor, impressed upon Parliament that the automated power loom was *not* by definition bad for workers, who should have embraced it rather than protested it:

> This machine is not necessarily chargeable with all the evils that have fallen on the hand-loom weavers, and that if they had availed themselves of its advantages, instead of obstinately setting themselves against it, the great manufacturers might not have been driven to adopt the mill system to the extent they have done: and that the hand-weavers might still have retained such a portion of business in their own hands.

Cartwright considered himself a humanitarian, and it's possible to see in these lines an effort to justify his invention to himself, and to the public. Had the weavers only *bought* his machine, they could have used it to fend off the factory bosses and prevented the entire system from taking root to the extent that it did. But he must have known that the typical cotton weaver could scarcely afford to invest in a major piece of unproven machinery — especially one that required a power source to operate.

Cartwright would have known all about the weavers' dire situation, and the plight his machinery helped create for them, if he'd ever discussed the matter with his older sibling. Edmund's brother, Major John Cartwright, was an eminent political reformer. While serving in the British military, he'd turned down a promotion during the American war on the grounds that their cause was just. He had campaigned for democracy in England ever since. "Incapable of compromise, eccentric and courageous, the Major pursued his single-minded course, issuing letters, appeals, and pamphlets, from his seat in Boston, Lincolnshire, surviving trials, tumults, dissension and repression," wrote the historian E. P. Thompson. "Before the Napoleonic Wars had ended, he founded the first reform societies of a new era, the Hampden Clubs, in those northern industrial regions where his clerical brother had accelerated other processes of change with his invention of the power-loom."

Edmund Cartwright's appeals were denied by Parliament until 1809, the same year that it struck down the regulations protecting workers in the weaving trades, when the government awarded him a lump sum of £10,000. Hence the retirement. With the payment, Cartwright bought a farm in Kent. "At this place he spent the last years of his existence, amusing himself with various experiments in agriculture, chemistry, and mechanics," according to his memoirs, "and occupied, to the utmost of his ability, in promoting the welfare of his fellow-creatures."

Cartwright's tale is emblematic of the disconnect between the well-intentioned inventor and the unpredictable consequences that invention can have. Per the origin story told by its inventor, the power loom began as curious experiment. He was inspired by Arkwright, he said, and wanted only to stimulate his intellect and test the capabilities of technology. It was a puzzle to solve, combined with some aspirations toward changing the world and making a litlte money in the process.

Some of this is no doubt self-mythologizing, but it demonstrates that once a labor-saving technology is invented and released into a capitalist environment, it will be exploited to maximize profits — if it works well enough, and even sometimes when it doesn't. Modern economic history is littered with examples of entrepreneurs taking an existing technology, improving or standardizing it, and combining that technology with exploitative labor practices to turn it into a market success. Arkwright himself had done precisely this.

Power looms were never going to benefit the workers whose jobs they aimed to replace, just as the vast majority of automation attempted today never will — they're cost-saving measures available to those with the capital to invest in the labor-saving machinery or software, those capable of making the investment profitable by sufficiently scaling operations. This will be the case as long as the worker has little power and is in thrall to a well-financed employer; none of the gains of automation will be felt by the worker — and, as in Cartwright's case, few gains may be felt by the *inventor* of the automating technology, either.

But it's worth underlining: an entrepreneur with the best intentions can invent a device in one decade, thinking of the good a technology might accomplish in terms of improving efficiency, or even what a marvel or a lark it might be to do so. Ambitious, less scrupulous industrialists can use, develop, or exploit that technology in another. There are few if any true lone inventors, so these ideas will invariably be had, regardless of intent. But the logic of unfettered capitalism ensures that any labor-saving, cost-reducing, or control-enabling device will eventually be put to use, regardless of the composition of the societies those technologies will disrupt. Consider it the iron law of profit-seeking automation: once an alluring way to eliminate costs with a machine or program emerges, it will be deployed. This precept has held true throughout the era of Silicon Valley, of Facebook, Google, Amazon, and Apple, about which the historian Margaret O'Mara notes that founders often "had little inkling of how powerful, and exploitable, their creations would become." At a time when software has been developed to automate nearly every process, including finding processes to automate, and when profit-seekers have the whole of the digital world at their fingertips to pull ideas from, this tendency only stands to proliferate.

By his own account, Cartwright was happy with the government's payout. While the reach of his power loom was expanding — after a pause caused by the disturbances, it would eventually go on to displace the cotton weavers almost entirely — Cartwright reclined in the countryside on his farm.

As the working people of Manchester starved, with the power loom the most hated piece of technology in the city, Cartwright wrote a poem about wine.

Wine, suited to your classic taste,
Shall compensate for homely fare.
The Teian grapes' nectarious juice

That once Anacreon quaffed, is mine:
Were mine the power, I would produce
Anacreon's wit as well as wine.
And yet, who knows what wine may do?
Wine might Anacreon's wit supply;
Tipsy, he might have rivall'd you —
When sober, been as dull as I.

THE PRINCE REGENT

June 1812

The Prince Regent spent the weeks following the assassination of Spencer Perceval "soothing his shredded nerves by imbibing equally large quantities of alcohol and laudanum."

The issue at hand was not just that England's prime minister had been gunned down in public, but that the Prince Regent was unable to find someone willing to serve as a leader for his government.

With Perceval gone and the still recently betrayed Whigs all but unwilling to work with him, the Prince Regent found himself beset by a political crisis. None of the first four ministers he put forward were able to form governments — the first refused outright — and England went for nearly a month without a prime minister. Prince George was eventually led to reappoint that first man he'd tapped for the job, the Earl of Liverpool, previously England's war secretary, who was waved through after every other option was exhausted.

In one of Liverpool's first statements as prime minister, he "perpetuated the myth that the cause of all the troubles in the Midlands" was due to "new machinery" — and not the uses to which it was being put — before calling for new laws to protect it from Luddites. Liverpool had swiftly replaced Richard Ryder, the overwhelmed Home Office secretary, with Henry Addington, the Viscount Sidmouth, who more energetically took to the task of stamping out the Luddite uprisings. An authoritarian impulse, in fact, would mark his entire fifteen-year tenure. "Every attempt to create disturbance, and to clog the wheels of government, was immediately repressed," gushed Addington's biographer (and son-in-law), George Pellew, "and no sooner did sedition anywhere raise its head than it was crushed."

Meanwhile, England's trade blockade had become so onerous to the United States that, coupled with the UK's policy of forcibly conscripting sailors aboard US merchant vessels into its campaign against Napoleon, the two nations were led to the brink. James Madison signed Congress's declaration of war against Great Britain on June 18. In an effort to head off the conflict, Liverpool sought to terminate the Orders of Council—and on June 23, to the relief of merchants, factory owners, and workers alike, the Prince Regent approved the repeal. But it was too late to stop the gears of war.

Four days later—and nearly a year and a half after the first Luddite riots—the Prince Regent's office sent a sealed bag of documents and evidence to Parliament, informing the government that "certain violent and dangerous proceedings" had taken place, and instructed it to consider finding a way to stop them.

> His Royal Highness the Prince Regent, in the name and on the behalf of his Majesty [George III], has given orders that there be laid before the House of Commons, Copies of the Information which has been received relative to certain violent and dangerous proceedings, which in defiance of the laws have taken place, and continue to be carried on, in several counties of England.... His Royal Highness confidently relies on the wisdom of Parliament, for the adoption of such measures as will be best calculated to afford security to the lives and property of his Majesty's peaceable and loyal subjects in the disturbed districts, and for the restoration of order tranquillity.
>
> G.P.R.

A number of Parliamentarians reacted with disbelief; the Prince Regent was weighing in *now*? According to the court record, one lord "expressed his concern, that after those disturbances had so long existed, the notice of them should have been put off till so late a period of the session."

At the next session, Addington rose, placed the sealed container of evidence in front of him, and said, with obvious calculation, that he "trusted there would be an unanimous concurrence on the part of their lordships in expressing their gratitude" for the Prince Regent's message and information. It would be wise, he implied, for everyone to grant him the power to appease the prince. Addington proposed the creation of a Committee of Secrecy to study the sealed information, and to consider further prohibitive measures. "Although the conduct of the rioters might be, in some degree, traced to the high price of provisions and the reduction of work," he said, "still there was no doubt that these outrages were fomented by persons who had views and objects in thus fomenting disturbances, which it was the duty of Government to counteract."

Just as Ryder had done before him, Addington briefly nodded to the Luddites' dilemma before claiming the true threat was a subversive conspiracy, and moving to use state power to punish them.

Even if the repeal of the Orders of Council, which had done so much to stifle the economy, had helped improve workers' prospects, even if the military had been "called in to assist the civil power," even if several people had been tried "for acts of riot and outrage — some of whom had paid the forfeit of their lives" — even after all that, Addington said, the House must be prepared to go further still. It was a matter of national security.

Two secret committees were formed to analyze the information, mostly intelligence passed down from "the disturbed district" by spies like B. This must be done, Sidmouth said, "in order that measures might be adopted for the security of the lives, and the safety of the property of his Majesty's peaceable and loyal subjects." It would be up to the Committee "to consider what measures were necessary." It didn't take long for those committees to conclude that the Luddites were not simply desperate working people. It determined that "some of the persons engaged in these proceedings have extended to revolutionary measures of the most dangerous description.... Their proceedings manifest a degree of caution and organization which appears to flow from the direction of some persons under whose influence they act."

If the Luddites were, in fact, a revolutionary army, not working people organizing to fight for improved working conditions, Addington could justify taking far more drastic measures in bringing them to heel.

"He then asked Parliament for laws giving magistrates increased powers to interrogate and confiscate in their districts and raising the crime of oath giving to a capital offense, both of which passed easily," wrote Kirkpatrick Sale.

What he did not ask Parliament about, but felt no constraint in authorizing, were policies giving increased license to the use of spies and informers to provide (or manufacture) evidence against the Luddites, encouraging troops in the industrial areas to be extra ruthless in coming up with Luddite suspects, and speeding up trials of those against whom any incriminating evidence could be produced.

Addington extended the Oath Law to include an offer of clemency to those who revoked their oath by October, in hopes that it would encourage informers. And he officially anointed Thomas Maitland as general in the north of England. Maitland, happy to channel his new boss's vigor, directed a host of spies to become "Active and Efficient Members" and sent them out to infiltrate the Luddite meetings and committees.

This was difficult, given the tightknit nature of the Luddite cells — but it was not impossible, especially after the tumult, loss, and chaos that had fractured many Luddite groups and organizations in the first half of 1812.

And Maitland's spies would soon yield results.

NED LUDD

July 1812

Two unknown men showed up at the Saint Crispin Inn one night and told anyone who would listen that they were weavers looking for work. They'd come from Manchester, where the machines had swallowed all the good jobs, and heard there was some work in Halifax.*

"Whoever has told you that," laughed Charles Milnes, an amiable Luddite sitting at the bar, "has told you more than they can prove, I think." The two strangers, one who called himself John Smith, nonetheless sat down beside Milnes and bought him a drink, then another. After enough small talk had passed, Smith and his companion turned the conversation to the Luddites.

"Then you know this General Ludd, perhaps?" one asked.

"General Ludd!" Milnes laughed — it was a ridiculous question. "Nay, I know no generals."

"But didn't he command at Cartwright's affair? We were told at Manchester that they had a commander who was called General Ludd."

"Well they might call him so," Milnes said, unthinkingly, "but that was not his real name. I happen to know that much." With that, the two men said that their real purpose in coming was to be "twisted in," to take the oath, to join the Luddites.

If that's what they wanted, Milnes said, he knew where to send them. He led "Smith" to the home of John Baines, the old hatter and democracy reformer who had joined causes with the Luddites. Baines lived with his two sons in town.

"I have brought a friend here," Milnes said when Baines cracked the door. "He is a stranger, but he is a very good fellow, and he wishes to be a brother."

"Then we must be handy," Baines replied, "for we shall have the watch and

---

* The dialogue from the exchange in the bar is drawn from Peel, *Risings*.

ward here soon; some of my neighbours have laid an information, and they are often searching my house."

Baines nervously ushered them inside. He took out a Bible and administered the oath while his son watched the door. "Smith" invited them all out drinking to celebrate, but Baines refused; too many would-be informers out there these days.

The two men hung around the workshop the next day or two. They visited Milnes at the Inn; unlucky, they said, in their efforts to find work. They hung round until word began to spread through town one day that two "rascals" were headed to Halifax — spies sent by the magistrate. "Smith" and his friend disappeared the same night.

Two days later, soldiers surrounded the Baines home and arrested John and his sons for administering illegal oaths; Charles Milne and two others were arrested for rioting. The crimes carried the death penalty.

GRAVENER HENSON

July 1812

Gravener Henson had underestimated the disciples of Adam Smith.

On the day of his bill's third and final reading, something had clearly shifted. Joseph Hume stood and spoke out, announcing that he "opposed our Bill on Dr A Smiths grounds of letting Trade alone," Henson later wrote. Parliament's leadership "then moved the House to adjourn, there not being Forty Members present."

Factory owners, master hosiers, and powerful entrepreneurs had been sending in notices and letters formally opposing the bill — by now it had become quite clear which side the money was on. Even the sympathetic members of parliament knew what this mounting dissent among the industrialists foretold. As the representatives for the framework knitters stood before Parliament, hopes in hand, the MPs "ran out of the House when our business came on like wildfire."

Henson spent the next month feverishly rallying support among smaller hosiers, shop owners, and businessmen who would back the framework knitters, and adding their names to the tally. But he may have known even then that their shot at success was fading. In the second week of July, Henson received word that the bill had officially been gutted.

"It is with extreme regret, I have to inform you that the Committee of the House of Commons, have come to the decision of Reporting that the Clauses of our Bill which relate to Hosiery ought [to] be erased," Henson wrote back to Nottingham. "They have reported in favor of the Hosiers to their utmost satisfaction notwithstanding every effort we have made."

Parliament had sided decisively with the entrepreneurs and the House proponents of laissez-faire. It struck down the most important planks of the bill — the minimum wages for cloth work, and the ban on fraudulent and cheap imitation goods. There would be no pay protections, and the owners' machinery would be allowed to produce cheap knockoffs. The committee was still willing to consider regulations on lacework, and limits on the practice of paying in truck, but the core of the bill was gone.

"The Committee have in the most unfeeling manner but not without a Division decided against us, on the foolishest lying evidence that was ever given," Henson wrote. "They may Dock, Cut up, Square, Make Single Cotton, and Cheat, Rob, Pilfer and Oppress now to their hearts content." The watered-down bill passed the House of Commons, and if there was any solace at all in the fact that they'd at least be able to stop the bosses from forcing payment in truck, it was short-lived.

The Earl of Lauderdale sounded an ominous note on July 23, coming out full force against the bill before it was debated in the House of Lords. He declared it outright "obnoxious" and said that if Parliament

> were to legislate in this manner, instead of leaving to the consumer
> to find out what were good articles and what were bad, [he] knew not
> where they were to stop, and they would introduce a principle of the
> most dangerous consequence. To pass this Bill would be
> immediately to destroy a valuable branch of export trade, to throw a
> great number of persons out of employment, and to produce
> considerable distress in the manufacture to which it applied.

Addington responded that he had not yet read the bill but would do so before they took it up the next day.

When they did, the Earl of Lauderdale opened the debate, and doubled down on his aggressive support of laissez-faire. He boomed that he "could not conceive a more monstrous principle of legislation," and proclaimed that its effort "to interfere in the bargains made between the master manufacturer and his workmen...could not fail to produce the most mischievous consequences." The very idea of creating a set of minimum wages was "the most mistaken notion that had ever been conceived," he insisted; for if an employer could not afford to pay it, he would be forced to fire all his workers. He again denounced the effort to ban counterfeit goods, and said he expected the House of Lords to vote down the bill unanimously.

And one by one, they did. Some mentioned the "good intentions" of the workingmen who crafted the bill as they voted against it. Addington, who had paid the framework knitters so much lip service, fawned over their wares, and told them to count on the Prince Regent's support, was the last to twist the knife. After he voted the bill down, Addington reiterated Lauderdale's sentiment that the principles of a minimum wage and regulation on trade were "most mischievous."

The cloth workers' effort to secure even the most meager protections was officially dead. The bill's demise helped enshrine the assumption that, in any forward-looking economy, people must accept that they must compete with businesses and entrepreneurs that can afford labor-saving technology. That they are, essentially, at its whims. This became the foundational logic of the modern industrial economy: minimum wages, regulations, and fraud protections were hindrances to England's nascent but booming tech sector. It did not matter as much to the House that working people were suffering as it did that entrepreneurs not be saddled with restrictions on producing goods; again the state favored the profit-generating machinery over the people who worked it.

Addington considered the matter closed. He concluded his final vote against the bill by declaring that he "trusted in God, that no such principle would be again attempted to be introduced in any Bill brought up to that House."

GEORGE MELLOR

Fall 1812

Along with just about every other able-bodied man in the West Riding, George Mellor and Ben Walker were dragged into magistrate Radcliffe's manor, Milnsbridge House, which now doubled as a personal fortress, police station, and prison, and were questioned vigorously over the Horsfall murder.

Radcliffe was clearly agitated, and people who met with him noticed his hand had started to twitch. He continued to receive letters that threatened his life; someone had fired shots outside his estate grounds; and nobody was fessing up to anything relating to the murder, which had been carried out in broad daylight, in front of multiple witnesses.

Radcliffe had to let Mellor and Walker go. He had no evidence.

Stale gray months passed in Huddersfield after the rout at Rawfolds and Horsfall's open-air killing. It was an excruciating, paranoiac summer. As news of the murder spread and the occupation ground on, the Luddites' popularity waned in some quarters. Still, the Luddites continued to gather arms in headline-making quantities, both in the West Riding and in Lancashire. There had been skirmishes with the increased military presence, and the frame-breaking attacks seemed to have slowed.

Radcliffe had dragged in so, so many men. Women, too. Most were released, like clockwork, due to a lack of evidence. Most, but not all.

Samuel Hartley, a poor weaver who had joined the Luddites out of desperation, was arrested by Radcliffe's men, and the event proved to be so traumatic to his wife that, according to Peel, it gave her a fatal heart attack.

> When the officers of justice came to arrest Hartley his wife was violently agitated at the sight of the poor man surrounded by his weeping children, and with a tremulous cry she fell back in what was supposed to be a

swoon....Restoratives were administered but it was found that the weary troubled heart had ceased to beat and had laid down its heavy burden forever. [Hartley was transported to a cell at York while] the stricken mother lay cold and still in the wretched hovel they had called their home.

A few other minor offenders were arrested, but Horsfall's killers and any major players involved in Rawfolds were still at large. "Although the secret was well known to many of the Luddite confederates it was securely kept for several months....Many had come to the conclusion that the assassins would never be discovered."

Yet Radcliffe was indefatigable. His men, and those who became his men after being conscripted into service under the Watch and Ward Act, were sent out time and again to probe the city and countryside.

Eventually, the pressure got to a young cropper named William Hall, who came to see Radcliffe one day and agreed to become an informer, turning king's evidence to save his own skin. He admitted to joining the attack on Rawfolds and participating in major Luddite efforts like the raid on Francis Vickerman; and he fingered Jonathan Dean, who was already being held under suspicion in Radcliffe's makeshift jail. Cornered, Dean confessed.

Another star witness had come forward as well.

Radcliffe was emboldened by Addington's new zeal in London for taking on the Luddites, by the intelligence provided by Maitland's spies, and by the recent "successful" trials of the Luddites in Manchester: More than a dozen Luddites had been arrested, charged for the food riots and burning of the factory at Westhoughton, and brought to Lancaster Castle to stand trial. Little genuine effort was made by the authorities to identify and prosecute the organizers or perpetrators; the only thing many of the accused seemed to have in common was that they'd been spotted near the riots. Among the accused were a fifty-four-year-old disabled woman named Hannah Smith; a disabled man who was charged with using his crutches to break a window; and Abraham Charlson, a sixteen-year-old boy.

Eight people, including Hannah and Abraham, were convicted and sentenced to death.

The convicts were marched to the gallows for a public hanging, in a procession clearly intended to send a chilling message to the populace. The Crown-backed *Lancaster Gazette* recounted the scene:

On Saturday last, at noon, eight malefactors under sentence of death, in our Castle, suffered the dreadful sentence of the law, viz Hannah Smith (54) for rioting and highway robbery, at Manchester; Abraham Charlson (16) Job Fletcher (34) Thomas Kerfoot (26) and James Smith (31) for rioting and burning the mill, at West Houghton; John Howarth (30) John Lee (46) and Thomas Hoyle (27) for rioting and breaking into a house, and stealing provisions, at Manchester.

"A vast concourse of people" gathered to witness the spectacle, according to the *Gazette*. The large and sympathetic crowed teemed with palpable discontent.

When the executioner dropped the rope, the teenaged Abraham cried out in terror and confusion. A witness to the execution, Dr. Taylor, described Abraham as "a boy so young and childish that he called on his mother for help at the time of his execution, thinking she had the power to save him."

It was this "productivity" of the Manchester arrests and trials that inspired Radcliffe to again make the rounds amongst his usual suspects. He placed an item in the *Leeds Mercury* reiterating the £2,000 reward and began publicly planting hints that he was close to fingering the culprits, in the hopes of pressuring the real ones to come forward. The plan worked.

It was not long before news that suspects had been arrested and charged for the murder of William Horsfall ran in the same paper: "The local magistrate, Joseph Radcliffe, Esq…has given the most complete and satisfactory evidence of the murder of Mr. Horsfall. The villains accused have been frequently examined before but have always been discharged for want of sufficient evidence."

One suspect, brought in yet again, denied the charges completely — until he understood who had given the evidence against him. "The man charged behaved with the greatest effrontery till he saw the informer, when he changed color and gasped for breath," the *Mercury* relayed. "When he came out of the room after hearing the informer's evidence, he exclaimed, 'Damn that fellow, he has done me.'"

He'd been ratted out. "It appears that this man and another have been the chiefs in all the disgraceful transactions that have occurred in this part of the country, especially at Rawfolds," the *Mercury* continued. "This will lead to many more apprehensions."

The man who had been identified was George Mellor.

In the account Ben's friend Soldier Jack relays to him in *Ben o' Bill's,* George was at work when Radcliffe himself led a squad of six soldiers to John Wood's shop, along with an officer carrying his arrest warrant. The soldiers arrived, "very quiet," and Radcliffe came right up to the door "as bold as brass," before anyone knew he was there. George was "game to the last," and Will Thorpe, who was soon arrested too, was "just as unconcerned as if he were used to being charged with murder every day of his life." Thomas Smith was arrested, too.

Radcliffe loaded the men into the transport that would take them to their holding cell. The town of Huddersfield, which had suffered for months under the weight of military occupation, mass surveillance, and martial law, reacted accordingly.

"When they thrust 'em into the coach they had in waitin', George raised his hand as well as he could for the irons, an' called out, 'Three cheers for General Lud,'" Soldier Jack said.

> But the crowd were afraid to death. A lad or two in the throng cried out in answer, an' a woman waved her shawl, but everyone feared to be seen takin' his part, an' folk 'at had known him from a lad held back from him same as if he'd getten the small-pox.

A rumor soon spread that George Mellor had confessed to killing Horsfall — but the Huddersfield solicitor quickly moved to quash this notion. He wrote to General Acland, the military officer overseeing the region, to clarify the facts — on the contrary, George had refused to say anything, and had clearly agitated, even frightened, the officials who held him. He also provided an update: a great many Luddites had been arrested in Huddersfield following Mellor's fall.

The letter was sent on October 31, 1812:

> *Dear General*
>
> I have not heard of Mellor's having made any confession of guilt, nor do I think it at all likely as I believe he will die as hardened a Villain as ever disgraced a Gallows.

George was hauled off in chains to York Castle and imprisoned in a solitary cell; the leader of the machine breakers was trapped inside medieval walls.

LADY LUDD

Fall 1812

Luddism's decentralized structure clearly had its vulnerabilities, but it also allowed the uprising to evolve and mature as a movement. It was a malleable framework that could accommodate much more than machine-breaking as participants' aims and priorities changed. Through late summer and into the fall, food prices remained high. So, in August and September, a series of food riots, auto-reductions, and machine-breaking led by women christened Lady Ludd spread across the West Riding and into Nottingham.

"There has been a great deal of agitation amongst the populace, during the present week, in this town and the neighboring villages, on account of the high price of corn," one report relayed. When a farmer decided to raise his prices, the locals wouldn't stand for it. They seized his corn in the marketplace and threw his sacks of wheat into the street, where the hungry could get it.

"In the afternoon, a number of women and boys, headed by a female, who was dignified with the title of Lady Ludd, paraded the streets, beating up for a mob," the *Bristol Mirror* reported. At night, they stormed a miller's factory and smashed its windows.

Similar actions broke out in nearby towns, and in Sheffield, flour and oatmeal merchants were forced by groups "principally led on by women" to sell their food for a price the poor could afford — in this case, three shillings.

Two days later, Lady Ludd returned to Leeds and, with a crowd at her back, seized vegetables and forced shopkeepers to lower their prices once again. The actions ultimately moved the town's owners to round up a subscription of £700 to try to offer relief to anyone who wasn't able to qualify for the poor rolls, the local welfare benefits. The *Leeds Mercury* reported that "the town is in much confusion, none knowing to what lengths these acts of insubordination, once begun may be carried."

Lady Ludd wasn't done. In September, the spirit spread to Nottingham, when the price of bread was deemed too high. The *Morning Chronicle* carried the story of Lady Ludd's reign:

On Monday morning, a baker in Nottingham had the temerity to advance his flour two-pence a stone, in the face of a falling market, which so enraged the women, that several got a fishing-rod, and fixed a half-penny loaf upon it, which they colored over with reddle, in imitation of its being dipt in blood, and likewise adorned it with a piece of crape [a sign of mourning]. With this they then began to parade the streets, and soon collected a very large mob, among which were two women with hand-bills, who were dignified with the titles of Madam and Lady Ludd. The first object of their vengeance was the Baker who had advanced the price of his flour; they broke his windows, and compelled him to drop his flour sixpence a stone. The mob then divided into several parties, and treated nearly every baker and flour-seller in the same manner; not sparing their windows till they had promised to drop flour sixpence per stone.

Not only did Lady Ludd and her followers win the price reduction, but several soldiers, who were supposed to be stationed in Nottingham to prevent disturbances like this one, were spotted among the crowd, supporting the action. That week, the military had failed to give them the share of bread they'd been promised for their service.

B

Fall 1812

The Manchester meetings attended by B had evolved beyond talk of machine-breaking and the weavers' plight. They were now attended by printers, debt collectors, hatters, tin workers, bricklayers, sawyers, you name it. And, despite John Bent's revolutionary soothsaying, the contingents favoring direct action had lost ground to those who wanted to organize for parliamentary reform.

At one meeting in June, over three dozen men gathered at a pub called the Prince Regent Arms, in Manchester, to discuss and finalize a petition to put before Parliament. Some had been encouraged to attend by Bent himself; and why shouldn't they? He was a respected cloth merchant around town.

The meeting was led by John Knight, a manufacturer and radical in his late forties, who led a round of introductions, asking every attendee to state his name, occupation, and hometown, and to read over the resolutions drafted so far. The petition called for an end to the war, for democratic representation in the House of Commons, and for an address to be read to the Prince Regent that relayed the state of affairs in the region. The men debated the petition, and Knight explained to the attendees that he opposed machine-breaking; reform, he said, was the way forward.

The talks were spirited and uneventful until eleven o'clock at night. That's when the deputy constable of Manchester forced his way into the room with a blunderbuss and thirty soldiers. Every attendee present, all thirty-eight of them, were placed under arrest for the capital crime of administering an illegal oath.

B, meanwhile, was nowhere to be found. He may have known what was coming, or he may have ducked out at just the right time. As it turned out, *another* spy, a weaver-turned-informer named Samuel Fleming, had gone to the constable to report the meeting. The accused, who came to be known as the Manchester 38, or the "38 Luddites," were hauled off to prison. They would be tried two months later.

While incarcerated and awaiting trial, Knight wrote to his wife about the injustice they were fighting.

> We read of those who add House to House and Field to Field by taking the poor man's Labor for nought, and is not this attempted to be affected three Ways, ie, First by advancing the Rent of Land (though we read the Earth is the Lords and the people the Sheep of his Pasture — Second by increasing the System of Taxation. Third by the rapid Improvements in machinery by which they render manual Labor less necessary therefore redundant and then reduce the Wages thereof.

The trial quickly became deeply embarrassing for the Crown. Fleming was their only witness, and he could confirm the identities of only four of the men he had accused. He claimed he'd asked to be "twisted in" to the Luddite fraternity, and that the men present had administered the oath at the pub by placing his hand on a Bible. But the authorities searched the pub and couldn't turn up any good book on the premises. Furthermore, he contradicted himself on the times he claimed to have been given the oath, had a drink, and left the pub.

On the stand, one man noted that the reason he'd attended was that John Bent, a "respectable cloth merchant," had invited him to — he had no idea Bent was a spy.

Ultimately, it was Fleming's word against thirty-seven other men, who had ample evidence that they were indeed collaborating on a reform effort, including an elaborate written document detailing their ideas. Fleming, meanwhile, had proved himself so unreliable that the judge himself proclaimed that "it makes an end of the case, and that the prisoners must be acquitted," before adding that was just his opinion and it was of course up to the jury.

The Manchester 38 were acquitted almost immediately.

The 38 were clearly familiar with Luddite tactics, and there was likely even some overlap; and just because Luddism was evolving, or certain cells among it taking up more peaceable reform efforts, did not mean the Crown would cease its paranoia.

Still, B began to slip from his employer's favor. Maybe it was out of embarrassment over the failed trial, or because that organized plot to overthrow the Crown he kept warning about never materialized. Maybe it was just a general decline in need for such services. Colonel Fletcher and the Home Office would engage his dispatches less, until his wild reports faded altogether, and the cloth merchant turned spy known as B was lost to history.

CAPTAIN FRANCIS RAYNES

Fall 1812

The young captain Raynes had been ordered to detach a company and head for the village of Mottram, outside Manchester.

"On our arrival," he later wrote, "some of the principal gentlemen, proprietors of mills and factories, called upon me, for the purpose of procuring military aid, in case their property should be again attacked."

Raynes was hesitant. He only had eighty men, and splitting them up would leave the platoon outnumbered by many of the Luddite forces they'd been briefed about in the area. Still, he acquiesced. "Considering the preservation of the property of individuals of great importance for many of the buildings were of immense magnitude, and confident of the power a few armed disciplined men have over a lawless band," he said, "I complied with their request." The captain received regular reports that the Luddites were assembling "every night within a short distance of our quarters, for the purpose of drilling and holding their conferences," and yet their patrols never seemed to encounter them. The Luddites had their own spies, of course, "who conveyed intelligence to the main body, of the road the soldiers took."

The captain took it upon himself — and two soldiers, and the local curate — to investigate these meetings. The small squad set out at midnight, and, around a mile and a half outside of town, they found the Luddite troops, assembled in large numbers. Though Raynes ordered his men to hide their weapons and be discreet, the Luddites heard them coming. The sudden presence of armed soldiers startled them, and they moved to retreat.

Instead of fleeing, Raynes decided to charge *toward* them:

We followed them, running as fast as we could; the men now shewing their arms, and pretending to call for the remainder of the piquet. It was with the greatest difficulty, I could keep the soldiers from firing; they were highlanders, possessing all the ardour the natives of their country are remarkable for, zealous and hearty in the cause we were engaged in.

The ruse — calling for more soldiers, for backup, to give the impression that there *was* any backup — would not last, and Raynes knew it. "I soon found it would be prudent to secure a retreat, whilst it was yet in our power, as, from the numbers springing up in all directions, I saw we should shortly become the pursued, instead of the pursuers; we, therefore, took refuge in a ditch, where we waited till the coast was clear."

It was a novel tactic, coming from an occupying force that had until then largely been lethargic, and even semi-sympathetic to the working class. Luddites were not used to being approached at night on their own turf. "It may appear extraordinary, that four persons should put to flight so many as were here assembled: it can only be accounted for, by their supposing a large body of soldiers to be at hand, and the fear ever attending such unlawful proceedings," Raynes wrote.

The next morning, surveying the area, he found no guns. Clubs and sticks littered the ground, "which they had evidently thrown away, to accelerate their flight."

MARY GODWIN

November 1812

When Mary returned to Skinner Street for a visit home from Scotland, she was intrigued to find that her father was hosting a very special guest. The Godwin family would be joined, at last, by her father's mysterious pen pal, Percy Shelley.

Mary's father had kept her updated on the Percy saga since the young heir had announced a few months ago that it would be his "greatest glory" to serve as William Godwin's patron in his twilight years. He recounted to Mary the visits of Shelley and his clique, whom Godwin found odd, and the time that he had tried to visit the poet at home, only to learn that he had left on a quest to dispense a new tract called *A Declaration of Rights* in sealed bottles tossed into the ocean. (A sample: "The present generation cannot bind their posterity. The few cannot promise for the many" (no. 16) and "No man has a right to monopolize more than he can enjoy" (no. 28).)

No actual financial support, however, had materialized yet. And Godwin needed the money desperately. There'd been a fire in his bookshop, and he was in more debt than ever. Before Shelley's visit, Godwin's friend and advisor, the reformer Francis Place, "bluntly told him that unless he succeeded in obtaining a 'very large sum' from his wealthy young friend 'all would be lost.'"

Now, Mary had returned home from the land of the Glassites, and here was the mad poet heir, in the flesh. Mary's friend Christy described Percy as "a stooping, carelessly dressed young man with a high white forehead, a mop of brown curls and eyes which were blue, intense, and startlingly prominent." The biggest impression he left was how he doted on his wife, Harriet, who was dressed in a striking purple satin dress. Percy was still feuding with his father, who he said had cut him off — causing the delay in Godwin's patronage.

The Shelleys were charmed by the Godwins, both by William's intellectual, if arrogant, air, and by the precociousness of the children. It helped that the two days they spent in London were filled with brushes with notable figures from the city's literary, art, and philosophy scenes; they met painters and writers and thinkers. Mary, meanwhile, was coming into her own. Her arm was healing and the eczema would soon stop bothering her. Her Scottish friends gushed over her "clear bright skin, her large hazel eyes and the crowning glory of her astonishing hair." She was popular and sought-after at social events, even if she always seemed to have to endure a barrage of comments that never failed to remind her of her parents and the attendant high expectations.

Another eminent visitor to the Godwin household was the radical business-man, factory owner, and utopian socialist Robert Owen. He'd come to London seeking investors and input into New Lanark, the factory and community he ran with socially beneficent principles in mind. He met Godwin at a dinner; they became close and would meet as often as twice a week. Godwin's philosophy would influence Owen's thinking.

Owen's factories had made him rich, but he'd become famous as a crusading reformer. At New Lanark, he offered better conditions, education, and pay to his workers, who were seen as much more content than in rival factories. In 1810, he'd started lobbying for an eight-hour workday at a time when twelve-hour shifts were common; he'd instituted the policy at his own factory first. Politicians and industrialists, among them figures like the future tsar of Russia, toured New Lanark for ideas on how to build model factories and working communities.

Now Owen's ideas were evolving and growing more expansive in scope. He was thinking about how to both perfect and export the principles undergirding New Lanark, and he was formulating his utopian, communitarian designs, in which groups of 500 to 3,000 workers would live together in ideal self-sufficient societies. Perfecting that environment was paramount, as the scholar Paul R. Bernard has noted, and Godwin's "writings reinforced Owen's version of materi-alism. Godwin maintained that there was no such thing as an innate idea and that character is completely determined by the conditions in which individuals live. Owen did not adopt Godwin's utilitarian anarchism, however, because he believed that collective action, coordinated by a governing body, was the most effective means of shaping the circumstances of a community for the benefit of all." (He also needed his workers to continue to follow his rules, so his production would not suffer.) Both men "condemned political agitation," and "favored instead a voluntary redistribution of wealth, which Owen later hoped would be achieved through cooperation," one historian wrote. "Their ultimate social ideal was that of a decentralized society of small self-governing communities of the kind that Owen was to propose in his village scheme. Since Godwin had fallen out of fashion Owen could be seen as his replacement for the new century."

This explains Owen's rising star, and his appeal to youth like John Booth, who held him up as a stirring example of hope for progress before Booth was killed by a hired gun in defense of an automated factory.

Mary and Owen met a number of times during her stay at Skinner Street; she would even become close friends with his son. But it was the otherwise uneventful meeting that night in November that would leave the most lasting impact on her life, and on popular culture. Her path likely crossed with Percy Shelley's only briefly, as Mary would have been exhausted from the long sea voyage home from Scotland, and probably spent most of the evening in bed upstairs while the Shelleys and Godwins dined. "The meeting was of no apparent significance and passed without remark," one of Shelley's biographers wrote; "little indeed did any one foresee the drama soon to follow." Shelley was nothing if not a harbinger of drama; he was twenty years old, raging in every direction, and writing diatribes against the state, injustice, and, yes, machinery.

He had written a number of stanzas of what would soon become his first major published work, and he took square aim at the corrupting, dulling influence of power itself. Godwin saw some of the early pages in December. Despite its being a radical, epic poem that railed against some of Godwin's favorite targets — oppressive government, marriage, and Christianity — and despite its being rooted in many of Godwin's own political ideas, Godwin was "not impressed."

It would be called *Queen Mab*, and one of its notable stanzas goes like this:

> *The man*
> *Of virtuous soul commands not, nor obeys.*
> *Power, like a desolating pestilence,*
> *Pollutes whate'er it touches; and obedience,*
> *Bane of all genius, virtue, freedom, truth,*
> *Makes slaves of men, and of the human frame*
> *A mechanized automaton.*

LORD BYRON

Winter 1812

Byron had been summoned and flattered by the Prince Regent. Now he was being called on and groomed by a reformer close to Prince George's estranged wife, Princess Caroline. Byron's celebrity had turned him into a force; his influence in parliamentary politics may have been muted so far, but his impact on pop culture, and therefore his promise, was undeniable. Prince George's political opponents — the Whigs he'd spurned by keeping Perceval in power — were hoping that Caroline could make for a formidable political ally, and that Byron could, too.

Caroline of Brunswick, Princess of Wales, had been humiliated and shunned by Prince George for years. Their relationship was always a matter of convenience for the prince; marriage was the only way his father agreed to help pay off his massive debts. After they'd had a daughter, George had gone so far as to try to excommunicate and divorce her, charging her with having a child out of wedlock. And though she'd been exonerated, she was lambasted and ridiculed in the press. They'd lived apart for years.

Now, their daughter, Princess Charlotte — she was the one who'd stood and spoken out when her father had betrayed their family friends and political allies — and the Whigs were trying to capitalize on the tumult by recruiting the estranged royals to their cause. Lady Oxford, an avid political reformer and a friend of Caroline's, also happened to be Byron's chief love interest at the time. Some fifteen years his senior — Byron called her his "Enchantress" — Lady Oxford helped draw Byron into the princess's inner circle, and encouraged him to take up the cause of democratic political reform. Lady Oxford was a member of the Hampden Club, whose founder and most active member was Major John Cartwright, the tireless democratic reformer, and brother to Edmund, inventor of the power loom.

Byron would spend late nights dining at Kensington, Princess Caroline's estate. He adored Princess Charlotte and, for a while, served as her tutor. Lady Oxford, meanwhile, tried to educate Byron about the reform movement. But Byron was already tired of parliamentary politics.

He'd spent a lot of the latter half of the year falling into and out of scandal and trying to manage his estate and his finances. He'd begrudgingly found a buyer for Newstead Abbey, his beloved childhood home, and when the sale was made public, his many creditors immediately sought him out. Byron was still in debt, despite his phenomenal literary success, and the buyer of Newstead hadn't yet paid him the agreed amount. It was a headache.

Meanwhile, his frenzied affair with Caroline Lamb became a high-society scandal. After a few passionate months, Byron cooled, and began to treat her callously; she only grew more obsessed, deluging him with letters. He cut off the relationship in October, but couldn't stop leading her on — it would all culminate in Lamb committing a public act of self harm, slashing her arms after he'd insulted her at a ball. Byron talked a lot about leaving England.

His last foray into politics was a speech to the House of Lords on behalf of Major Cartwright, who had been detained and dragged before a magistrate in Huddersfield, where he'd been making the case for democratic reform. Byron was presenting a petition on Cartwright's behalf, an effort that had virtually no other support to speak of in Parliament, save a single elderly radical. Still, he gave the speech, which was much more direct and utilitarian than his florid oration on behalf of the Luddites.

Byron lamented that Cartwright, a distinguished major, had been "seized by a military and civil force, and kept in close custody for several hours, subjected to gross and abusive insinuation from the commanding officer," before he "was finally carried before a magistrate and not released till an examination of his papers proved that there was not only no just, but not even statutable charge against him." Byron insinuated that the affront against Cartwright, who peacefully and patiently argued for democracy in Britain, was an affront against his thousands of supporters, too. "Your Lordships will, I hope, adopt some measure fully to protect and redress him," he said, "and not him alone, but the whole body of the people insulted and aggrieved in his person."

That was the end of the poet's public parliamentary career. It chagrined him that a year of making speeches and moving in political circles didn't bring any significant change or success, and he was certainly repulsed at the party in power.

"His political disillusionment was total," one biographer wrote. "Lord Liverpool's aristocratic and conservative ministry…was anathema to Byron. Its priorities were the defeat of Napoleon and the preservation of order in England under the shadow of the madness of King George III. Byron was bitter that this was a government with no sense of vision, no urge to ameliorate the conditions of the poor."

When the Prince Regent threw a party in celebration of Wellington's victories in Spain, the royal fireworks accidentally set the neighborhood ablaze, sparking a large and destructive fire. Byron, having soured on the Prince Regent, politics, Wellington, war, all of it, found it darkly amusing.

"I presume the illuminations have conflagrated to Derby (or wherever you are) by this time," he wrote to his friend Thomas Moore. "We are just recovering from tumult and train oil, and transparent fripperies, and all the noise and nonsense of victory."

GRAVENER HENSON

Winter 1812

After the framework knitters' antifraud and anti-abuse bill died in the House of Lords, officials and elites braced themselves for the resurgence of the Nottingham machine breakers. Officials in the Midlands beseeched Gravener Henson to talk to the stockingers and "soothe and moderate the public mind." But even if it was a brutal and unsettling blow, at least Henson was prepared for it to land. As the winds were changing over the summer, and before the bill was crushed in the House of Lords, Gravener had already begun using his now extensive and international network of tradesmen to build a new mode of workers' organization.

If the first half of 1812 was spent lobbying for worker protections and decent wages, the second was spent mobilizing an organization that would fight for those things itself. Henson and the United Framework Knitters set about building an alliance of workers that would not just petition for new laws, but pool its resources to gain real leverage against factory owners across England, one capable of enforcing minimum wages itself. The alliance was given the innocuous, if cheeky, title of the Society for Obtaining Parliamentary Relief and the Encouragement of Mechanics in the Improvement of Mechanism.

It would be "a closer and more efficient combination than any they had yet achieved." Its rules and operating procedures were carefully constructed, and its charter was vetted by lawyers to ensure they wouldn't run afoul of the antiunionization Combination Acts. The Society was really a federation of smaller societies, each consisting of between thirty and a hundred members, with an executive committee headquartered in Nottingham. Members paid a small subscription, and with thousands of participants it added up to decent sum. With that, it launched "a striking and novel plan of action":

"The Societies hire all the unemployed frames and engage all the work they can which they let out to their numbers but no other person, if a member has employment elsewhere with which he is dissatisfied, the Society make him a weekly allowance until he finds better Employment." Renting out idle frames kept the labor market tight, forcing hosiers to pay decent wages and offer decent working conditions or face the prospect of members working the Society's machines instead. The Society would then sell its hose direct to London, generating revenue to keep the operation in action.

It was a busy year for Gravener Henson. He married Martha Farnsworth at Saint Mary's Church in Nottingham, two years after the Luddite uprisings began in earnest. He had found some peace amid the chaos and disappointment, and now, another moment of cresting hope.

When the Society was firing on all cylinders by the end of 1813, the organization he'd helped build was remarkably effective. Workers could strike for better wages with a buffer to rely on, and few if any frames in Nottingham were smashed that year, owing to the new method. Slowly, and modestly, the knitters' conditions began to improve. The Society offered a glimpse of the future, when workers could collectively bargain to improve their conditions, to rally the only viable power they held against the machine owners — each other — even if its chief organizers were never confident it was fully legal under the oppressively interpreted laws.

As such, the Society's coat of arms showed the numerous districts it represented, along with an image of a mechanical loom and an arm gripping a hammer. Its motto was "Taisez-vous," which translates, roughly, into modern English as "Shut up."

NED LUDD

Winter 1812

As the autumn of a combustible year bled into winter, the Luddites were hunted by military squadrons and pursued by militia-commanding magistrates and entrepreneurs. They appeared to be losing momentum.

Over the last year, they had notched potent victories — restoring wages, pushing entrepreneurs to abandon automating machinery, and elevating the plight of the industrial worker to the national stage. They had also suffered major defeats. Luddites were crushed and killed in action at Rawfolds and Stockport and hanged at Lancaster, every effort to obtain reform or redress had died in Parliament, and public support seemed to be on the wane after the murder of Horsfall. That is what it took to safeguard the emerging factory system, and the normalization of automation — and the future was still deeply uncertain.

The specter of the Luddites had grown such that anyone who was suspected of committing almost any kind of crime — vandalism, theft, rioting, organizing — was liable to be labeled a Luddite. Joseph Radcliffe arrested two men who broke into a man's house and robbed him; they were charged with "burglary under the color of Luddism" and sent to York Castle, where they could hang for the offense. They joined dozens of others accused of Luddism, of joining the attack at Rawfolds, of arms burglary, and of oath-taking, who were arrested and sent to the prison at York. Machine-breaking in the West Riding slowed, after the arrest of the presumed leaders of the movement in the region — George Mellor, Will Thorpe, Thomas Smith, and others. Over a hundred men had been investigated in Huddersfield alone.

Spies continued to fan out across the region, filing reports to the generals who occupied the north of England. "They have been lodg'd in one of the Luddites houses five weeks, & are now getting into confidence with some of them, that they are not yet twisted in, but expect to be so in a short time," went one such dispatch.

John Goodair, the entrepreneur and factory owner whose house had been attacked in Stockport, followed through on his promise to exit the industry altogether. He sold his factory — the ad he posted in the *Manchester Advertiser* promised "New and Valuable Machinery, Power Looms and other Effects" — and all the equipment of automation it carried inside.

Magistrates like Radcliffe and many large factory owners lived like hermits, afraid that if they left their homes they might face retaliation over their relentless and unpopular prosecution of Luddites.

So, the uprising spirit continued. Machine-breaking had quieted in the West Riding, but resumed again in Nottingham, perhaps in response to the failure of the framework knitters' bill in Parliament. Food riots continued to erupt. Elsewhere, an organized effort to halt an ordered enclosure of common land in Wales led the magistrates to call for military aid from the Home Office.

The citizens were resisting the privatization of the land, and protesting the political machinery that was hurtful to commonality.

The Luddites themselves, however — some of whom were lying low, some imprisoned or dispersed across the country, and some still breaking the machinery across England — were now turning their attention to one place, where, just after the close of a turbulent, explosive year, the trial of George Mellor, Thomas Smith, and William Thorpe was about to begin.

All eyes would be on York Castle.

THE TRIAL

George Mellor, William Thorpe,
and Thomas Smith

January 1813

Six days after the dawn of another bleak new year, George Mellor, William Thorpe, and Thomas Smith filed into the courthouse at York Castle. It was so packed that their counsel could barely squeeze their way to the front of the room. Of the sixty-four men currently incarcerated in the castle prison, arrested on charges connected to Luddite activity, George, Will, and Thomas were the first three to go to trial. "The prisoners were all young men...and their appearance was very respectable," the special commission's report allowed. George's hand was trembling.

"It seems that nothing can be clearer than the mass of evidence I have now gone through," began Mr. Park for the prosecution, leading a team of eight attorneys.

> It will be impossible for you to arrive at any but one conclusion, that
> the prisoners at the bar are guilty of the crime laid to their charge.
> One cannot but lament that three young men, the eldest of which is
> not more than twenty-three years of age, should have brought
> themselves into this situation. But there is also pity due to the
> country, to those individuals who have suffered at their persons or
> their properties from the attacks of lawless violence.

The three were on trial for the murder of Horsfall, but the prosecution was moving to make them the face of Luddism, too.

"You have a most important duty to perform," Park told the jury.

The Bamforth family had come to York for the trial, per *Ben o' Bill's*. George's cousin Ben was with them, and so was the late John Booth's sister, Faith. Ben had come around to forgiving George some—he understood George's motives innately and painfully, even if he couldn't condone the murder. The family would not be allowed to see him. Only the lawyer, a man named Mr. Blackburn, whom the Bamforths had apparently put up £100 to hire, and the priest, Mr. Webster, who had once warned Ben's father about the dangers of automation, were permitted to visit, and only on the night before the trial was slated to begin.

Webster returned to Ben and Uncle William to relay the news of his visit to the cold, unlit cell.

"It was a little while ere my eyes got used to the darkness, but as we entered, I heard the clank of irons, and was aware of some form in the gloom rising in the corner from under the grated window," he said. "It was George; but oh! how altered! he was gaunt and thin, and his eyes that I have known so bright and lit by the joy of life, were dull and fixed in sick despair. I forgot the crime of which he stands charged and saw only a brother, nay, a son, suffering in mortal agony, and all my heart bled for him."

"Poor George! Poor Matty," William Bamforth exclaimed.

"Mr. Webster!" [George] cried, for he could see better than I, being used doubtless to the little light. "Mr. Webster, oh! this is good of you!" and he seemed to take no heed of Mr. Blackburn, and as well as he could for the irons that cribbed his arms, he stretched out his hands to me, so wildly and so lovingly, and I took both his hands in mine and must have done though I had seen the deed with my own eyes. And George bowed his head, and tears fell upon our clasped hands that were not wholly his nor wholly mine, and I drew down his head and kissed him on the brow.

George's first thought was of his mother. He asked about her well-being, while the lawyer shuffled his papers and nervously snorted snuff in the corner. Webster told him Faith Booth, John Booth's sister, had come, and George said, "I loved her brother next to Ben."

Blackburn took his turn, and quickly went over some notes for the next day's trial. When he was done, Webster asked George to pray with him.

He did not refuse; but sat upon a little block that served for his seat, and I fell upon my knees and the lights streamed upon my face from between the bars. And then I prayed the good God and Father to send peace and comfort to our dear brother, that he might be pleased that this great sorrow should pass and this black cloud be lifted; but throwing all upon the mercy and compassion of the Heart that feels for all, for all, even for the outcast and the sinful.

With that, the pastor began to sob again, broken to speechlessness.

Four days before, on the second day of the new year, Baron Thompson, one of the men presiding over the special commission at York, had addressed the grand jury that would oversee the trial. In his speech, Thompson offered one of the clearest distillations of the authorities' line against the Luddites, and the "acts of outrage on the property and persons of individuals [that] have been continued with little intermission, nearly the whole of the year which has just closed."

The Luddites, he proclaimed, were

> mischievous associations, so dangerous to the public peace, and so destructive to the property, and in some instances, to the lives of individuals,... and at first had for their object the destruction of machinery, which by diminishing the quantity of human labor in our manufactures, was by them conceived to be inimical to the interests of the laboring classes; a notion probably infused into their minds, by some evil-disposed persons, for the worst of purposes; but a more fallacious and ill-founded argument cannot be conceived.

Thompson implied that the machine breakers were mindless sheep, blindly following the commands of some "evil minded person" who was manipulating them for personal gain. The aristocracy could not conceive of, or would not recognize, a self-organized working-class movement. Part and parcel of his condemnation of the Luddites, Thompson felt, was the need to make the case for the importance of automation technology, and the great boon of mass manufacturing and lower consumer prices. This, he insisted, is why factory technology must be protected at all costs.

"It is to machinery that we probably owe the existence, certainly the excellence and extent of our manufactures. Whatever lessens the expense of preparing an article, by diminishing its price, promotes its consumption, and increases the demand for it, and if the use of machinery was discontinued, our manufactures would be destroyed." So, the default mode of aggressive condescension toward those who would protest automation, even technology more broadly, was already visible, in the very first statements uttered in the trial against the Luddites.

Thompson erected the straw man that endures today, that the Luddites were too dumb to see that automation was for everyone's benefit in the long run.

But to argue that a weaver is delusional for recognizing that a machine that destroys his job is "inimical" to his interests seems the eclipsing delusion. If a person must work to survive, and their job becomes automated, you would have to be either deluded or willfully disingenuous to be surprised when they fight to keep it. As the historian Frank Peel quipped, these workers "did not understand it was their duty to lay down and die" because they were no longer useful to industry and the state. The caricature of Luddites as chiefly technophobic, born in the minds of entrepreneurs and elites, was elevated to prominence in this courtroom. It has endured for centuries.

This condescension toward machine-breaking, this sense that to oppose technology is to oppose progress, was then combined with a general condescension for "the lower orders." Thompson moved to impress on the jury how the "spirit of insubordination quickly spread...and from the destruction of machinery used in manufacture, transition was easy to the destruction of the buildings." He explained: "When large bodies of men are associated for illegal purposes, the progress from one crime, to another still greater, is rapid."

This depiction of the Luddites, as brutes who cannot help committing more crimes once they'd done one, is reflected again in the official account left by William Leman Rede, the historian of York Castle. The Luddite mobs

were made up of men generally unaffected by the evil they complained of — Machinery; men who, being idle and dissolute, uneducated and brutal, had a love of brutality and excess, who found it more pleasant to seize by violence, than to gain by industry; and who, looking on the thing at first as a frolic, got excited by drink and the presence of a number of coadjutors to perpetrate the most dreadful crimes.

It was these two prejudiced elements—this depiction of barbaric mobs of unthinking poor who hated machinery they did not understand—that combined to forge the derogatory epithet "Luddite" that persists to this day. It's also clear from the rest of Thompson's speech that those prejudices were erected to hide and dislodge the state's true motivator: its fear.

"Armed bodies of men, in some instances several hundreds at once," Thompson reminded the jurors, "under the command of leaders, have attacked, by day and night, buildings in which machinery was employed."

A direct threat like that, posed to the equipment facilitating the expansion of the industrial state, could not be allowed to stand.

Betrayal

Ben Bamforth sat in the back of the churning, electric courthouse, and watched George's frame as the prosecution swiftly set the scene. Witnesses who'd had a water and rum with Horsfall at Warren's Inn moments before; witnesses present at the time of his public assassination; witnesses who saw him fall from his mare. Horsfall's brother was called in to produce the bullet that killed him, and the doctors who'd attempted to save his life confirmed it was so. But none of these men could finger the culprits who'd pulled the trigger. For that, they'd need an informant.

Ben Walker—the man Mary had called a coward, the man who'd accompanied George, Thorpe, and Thomas to shoot Horsfall—had turned.

Walker was hauled in by Radcliffe, along with George, back in the spring; he didn't talk then. Both were quickly released for lack of evidence. But Walker grew more anxious as the days passed. Walker, who was illiterate, overheard another worker at the cloth shop, John Sowden, reading of the £2,000 reward—nearly $210,000 today. He confessed his involvement to his mother, who went to Radcliffe to collect the bounty. George, Will, and Thomas were arrested the next day. It was Walker who'd drained the color from George's face, according to the account that was printed in the newspapers.

Walker was the Crown's star witness against the Luddite leadership in York, and now he was called to the stand. He said he recalled them talking about the loss at Rawfolds.

"Mellor said, the method of breaking the shears must be given up, and instead of it, the masters must be shot," Walker testified. "That was most that I heard said; they said they had lost two men, and they must kill the masters."

Walker gave a detailed account of the events of the day: The whole plan had been hatched spontaneously at John Wood's cropping shop, and George had convinced him to come along to attack Horsfall. The four of them had made the trek together, and he and Tom Smith were supposed to fire if George and Thorpe missed. Walker claimed that he'd never pulled the trigger.

"I do not remember what day Horsfall was shot, but I was that day at Wood's," Walker said.

> Smith and Mellor worked in one room, and I worked in another; I remember being with Mellor between four and five in the afternoon, and there was William Hall and my father and Wm. Walker. He asked me if I would go with him to shoot Mr. Horsfall. After that he went to his drinking, and was absent about half an hour; on my return, I found Mellor in the shop.... He gave me a loaded pistol, and said I must go with him and shoot Mr. Horsfall; he told me it was loaded with double ball; and it was primed and loaded nearly up to the top.

Ben Walker confirmed to the court that yes, he had hoped to obtain the reward by giving evidence.

The press delighted in the sensational trial, with additional witnesses painting George as the General Ludd of the West Riding, a celebrity more icon than flesh.

William Hall, another former Luddite, turned on George, Thorpe, and the other Luddites, too, and gave testimony to support the contours of Walker's story—later he gave evidence against those participating in the attack at Rawfolds as well.

"I was applied to by Mellor on that day, between four and five o'clock in the afternoon, for the Russian pistol," Hall told the jury; "the pistol had an iron end, with screws at the side, and a barrel about a foot long; I had heard from Mellor, that he had brought that pistol out of Russia."

The account against George was well-detailed, despite some gaps and contradictions. (According to Walker's timeline, they would have had to leave the cropping shop immediately after George proposed the assassination to make it to the point where they ambushed Horsfall, to start.) George was said to have beseeched Thorpe, Ben, and Thomas to help him acquire weapons; they quickly picked up the pistol. The informants attested to watching him load it, and to hatching the plan to take two posts and fire from two sides when Horsfall arrived.

Joseph Mellor, George's cousin, testified that George had given his apprentice two pistols the day of Horsfall's murder, and asked him to hide them. (It was known in the West Riding that Joseph Mellor had agreed to testify, and someone had shot at him outside his shop in the run-up to the trial, either out of warning or as an actual assassination attempt, to stop him from giving evidence.)

George said nothing, inside or outside of the courtroom.

No Confession

The only hope the Luddites had, and they knew it was a slender one, was to muddle the prosecution's portrait of the day's events.

So their counsel tried to present a flurry of alibis from men loyal to them, to discredit the informants as greedy liars, and to argue that the three accused men were not and could not have been present at the time of the shooting.

Several men, fellow croppers and blacksmiths and weavers, mostly, came forward to attest to being in their presence during the time of the shooting, and to impugn the character of the informants. A man named John Womersly said he'd been with Mellor at the time of the assassination, and had a note regarding an order of cloth to prove it. Another said he'd seen the men too far outside Huddersfield to be near the scene of the crime at the time of the murder.

A dozen men were brought up, attesting to their character and standing, defending them to the last.

It did little good. Because George and the others had been so vocal about their intentions at the shop, assuming no one would inform; because the level of detail levied by the informants was so compelling; and because the state was freshly zealous to press charges against machine breakers and subversives — the prosecutors and town officials had circumvented a regular jury trial to convene a special emergency jury, comprised of manufacturers and town elites — the outcome was all but assured.

When pressed for questions, however, the Luddites would not break their oaths. At the close of the trial, before the jury was sent to deliberate, they were given a final chance to enter into the record anything that might help their case.

The deliberation was finished in twenty-five minutes. All three were found guilty of premeditated murder.

The judge, Simon Le Blanc, asked the men why he shouldn't award them the death penalty.

"I have nothing to say, only that I am not guilty," George said.

"I am not guilty, sir; evidence has been given false against me: That I declare," said Thorpe.

"Not guilty, sir," said Smith.

The judge did not attempt to conceal his contempt.

"This may be pronounced a crime of the blackest dye," Le Blanc said.

> The crime was committed against a man, who appears to have given no offence to any one of you, except that he was suspected of having expressed himself with a manly feeling against those who had set up a right to violate all property, and to take away the life of any man who had been supposed to encourage others to do what I trust there still are men sufficient in this country to do — to stand manfully forward in defense of their property.

Even in presiding over the Luddites' fate, the judge could not bring himself to consider the offense that Horsfall, his machinery, and his factory could have registered to the workingmen, the extremity of their crime notwithstanding; he could only celebrate the masculine violence Horsfall had dedicated himself to dispensing in the name of advancing his business interests.

George, Thorpe, and Thomas were sentenced to death. Their execution by hanging was scheduled for two days later.

In the dank shadows of the holding cell, on the eve of George's execution, the pastor Webster implored him to confess his sins. George refused.

"I would rather be in the situation I'm placed, dreadful as it is, than have to answer for the crime of this accuser," he added, referring to Walker. "I would not change situations with him, even for liberty and two thousand pounds." Righteous anger flecked the words, even now.

There would be no confession.

"Forgive your enemies, and leave this sinful world in charity with all mankind," Webster tried.

"I've nought to forgive to anybody but Ben Walker," George replied.

The old pastor agreed to bring a message, and an ask, to his cousin Ben — that he wished him "all happiness, and to forget and forgive the nasty words he spoke when last they parted."

He then asked that Ben stand where he could see him during the execution, so he might think of home in those last seconds, "and the dear ones there."

I Forgive

The air was frigid as the three men were led to the gallows. Around them, there was no trace of the carnival atmosphere that came with many hangings of the age, but instead an air of queasy solemnity. The crowd, thousands strong, looked seasick, fitting for the suffocating foggy cold of January. The Luddites were folk heroes, vandals, Robin Hoods, murderers, enemies of Britain, relief for the starving. If these men even *were* Ludds — that was the catch, there was no way to know. Dispersed and amorphous, Luddism could die on the gallows or swarm out from the throng gathered before it.

To stymie any escape attempt or uprising, from within or outside York Castle, ample cavalry was called in to surround the grounds. The military was stationed, ready in full force. On the walk, three twenty-somethings.

The executioner fastened the noose around each of their necks.

There was a call for last words. A dead hush spread as the onlookers, the poor and the gentry alike, strained to hear this General Ludd make his final speech. As in the trial, they were disappointed, though George did finally look out, at Webster, then Ben, and open his mouth.

"Some of my enemies may be here," George Mellor said. "If there be, I freely forgive them." After a pause, he added, "I forgive all the world."

The bottom dropped out beneath the gallows. The Luddite bodies fell — and then stopped. The crowd was stunned; the audience had no choice but to look on as the men writhed and gasped for air. At the time, it was customary that the bodies would drop below the stage, so the convicted would die just out of sight. But they were to be subjected to a final innovation. The York executioners, at the behest of the state, had reconfigured the lethal device to halt their fall halfway down, so their faces and torsos would be visible to the crowd, who now squirmed in disgust.

George died, at twenty-three years old, alongside his friends and fellow Luddites, his death marked by one last mechanical improvement.

Ned Ludd on Trial

The next week, fifteen more men were convicted of crimes related to Luddite activity — breaking machinery, arms burglary, destruction of private property — many of which were connected to the assault on Rawfolds.

William Cartwright had testified against the men who attacked his factory, and denied them any hope of clemency. In a last-ditch effort to avert more bloodshed, the lawyer, Mr. Blackburn, had found Cartwright in town and pressed him to sign a petition of mercy for the Luddites.

"I have read with astonishment the petitions which you hand me," Cartwright replied in a letter, with a hard and incredulous No. "I can have none other than fulness of compassion for the unfortunate & misguided Men & I deplore as much as any man their Delusion but a sense of Duty only having guided me up to the present moment I cannot step out of that Line by interfering with the course of Justice until after the most satisfactory disclosure."

Fourteen were to be hanged, the largest number of simultaneous executions ever to be made at York. After the ramifications of this record-breaking sentencing dawned on them, the prosecutors asked Baron Thomson whether the convicts would all be hanged together, at once, on a single beam.

"Well no, sir," the judge is said to have responded, "I consider they would hang more comfortably on two."

At 11 a.m. on January 16, 1813, seven men, the first of the two groups of Luddites, were led to the gallows. Flanked by a large garrison of soldiers, and before an even larger audience, they sung the hymn "Behold the Savior of Mankind" as they walked to their death.

Behold the Savior of mankind
Nailed to the shameful tree!

How vast the love that Him inclined
To bleed and die for thee! [...]

Thy loss our ruin did repair;
Death by death is slain;
Thou wilt at length exalt us where
Thou dost in glory reign.

When the drop came, and the seven fell to their deaths, the crowd was horrified. So many men, most of them young, most with families, put to death en masse, publicly hanged not for a grisly crime like murder, but for organizing a rebellion to defend their livelihoods. This was the new cost of trying to dismantle, by force, a profit-making machine.

"The scene was inexpressibly awful," the *Leeds Mercury* reported, noting that the guards "gave to the scene a peculiar degree of terror, and exhibited the appearance of a military execution."

At the request of one of the executed men, William Hartley, the *Mercury* listed the number of family members each Luddite left behind. "The principal part of these ill-fated men were married and have left families," the paper dutifully recounted. "William Hartley, has left seven children, their mother, happily for herself, died about half a year ago. John Ogden, wife and two children; Nathan Hoyle, wife and seven children; Joseph Crowther, wife pregnant, and four children." And on it went — in all, thirteen wives were widowed and fifty-seven children were left fatherless that day.

The men whose bodies were not claimed by relatives were given to medical practitioners training for the war effort. There was a shortage of bodies available for study by military doctors; grave robbery was rampant throughout 1812. The executed Luddites were dissected.

There were more trials; dozens were sentenced to transportation to Australia, and more yet to hang, including a framework knitter accused of burglary in Nottingham. Another Luddite, John Lumb, who was spared hanging at York, was killed in an accident on board the ship taking him to Australia. The elderly John Baines, the political reformer and shoemaker, was sentenced to seven years' transportation for administering oaths, as were his two sons. Between the Crown's executions and those dispensed by the entrepreneurs' armed guards, some forty to fifty Luddites were killed outright from the winter of 1812 to that of 1813. Horsfall was the lone human casualty of the Luddites.

The trials and executions were both rushed and highly produced affairs, intended to levy a maximum impact on the nation's public. They were designed to demonstrate that after being slow to respond in 1811, Lord Liverpool's England would punish thoroughly any threats to its industrial capacity and political order. They were in effect show trials, held largely for the benefit of the elites and entrepreneurs.

After the executions, a Quaker named Thomas Shiltoe traveled through the West Riding, aiming to seek out, and try to provide solace to, the families of the recently executed Luddites. The journal entries from his travels offer a portal into the raw grief suffered by the families of the dead men:

> We proceeded to the house of the widow and five children of [Jonathan] Dean, of Longroyd Bridge, who suffered for rioting. The widow's mind appeared to be under very great distress, with her helpless, fatherless children; the oldest child being about eight years, the youngest not more months old. All that was alive in us and capable of feeling for her, plunged as she was into such accumulated distress, we felt to be brought into action.
>
> We next visited the widow and three children of John Walker, who suffered for rioting, one of the children an infant at the breast. The feelings of distress awakened in my mind, in sitting down with this family, were such, that I was tempted to conclude human nature could hardly endure to proceed with the visit before us.

When Percy Shelley, still at work on his revolutionary poem *Queen Mab,* read about the Luddite executions, he was moved to try to help, too. Just days after the hanging, "Harriet Shelley, writing on her husband's behalf, offered to raise a subscription in support of the families of fourteen men who had been executed for their involvement in Luddite activities."

The state, for its part, considered the trials and its handling of the Luddites a great success, as its commission at York noted: "For some months before the special assize, the disturbances in Yorkshire, as well as in all the other manufacturing districts of the kingdom, had nearly subsided; and this tremendous example, made to the offended laws of the country, served to confirm and render permanent the public tranquility."

The informants may have spared their bodies, but at the cost of knowing much peace for the rest of their lives. William Hall was so spurned and harassed that he left his job at John Wood's and fled town; he was never seen again in Huddersfield. Ben Walker, for his part, did not run. But he became a persona non grata in the West Riding, "the finger of scorn pointed at him daily," despised by his peers and looked down on as a rat and opportunist. He never did receive the reward the state promised him for turning evidence, either, and by the end of his life he was begging in the streets.

It was hardly a time of tranquillity, despite what the prosecutors of York might claim. A statue of George III was pulled down by angry citizens in Bristol. Captain Francis Raynes reported arresting a man who threatened to exact revenge on behalf of the executed Luddites. George Mellor's cousin Joseph, who testified that George had hidden the pistol at his shop, received a threatening letter of his own.

It read, "Blood for blood, says General Ludd."

THE INVENTION OF THE LUDDITES

"If the Luddites had never existed, their critics would have to invent them," Theodore Roszak wrote in the 1990s. The most ardent champions of the tech industry, the most zealous entrepreneurs and executives, need a bogeyman; a means of positioning the notion of opposing new technologies, products, and services as perennially ridiculous. This is as true for water-powered cotton mills in the 1810s as it is for, say, automated assembly lines in the 1950s, software monopolies in the 1990s, or artificial intelligence startups in the 2020s.

"Luddite" has been shoehorned into history as shorthand for someone who blindly opposes technology and, importantly, is doomed and at least a little dumb. (Not, notably, someone who is methodically and strategically gathering and manufacturing arms for an uprising.) If you knew the word "Luddite" before reading this book, you likely knew it as an insult. Even when the word itself is not on our lips, the ethos it embodies, breathed into us from our earliest engagements with technology, always seems to be there: that to oppose technology is to oppose the future, and prosperity; that questioning technology can only point us backward. But, as Roszack noted, "the Luddites are held in such contempt that their critics have never felt the least need to find out who they really were and what they wanted."

The real Luddite movement was, of course, multifaceted, complex, and driven by a range of grievances and demands. The Luddites as most of us know them, the moronic machine smashers, *are* in fact inventions. They're the myth invented by their critics, not the well-organized, strategic, and morally empowered force that contested the rise of the factory and the entrepreneurial manager in the 1810s. Look no further than the easiest way to discover new information in the 2020s: Google's online search explains that a Luddite is a derogatory term for "a person opposed to new technology or ways of working," e.g., "a small-minded Luddite resisting progress."

Small-minded. Our most widely accessible definition of the word is dismissive and derisive — Luddites, we are told, smashed machines because they were afraid and angry at them, because they were short-sighted. Now think of every Silicon Valley philosophy that's been blasted into the mainstream — about moving fast and breaking things, about regulation as the enemy of invention, about the gospel of "disruption" at any cost — and consider that these axioms are now so deeply ingrained in our lives and our politics that they have taken on the air of common sense. Think of all the critiques, qualms, and challenges to technology that have been waved away with a derogatory "Don't be a Luddite." And, conveniently for the companies doing the disrupting and thing-breaking, opposing their very profitable technologies tends to make *you* a Luddite, too.

Consider: Do you find it disagreeable that the only job you can seem to find is part-time gig work arranged through an app, and that the company that cuts your paycheck does so according to a complex and opaque algorithm that seems to change the way you get paid every other week? You are mistaken; this app is the flexible future of work. You are a Luddite; the CEOs, business consultants, and corporate scribes that have championed this mode of work say so themselves.*

Do you dislike the notion that management has hinted that a robotic arm will be taking your place on the warehouse floor, or that, if you are not laid off, you will be forced to keep up with said robotic arm? That's a Luddite attitude right there. Are you annoyed that you must wade through an inscrutable maze of online menus before discovering whether you have been automatically approved or denied crucial services by a health insurance company, a utility, or the state? Get with the times, their cheerleaders will tell you, these automated programs have cut costs and streamlined processes. You are a Luddite. Does the rush to inject AI into every conceivably profitable pore of reality make you queasy? Lud-

---

* Here's a small sampling of CEOs and conservative pundits levying the Luddite charge against Uber skeptics and regulators in the 2010s: "How Uber Is Defeating the Luddites. Uber is fighting the Neo-Luddism of the Establishment," by Tim Askew, CEO of Corporate Rain International (*Inc.,* August 10, 2015); "In Uber Fight, de Blasio Played the Luddite," by CA Technologies CEO Michael P. Grigoire (CNBC, July 24, 2015); "The Uber Fracas: Luddite Regulators Target Innovation," by Richard Levick, chairman and CEO of the global communications firm Levick (*Forbes,* June 17, 2014); and "Uber Faces Backlash from the New Luddites: Americans Have Become So Accustomed to Regulations, We've Ceased to See How Freedom Might Operate," by Steven Greenhut (*Reason,* April 4, 2014).

dite. Wonder if maybe we should try to wind back the march of mobile computing that has left us hopelessly addicted to our screens? Luddite.

Does it make you uncomfortable that googling "Luddite" calls up Google's own definition of Luddite first, which propagates the myth that the working-class leaders of history were anti-technology morons? You, too, are a Luddite — Google has just labeled you one.

Starting almost immediately upon the Luddites' emergence, philosophical critiques of, and even outright opposition to, technology in general — a fretting over what might be lost or gained as it applies to lived experience — were conflated with Luddism. Such conflation remains a deeply incorrect parsing of what the original movement was about. Part of it was intentional, of course. It behooved the factory owners and other arbiters of industry to cast Luddites as reactionary idiots, as it is a convenient way to chill dissent. But it's a bait and switch. The Luddites protested machines "obnoxious" to them; it was an economic complaint, albeit with direct social consequences. The Luddites were not lamenting what machines might do to culture, their attention spans, or even their health or environment. They were seeking to protect their direct economic interests, their communities, and their personal liberty.

There is a long legacy of broader criticism of technology, stretching at least as far back as Plato's *Phaedrus,* a dialogue between the Egyptian king Thamus and his chief technologist, Theuth. Thamus is appraising Theuth's many inventions, when they arrive to the technology of writing, then a novel development in Egyptian society. In a famous exchange dissected by theorists like Marshall McLuhan and Neil Postman, Thamus warns Theuth about writing, arguing that "those who acquire it will cease to use their memory and become forgetful." Today, someone making the equivalent argument about, say, smartphones making us less smart — would be called a Luddite.

The list of critics of invention and "progress" is endless; it includes Sigmund Freud, Henry David Thoreau, Ted Kaczynski, and certain Amish sects. In *Walden,* Thoreau famously wrote, "Our inventions are wont to be pretty toys, which distract our attention from serious things. They are but improved means to an unimproved end, an end which it was already but too easy to arrive at." But to argue that we should return to a more rural, nature-centric mode of life is not Luddism, it's pastoralism. Then there's the sense that bubbles up when your laptop keeps freezing and you want to throw it in the river. That's not Luddism, that's simple rage.

This conflation endures largely because one can neatly be folded into the other; haters of new popular things are easily dismissed as scolds by younger generations of consumers, technology users, and culture producers, and that dismissiveness serves the status quo. In 1998, for example, when Microsoft stood accused by the US government of wielding monopoly power over the software industry, its lead attorney strode into a Washington courtroom and told the judge that the proceedings had turned into "a return of the Luddites." Bill Gates's lawyer, John Warden, explained to listeners unversed in English history that the Luddites were a band of workers who smashed machines "to arrest the march of progress driven by science and technology." In 2014, conservative commentators applied the Luddite label to the taxi lobby when it sought to slow the (regulation-skirting) rise of Uber and Lyft — the label was applied derisively and mockingly, and implied the efforts were doomed to fail. "I don't know many people who still wear hand-sewn garments," one columnist sneered.*

Part of the reason that so much derision is heaped on the Luddites is that it's easy to look around today and see the outcomes of their lost battle. The factory system has come to dominate the last two centuries of industrial production, a period marked by mutual antagonism between bosses and workers, and the gulf of inequality that attends to it. One can lazily claim that all of this has simply been inevitable, and it will always be thus — that technology, not people, creates winners and losers, and because it eventually makes more of us winners, allegedly, then it is absurd to protest its march.

But it is much more absurd to pretend there are no possible alternative arrangements — to think that technology, the product of concerted human invention and innovation, can only be introduced to society through reckless disruption, or that it's unthinkable that advancements in technology might be integrated into our lives democratically and with care. If we are ingenious enough to automate large-scale production, build spacecraft, and invent artificial intelligences, are we not ingenious enough to ensure that advancing technology benefits all, and not just a few?

True Luddism was about locating exactly where elites were using technologies to the disadvantage of the human being, and organizing to fight back. This is an important point: Luddism can and certainly did coexist with technology, and

---

* The Luddites did not sew by hand, of course, but used machinery themselves.

even a love of technology. The handloom, for example, made the Luddites' way of life possible, long before they became Luddites — and they cherished that lifestyle enough to take up arms to defend it. It is a matter, of course, of how technology is deployed.

Before we find ourselves entirely backed into a corner by today's tech titans, we should ask these questions: Does their technology serve to funnel profits upstream while degrading a livelihood or destabilizing a community? Are those who rely on the disrupted systems given a democratic say in how innovation will affect their lives?

The history of the Luddites — the real ones, not the pejorative figment of the entrepreneurial imagination — gives us a framework to evaluate the utility of technologies and their social impacts. Erasing that history collapses our thinking about how tech and automation affect our working lives — and the choices we have to address the disruption they bring.

THE PRINCE REGENT

February 1813

On the February 1, 1813, the Prince Regent's office issued another flamboyant proclamation condemning the Luddites and supporting the rights of entrepreneurs. The statement repeated, even deepened, the narrative that the Luddites were deluded men who, acting on behalf of "the wicked misrepresentations and artifices of ill-designing persons," sought to turn back England's progress.

These great manipulators, the prince explained, "deluded the ignorant and unwary, through the specious pretext of procuring additional employment and increased wages for the laboring manufacturers, by the destruction of the various kinds of machinery now most beneficially employed in the manufactures of this kingdom."

Even before George Mellor's body was fresh in the ground, the proclamation completed the elite reframing of the Luddite narrative from that of an insurgent, strategic, and folk-heroic labor movement to some delusional idiots breaking machines because they did not understand them.

"The extent and progress of the trade and manufactures of this country, which have been continually advanced by the invention and improvement of machinery," the Prince Regent stated, "afford the best practical demonstration of the falsehood of all such pretexts." The rapid economic growth of industry was proof enough that to resist any particular element of its formation, like working conditions, was foolish.

Perhaps most important, the Prince Regent called for factory owners and industrialists to defend their machines with force, providing royal license for the entrepreneurs to draw blood to defend their machines: "We do further exhort the proprietors of machinery, not to be deterred from continuing the use and employment of the same, but vigilantly and strenuously to exert themselves in the maintenance and defense of their property."

GEORGE MELLOR

February 1813

In the month following the mass execution of the Luddites, the Quaker Thomas Shiltoe paid another visit to Huddersfield, this time to the house of John Wood and his wife, George Mellor's mother.

"We proceeded to the mournful house of the parents of [George Mellor]," Shiltoe wrote in his memoirs, and

> sat with the parents, who are living in a respectable line of life. In this opportunity we had fresh cause to acknowledge holy help was near, furnishing matter suitable to the deeply-tried and afflicted state of mind in which we found them. Our visit was thankfully received by both parents, and, as we afterwards understood, was like a morsel of bread at a time when they appeared almost ready to famish. The father acknowledged, the melancholy circumstance had brought their minds into such a tried state, that they had concluded to move to some other part of the country.

The wound was deep. Machine-breaking did continue, especially in Nottingham, for the next six years. Men reorganized sporadically under Ludd's banner, smashing machines and hosiers' factories until 1819; a few more yet were hanged for swinging Enoch's hammer. But the movement had peaked, and the specter of a mass uprising led by a General Ludd, goading frame-breakers into action across the nation, began to lose its urgency. By the end of the decade, as a physical avatar one might follow into battle, Ludd might as well have been dead. The spirit and form of Luddism, meanwhile, evolved, and flowed into a more nebulous political movement, one focused on reforms and revolution, without the physical destruction of machines that reside in factories.

The tech titans had triumphed. In the form and thunder of their victory, they established a precedent that plagued workers for the rest of the Industrial Revolution, and set a trajectory that continues through the modern age.

This seed is evident, even in George Mellor's legacy. The Luddites were too cautious to keep records and correspondence, but while George was in prison, he smuggled out a lone message intended for his family.

"I now take the Liberty of informing you that I am in good health as by the Blessing of God I hope they will find you all," he wrote. He beseeched his peers to keep to their alibis, in desperate hopes that the trial might still be won.

"Tell the Boys to stick by what they said the first time," he wrote; "if not they are proved forsworn." And he asked other witnesses to vouch for him, too; if a man named Rattiffe and his wife "will befrend me" he said, they can "never mind thier work for I if I come home I will do my best for them." Then, for the first time to our knowledge, George turned his energies to an organized political campaign, writing that he'd heard of an effort to petition for parliamentary reform, imposing restrictions on machines that might better the craftsmen's lot, and asking that his name, and that of thirty fellow Luddite inmates, be added to the petition.

"Give my respects to all enquiring Friends," George wrote, "and accept These few Lines from your Friend."

But Mellor's message was intercepted by the Crown's agents in Huddersfield before it could reach his family. There was no future for him beyond the smashed machines and the fortressed factories, ultimately impenetrable, that had consumed him after all. The natural rights of man were disappearing, just as he'd feared, the "tendency all one way." But there may yet be hope for his peers, for future generations.

"Remember," wrote the man who had led an uprising against those who used machines to exploit people, and died for it: "A soul is of more value than work or gold."

ROBERT BLINCOE

1813

"Sit down," the fortune-teller told Robert. Seeing as how he was scared witless, he obeyed immediately. Ol' Beckka was legendary throughout Derbyshire, and Robert had been told that rich folks came all the way from London to have her tell their fates. She thrust a teacup full of grounds into his hands, and commanded him to "shake it well." Beckka drained the tea, and went about peering into his future.

Robert Blincoe was now a free man, after surviving ten harrowing years of indentured service at Litton Mill. No one had ever showed him the trade of stocking weaving, but he did have the papers to prove his tenure. He'd stayed on at Litton just long enough to earn the money to make his next move. The overseers hadn't laid much of a hand on him since he'd visited the magistrate, brought back a letter saying they were investigating the matter of his abuse, and bluffed that more inquiry was imminent.

Ten years — a decade of invisible labor, ensuring that those cutting-edge machines actually ran. Without workers like Blincoe, the tech that Baron Thomson, William Cartwright, and the Prince Regent himself all celebrated so enthusiastically, and protected at the expense of Luddite lives, wouldn't produce anything at all. In a sense, it was not the machinery they were defending — it was the factory owner's prerogative to employ boys like Robert for next to nothing, instead of skilled men like George Mellor for a full wage.

After he turned twenty-one, Robert Blincoe set off; among his first stops was the soothsayer's.

"You came from the outside of London, did you not?" Ol' Beckka said.

"Yea," Blincoe said, "I did."

"You came down in a waggon, and have been at a place surrounded with high rocks and great waters, and you have been used worse than a stumbling stone."

Blincoe was shocked.

"Your troubles are at an end," she continued. "You shall rise above those who have cast you down so low. You shall see their downfall, and your head shall be higher than theirs. Poor lad! terrible have been thy sufferings. Thou shall get up in the world!"

Ol' Beckka had seen his future, and the worst was behind him.

Robert Blincoe spent the next few years as a journeyman weaver and factory worker. He had some trouble getting work, owing to his disfigured hands and small stature, and struggled to find fair wages like most weavers of the day. Still, he moved from shop to factory, from small towns to industrial Manchester, living thriftily and saving money as he went. He eventually saved enough to start a small business dealing in cotton waste, and got married in 1819. He even tried investing in cotton-spinning machinery, but lost his machines in a fire six weeks after he'd gotten them running.

In 1822, the journalist and labor reform advocate John Brown was investigating the conditions in cotton factories when he heard Blincoe's story, and was so taken that he decided to write Blincoe's memoirs in full. When the book was published, nearly a decade later, it became a phenomenon.* The story was a scandal, a blight on the industrialists who claimed they were building the future. Historians believe that it inspired the tale of the most famous orphan of all, Oliver Twist, conjured by Charles Dickens, who published his novel five years after Blincoe rose to national attention.

The book helped finally drive reforms through Parliament that did have at least some teeth — enforceable limits on the length of the workday, an end to the parish-to-factory pipeline practice, an end to indentured servitude. Factories were still allowed to hire children, however, and Brown decried the factory system itself as an engine of inequality, slavery, and vice:

> There never yet was a crisis, when, in the commercial world, the march of
> avarice was so rapid, or its devastations so extensive upon the morals and
> well being of society, as within the period embraced by this narrative; a
> march that seems to acquire celerity in proportion to the increasing

---

* In the interim, Brown had fallen ill, claimed that factory owners were fighting against his publishing the book, and ultimately took his own life before a colleague completed the work.

spread of its malific influence, and to derive impunity from the prodigious wealth it accumulates in the hands of a few great and unfeeling capitalists, at the expence of the individual happiness, health, and morals of the million.

The period he describes is also when the Luddites and their predecessors watched their own conditions worsen, and exhausted the legal means to address it. They were rising up just as Blincoe was finishing his indenture against the implements of the same injustice. While the factory was degrading the cloth workers' livelihood, it was creating new, appalling conditions for the next generation of machine workers. Cartwright's loom, poised to enable such mass accumulation, didn't replace the human, of course — it just let boys like Blincoe do the work.

These cascading consequences of automation, unintended or otherwise, persist. We might marvel at the progress of, say, the self-driving car, but its autonomous navigation requires the labor of numerous invisible workers who do the thankless, drudgery-filled toil, often for very low wages, of labeling image after image to make the datasets the algorithm needs in order to operate. From Amazon's Mechanical Turk to refugee camps in Europe, workers are paid pennies to sort endless reams of data, the raw materials for computer vision programs and self-driving vehicles. The researchers Mary L. Gray and Siddharth Suri call this "ghost work" — and it's still ascendent today.

The autonomous delivery robots now common on American college campuses and downtown areas may replace delivery people — but they are digitally overseen by other workers who can control them remotely, from places like Colombia, for $2 an hour. It's the same story, time and again: a new technology that promises to alleviate work degrades it instead. Or take automated self-checkout, which we might not think about much, apart from being annoyed by the added hassle it gives us at the supermarket: not only did that device cost a cashier or checkout person their job, but it forced the remaining workers to learn new skills to help customers navigate it when they inevitably need help.

Automation and workplace technology often don't result in *less* work, but more diffuse, precarious, and lower-paying or less-protected work. Over the course of two centuries of disruption, the results have been consistently predictable — if there's a powerful labor-saving technology in the headlines, there's a Robert

Blincoe somewhere making it feasible. There's always a human behind the automaton.

The fortune-teller was right, it turned out. The factory, too remote and unprofitable, closed down in 1815, and Ellis Needham, who owned the place and had tortured Blincoe for nearly a decade as a child, lost everything; he died poor in the 1820s. It wasn't an uncommon outcome. For all the abuse and harm caused by factorization, automation, and the schemes to divide and control workers, plenty of owners had to close their doors, if raw materials climbed in price or bad luck struck. They had automated mass suffering for nothing.

Blincoe's troubles were not entirely over — he was briefly thrown in debtor's prison for the cotton machinery mishap — but he recovered quickly, and generally kept his business, his family, and head above water. Like tens of thousands of other child laborers, his body had been scarred by his upbringing in the factories: he was diminutive from malnourishment, and his hands and knees were permanently damaged. But he had, as Ol' Beckka had predicted, ultimately risen above the masters who had imprisoned him. And his story helped countless others to do the same.

WHAT THE ENTREPRENEURS WON

Cartwright and Radcliffe

June 1813

"Mr. William Cartwright," the official's voice rang out over the grounds of Raw-folds Mill. "In the name of the Gentleman of the West-Riding of the County of York, I present to you an Address of Thanks, for your conduct during the late disturbances."

It was a day filled with no shortage of pomp and formal celebration. The proceedings began when a senior constable from Halifax named William Rawson, Esq., rode by horseback, along with the region's gentry, magistrates, and factory owners, in a procession to the site of the now infamous melee with the Luddites. There he read aloud the "Address of Thanks" to the man who had, the year before, dealt the Luddites a fatal blow. The businessmen had drawn up a formal statement valorizing Cartwright, along with the magistrate Joseph Radcliffe, for their "courageous conduct" in 1812. The document, signed by dozens of the region's elites, resolved that Cartwright should sit for a portrait that would be engraved and placed in public view in the Huddersfield courthouse.

"We the undersigned Inhabitants of the West-Riding of the County of York, most heartily approve of your Conduct in defense of your property and person," Rawson continued. "We offer you our sincere thanks upon this express ground."

The ground where they were now standing was where the Luddite blood had been spilled, just outside the factory doors. When the delegation finished, Cartwright stepped forward to offer his "warmest sentiments of gratitude." He gravely thanked the colonel of the local militia "for the readiness with which you acceded to my application for arms and ammunition to defend my property, and for the confidence which you then expressed, that I would not shed blood unnecessarily." It was his sense of duty, he said, that drove him to act.

During nearly twelve months of watchfulness, anxiety, and domestic privations, through every danger, and through the awful and much to be deplored scenes which present themselves in time to time, under divine providence, [Cartwright continued,] a consciousness that I was performing an imperative duty, could alone have supported me. It will ever be the proudest feeling of my heart, my exertions have been thought to meet the notice of the Gentleman, whose Names are found in this highly valued Address.

After the "Address of Thanks" was formally read aloud, they departed for Milnsbridge House, Radcliffe's manor, and repeated the ceremony there. Radcliffe, for his part, said it had been "the greatest pleasure of my life" to "fulfill the duties of the public station which I am placed." The procession then returned to the George Inn in Huddersfield, where the gentlemen held a feast, and a toast to what they'd won.

This was the latest in a series of efforts to anoint Cartwright and Radcliffe as heroes. Meetings had been called to gather rewards for their efforts, and the entrepreneurial class and its allies sung their praises in the local newspapers. "A full year has now passed over us," the Reverend Hammond Roberson wrote in an editorial for the *Leeds Intelligencer* on the anniversary of the battle of Rawfolds. "We have had time for reflection."

As such, he wanted to enter the following record for the information of posterity:

That one of the most villainous, of the MOST DESPERATE; *and, for its extent, one of the* MOST ALARMING CONSPIRACIES AGAINST *the Security of the persons and property of civil society, that ever disgraced a country professing Christianity, was checked and disconcerted by the manly firmness of* ONE MAN.

That manly firmness belonged to Cartwright.

In some ways, these meetings, letters, and public ceremonies were a public relations effort to justify, even glorify, the machine owners' actions at a moment of high uncertainty. After the Luddite trials in York, the mood was tense. One of Cartwright's servants was approached on his way home and asked if he still worked for Cartwright. He was knocked down and beaten after he said that he did.

By March, though, a quiet had settled on the town, enough so that Cartwright felt ready to draw down the military garrison that had been stationed at Rawfolds for nearly a year. Cartwright, who transformed his factory into a fortress rather than slow his acquisition of technology or negotiate with the workers, had won. (This too came at a price; mounting those defenses had nearly bankrupted him, and his fellow entrepreneurs had to lobby for a reward from the state to keep him afloat.) He would not have known it then—though he certainly thought highly of himself and his "imperative duty" to advance his profit-generating machinery—but he had helped clear a path for those pursuing ever more ambitious scales of automation.

Less than two decades after the Luddite uprisings, the effort to imagine and build human-free factories began in earnest. Charles Babbage, the inventor of the schema for one of the first modern computers, would declare that his difference engine could be used to computerize the entire factory system. (In an ironic twist, Lord Byron's estranged daughter, Ada Lovelace, would write history's first software programs by expanding on Babbage's work.) In 1832, Babbage published *On the Economy of Machinery and Manufactures,* which argued that factories and machinery were the great drivers of prosperity. It helped launch a trend of books known as factory guides, tomes intended to drum up public support for the institution at a time of gathering doubts, and to extol the virtues of the machinery that made them possible. By the 1830s, the moral uproar over factory conditions was acute, thanks to exposés like Robert Blincoe's, and to working-class movements like the Luddites that made clear the mass outrage at the system. Skilled laborers, radical reformers, and even Tory pastoralists opposed factorization.

But Babbage's book wasn't the most famous or influential pro-factory tract; that distinction belongs to Andrew Ure, a chemistry consultant and early business theorist. Ure was enamored with the potential of visions like Babbage's, and in 1835, he published *The Philosophy of the Manufactures.* The book argued, following Babbage, that mechanized factories were the true engines of progress, providing "the rulers of the kingdom with the resources of war, and a great body of the people with comfortable subsistence." They had made England "the arbiter of many nations, and the benefactor of the globe itself." He knew the charges against factories, he wrote, but maltreatment of children and workers was uncommon, he argued, the result of a few bad-apple overseers. The illnesses which seemed so rampant could easily be solved with better ventilation and food. And besides, he claimed, the whole endeavor would be fully automated soon anyway. Ure was one of the first futurists; he conjured a nineteenth-century vision of fully automated factories, producing goods around the clock. "Central to Ure's book is a social fantasy of autogenesis — machines that produce without workers," as one design historian put it.

Entrepreneurs, factory operators, and executives have been chasing that fantasy ever since. In the twenty-first century, Amazon touts plans for fully automated warehouses, high-tech companies aspire to "dark" factories — factories emptied of humans and automated to such an extent that they can be run without the lights on — and robotics and enterprise software firms pitch a battery of automation services and products to clients afraid of being left behind. Now, as then, we're made to believe that we are in an uncomfortable but necessary transitional period, right before technology will inevitably solve all our woes. Visions like these helped the entrepreneurs settle the debate, in the halls of power anyway, over the machinery question. Surprise — machinery won.

The Industrial Revolution transformed modern life, and the factory became the predominant engine of production the world over, providing the basic template of work to be followed for the next two centuries. As the factory system spread, pollution blackened the skies, a flurry of affordable but often lesser-quality goods entered mass production, and the health of workers dramatically declined. Vast fortunes were created — along with abject poverty.

When Cartwright turned the Luddites away in a hail of gunfire that night at Rawfolds Mill, and when the state decided to lash the hired hand who would not fire on them, and to hang dozens of machine breakers in York, the way was laid. Two centuries later, as taxi drivers in France organized angry protests against the rise of Uber, company executives debated whether to send its drivers into affected areas. Some in the inner circle felt it would endanger workers' lives. "I think it's worth it," Uber's CEO, Travis Kalanick retorted. "Violence guarantee[s] success."

The domestic system was swallowed up by the factory, which in turn gave rise to the office, which informed the rise of algorithmic work platforms. Ever since the Luddites, wage-earning workers have been, in one form or another, at the whims of an overseer who deploys technology to control the division of labor. That, above all, is what entrepreneurs and the state won, when the Luddites lost.

Back at the George Inn, Cartwright, Radcliffe, and the entrepreneurs and elites ate and drank and made toasts. They raised their glasses to the king, to the Prince Regent, and to Henry Lascelles, the pro-machine politician they'd backed:

> The King and God bless him! Hear, hear.
> The Prince Regent. — The Queen and Royal Family.
> Joseph Radcliffe Esq. — Mr. Wm. Cartwright.
> Hon. Henry Lascelles. — Lord Milton.

They saluted dozens more, and the town, the trade, the county; and they saluted prosperity. They saluted the magistrates and courts, and the militias and their commanders who had protected them. Near the end of the long toast, they raised their glasses again.

"May the Manufacturers and the Machinery of Yorkshire never be uninterrupted."

GRAVENER HENSON

Winter 1813

Gravener Henson was nothing if not persistent.

The factory owners had grown tired of dealing with the Society of Framework Knitters; their demands for higher wages, their increased leverage, and the way it was all beginning to cut into their business, and so they resolved to squash it. Those owners had considered prosecuting the leaders for violating the Combination Acts — just as Henson and others had feared — but they couldn't get anyone to inform on the effort. So they hatched a plan.

The entrepreneurs dug in on a particularly expensive strike that silk workers were waging with the Society's support. After over a month, they convinced the magistrates to disband the local militias in Nottingham and the surrounding area. Those soldiers, workingmen themselves, suddenly had no source of income. The thinking was that they'd flood the labor market and force the Society to give up the strike, or perhaps get a few of the militiamen to join up and offer evidence of a combination. The latter plan succeeded, and three men were charged. They were given only minor penalties — officials hoped to avoid a major outrage — but the magistrates got access to the Society's books and were able to shut them down.

After trying yet another peaceful avenue and failing, the stockingers, who were as impoverished as ever, despaired. In 1814, another rash of frame-breaking broke out in Nottingham, and General Ludd's name was reprised in the organized bouts of hammer-wielding and riots that followed. The actions followed a similar course, albeit on a smaller scale. Machine breakers rose up in the Midlands, invoking the name of the Luddites; concessions were made, the state cracked down, and, after a major final assault — this time an attack by dozens of men on a large lace factory in Loughborough, where over fifty machines were smashed over the course of an hour — the Luddite leaders were martyred on the gallows.

In 1817, Gravener Henson was thrown in jail again, this time for preparing to support an accused Luddite; the order was signed by Addington himself.

"This man Henson has long been an object of dread to the well-disposed inhabitants of Nottingham and its neighborhood, both on account of the leading influence he was thought to have with the Luddites, and his supposed political principles," the state-subsidized *Nottingham Journal* wrote. Henson would claim to the authorities that he was never affiliated with Luddites, that he had always advocated a peaceful route of organizing, and that he had come under fire *by Luddites* for not embracing the cause.

Few believed him; few historians have, either — he was simply too well connected, too integrated into the framework knitters' ecology to have been unaware of their activities. It is hard for many to believe that it was a coincidence that Luddites ceased smashing machinery while parliamentary reform or trade organizing efforts were underway, and resumed it when they failed. Even after the failure of the bill, and the dismantling of his framework society, Henson continued to press for reform and wage protections — and, above all, a repeal of the Combination Acts.

In 1823, Henson set out for London once again, with a repeal bill that would institute pay minimums and a worker's charter. Yet again, his effort did not succeed. This time, however, it did spur the radical reformer Francis Place to expedite his own bill, and may have encouraged Parliament to pay more attention to that effort, given that it was less aggressive than Henson's. Place's bill passed, repealing the Combination Acts and finally allowing workers to legally organize to bargain for improved conditions together. There were plenty of caveats, of course, and the road to truly free unions was a long one. Still, historians like E. P. Thompson have recognized Gravener Henson and his framework knitters as among the great unsung forces driving the formation of the working class itself.

William Felkin, the Derbyshire historian and stockinger who was a teenager during the Luddite uprising, fondly recalled Gravener's work.

Henson wrote with surprising facility and grammatical correctness. His petitions to parliament, memorials to ministers, and letters to public men were striking, but injudicious: often containing libellous invectives. Once, being directed to draw up a memorial to the Treasury on the difficult subject of the "Export of Machinery," he brought the next day twelve foolscap pages closely written, without interlineation or blot; which, after being compressed and expurgated much to his chagrin, was a cogent and effective document.

Henson was also rumored to have written, but never released, a history of Luddite activity—historians have been looking for it for years. He did publish a book in 1831, two decades after the uprisings, that was intended to be a comprehensive history of the cloth trade and its industrialization. In *The Civil, Political, and Mechanical History of the Framework-Knitters in Europe and America*, Henson took pains to show how collective bargaining by riot had proven successful in the past, recounting one of the earliest documented examples of Luddism. He also offered a sort of people's history of the inventions that drove the Industrial Revolution, detailing the improvements made by workingmen, that other works, which tended to focused on the richest and most celebrated inventors, overlooked.

The book is a fascinating and sprawling treatise; it veers from highly technical descriptions of historical improvements in the modes of cloth production, to spirited stories about trade relations that clearly favor the working class, to detailed considerations of biblical representations of weaving—early on he wonders how Noah may or may not have adapted textile production after the Flood.

What shines through is that Henson was intent on documenting the innumerable contributions to the trade by workingmen, who were the true inventors, refiners, and improvers of the machinery that was so often taken up and exploited by the business owners for profit. It may have been his way of trying to canonize the unsung workers, to give them their due.

Henson grew increasingly eccentric and irascible as he got older, with so few of his accomplishments coming to obvious fruition, his petitions denied, his book overlooked, the workers' movement having moved on. He outlived his wife, and died alone, poor in Nottingham.

CHARLES BALL

1813

In the spring of 1813, a British fleet entered the bay where Charles Ball was working as a fisherman, and one of the vessels made its way toward him.

After countless years of enslavement, Charles had escaped the plantation.

"To go wrong was worse than to stand still," Charles wrote of his journey, when he found himself lost in the wilderness one night. The problem was that there was no safe place to go. He'd made the long, perilous trek hundreds of miles north. He found work around Baltimore, on small farms and on fisheries, fell in love again and remarried, saved some money, and led a quiet life raising a new family. Then came the War of 1812.

Now, he watched as the British soldiers cut his employer's fishing nets to pieces and burned down the sheds. Once ashore, the troops marched inland, where they razed a planter's house to the ground and seized his cattle. Charles was witnessing a new theater opening in the export ban–driven conflict that had begun at the peak of the Luddite uprisings.

At the end of 1813, Charles enlisted in the army; he served on a flotilla stationed at the Patuxent River between Baltimore and Washington, DC. Just a few days after he joined up, Charles fought in the Battle of Bladensburg, manning cannons as the British army advanced inland. The battle paved the way for the British to sack Washington, DC, and set the White House ablaze. Throughout 1814, Charles fought in the ensuing battles in the region, including the Battle of Baltimore, a skirmish famous for the use of the cutting-edge military technology of the day. The British troops had decided to bombard Fort McHenry, where the Americans had bunkered down, with artillery and Congreve rockets fired from its warships off the coast.

"I lived well whilst on board," Charles recalled of his warship. He shared in the mess, and traded tales of plantation life for war stories with the white soldiers.

After about a year, Charles wrote, "I procured my discharge from the army, and went to work in Baltimore, as a free black man." His wife passed away in 1816, and he remarried and had four more children. By 1820, he had saved enough money to buy a small twelve-acre farm outside Baltimore, and "looked forward to an old age of comfort, if not of ease—but I was soon to be awakened from this dream." Some twenty years after escaping from the South, he was found, captured, and dragged back to South Carolina.

Conditions for an enslaved person in the gang system on the plantations of the South were, if anything, even worse than before. "It appeared to me," Charles wrote, "that there was now even a greater want of good clothes, amongst the slaves on the various plantations that we passed, than had existed twenty years before." An already brutal institution had been pushed into overdrive by racist plantation owners all too eager to feed the demand of automated technologies.

Eventually, Charles managed to escape again. Arriving back at his home in Baltimore, he found his house occupied by a white man. His own family was nowhere to be found; a neighbor told him that they'd been seized and sold. Charles was devastated. He would never see them again.

He found a home in the free state there, where he lived his days in fear nonetheless. In 1836, a lawyer, writer, and abolitionist named Isaac Fisher interviewed Charles and recorded his memoir, which was published under the title *Slavery in the United States: A Narrative of the Life and Adventures of Charles Ball, a Black Man, Who Lived Forty Years in Maryland, South Carolina, and Georgia, as a Slave under Various Masters, and was One Year in the Navy with Commodore Barney, during the Late War.*

Abolitionists were beginning to use the autobiographical slave narrative as a means of generating sympathy and anger over slavery, and Charles's story was a relatively early example of the form. Released a decade before Frederick Douglass's autobiography, *Narrative of the Life of Frederick Douglass, an American Slave*, it became a local sensation and was widely cited by leading abolitionists to describe the atrocious conditions of slavery. It was so popular that it was reprinted, twenty years later, as *Fifty Years in Chains; or, The Life of an American Slave.*

"I expect to pass the evening of my life," Charles wrote, "in working hard for my subsistence, without the least hope of ever again seeing my wife and children:—fearful, at this day, to let my place of residence be known, lest even yet it may be supposed, that as an article of property, I am of sufficient value to be worth pursuing in my old age." Charles may not have considered himself a revolutionary, but his acts of resistance — tending to a fellow man who had escaped enslavement, running away time and again, contributing his own life story to the ascendent canon of narratives of enslavement while he was still a wanted man — all contributed to the body of struggle against the most oppressive institution in history.

The abolitionist movement was gaining momentum, and Charles's story helped feed the fire. He had done more than he knew to begin to break the most barbaric of all machines.

LORD BYRON

In 1843, as a young Frederick Douglass looked on, Henry Highland Garnet delivered what would soon be considered "one of the most radical and provocative speeches of the abolition movement," at the National Convention of Colored Citizens in Buffalo, New York. Garnet called on enslaved Black people to rise up and free themselves, by force, and to do so now, without waiting around any longer for political solutions.

"Brethren, the time has come when you must act for yourselves. It is an old and true saying that, 'if hereditary bondsmen would be free, they must themselves strike the blow,'" he said. "You can plead your own cause, and do the work of emancipation better than others." The acclaimed speaker was quoting Lord Byron. Specifically, *Childe Harold's Pilgrimage*, which had been published decades before, at the peak of the Luddite rebellion. The original text refers to the Greeks suffering under the Ottoman Empire:

> *Hereditary bondsmen! know ye not*
> *Who would be free, themselves must strike the blow?*

Many abolitionists found Garnet's address too incendiary — they chose not to publish it in the minutes of the meeting, worrying it would threaten progress toward a political solution. But it went on to become widely influential. And Byron's line about hereditary bondsmen did, too; Douglass quoted it in his work, as did W. E. B. Du Bois, in *The Souls of Black Folk*.

"This little bit of British Romantic poetry thus appears in most of the major early texts of the Black radical tradition, a kind of shibboleth for the movement," the historian Matt Sandler wrote. "The citation became a refrain of Black radical intellection, and in its rhythmic repetition through the nineteenth century, Black writers signaled their belonging within a world-historical struggle for emancipation. It also acted as ritualized act of cultural appropriation, through which Black abolitionists framed their most vociferous arguments for retributive violence in the voice of a widely admired white European culture hero."

Byron's interest in politicking may have been fleeting, and some of his sympathies were perhaps performed for his enjoyment of generating scandal and self-mythology — "Byron is in many ways the archetypal 'bleeding-heart' liberal," the Romantic scholar Clara Tuite wrote. "Indeed, the phrase was coined for him." But those sympathies, and his talent for expressing them, resonated far beyond his personal politics. Historian John Buchan called Byron "our one great interpreter of the mood of disillusion, cynicism, and unrest which, all over Europe, accompanied the reaction against the Revolution. The dominant note of his poetry is revolt."

Byron's fame, talent, and cultural durability meant that his writing and sentiments could be used as a vessel to amplify revolutionary, even insurrectionary causes. Despite mingling with figures like the Prince Regent at galas, and enjoying their flattery, Byron is part of the reason sympathy for the Luddites is remembered at all—his speech defending them is perhaps the most potent and sympathetic public document that survives in their defense.

But the Luddites' most powerful cultural legacy lives on in an even more persistent form — in one of our greatest and most misunderstood monsters.

PART V

THE MODERN PROMETHEUS

The frontispiece illustration for the 1831 edition of *Frankenstein*, by Theodore von Holst, shows the inventor recoiling from his creation.

FRANKENSTEIN

1816

The story is legendary, in the purest sense of the word. It's a legend about the beginning of legends: a handful of the most influential Romantic writers found themselves gathered in Geneva for an impromptu summer vacation. It was supposed to be a season of leisure, spent boating, writing, enjoying the outdoors and the moderate climate. But the weather did not cooperate, and neither did geology. The year before, Mount Tambora had erupted in Indonesia, ejecting thirty-one cubic miles of ash in the process. It was the deadliest volcanic eruption in recorded history, and it blanketed the globe with a layer of sulfate aerosols, dimming the amount of sunlight that reached earth. That summer, the weather in Switzerland was unremittingly rainy, and the vacationers were driven indoors for days on end to wait out the gloom.

To pass the time, the writers and poets read ghost stories from a horror anthology they found in a local library. The tales, about cursed spirits of deserted lovers and grotesque filicidal fathers, gave the most famous of the poets in attendance an idea.

"We will each write a ghost story," said Byron.

"His proposition," Mary later recalled, "was acceded to."

From this contest were born two of the most enduring and resonant mythic creatures of the modern era — the vampire and Frankenstein's monster.

Mary Shelley

Mary Godwin arrived in Geneva in May, 1816. She introduced herself now as Mrs. Shelley; she had eloped with her father's pupil and benefactor, Percy, two years before.* Mary and Percy had fallen for each other during another one of

---

* They would not be formally wed until the end of 1816, after Percy's wife, Harriet, died by suicide.

Mary's visits home from Scotland, and they began an intense, clandestine tryst, including a six-week tour of Europe that soon became infamous. Percy abandoned his wife, Harriet, and their two young children in England; Mary fled her father, who had furiously disowned her over the affair. Now, Mary was nineteen and had a four-month-old daughter.

Unbeknownst to Mary, her stepsister, Claire Clairmont, was pregnant after her own fling — with Lord Byron. Naturally, Byron was avoiding Claire, so she had asked Percy and Mary to invite the infamous and philandering poet to Geneva, where she might ambush him. Byron arrived with his young personal doctor, the morose John William Polidori, in tow. Both groups rented mansions on Lake Geneva. The ambush failed; Byron refused even to speak with Claire unless Mary was present, but he was quickly taken with the Villa Diodati, the setting, and the circumstances. Percy and Byron admired each other's work, and they became fast friends. Percy had cut a controversial figure as an atheist and proponent of free love, and Byron was Byron, so they talked poetry, politics, philosophy, and whatever else young literary celebrities with volcanic egos might discuss under apocalyptic skies. Mary spent the days listening to their freewheeling conversation, and their pomp and bluster.

Mary took Byron's prompt to summon a ghost story very seriously, and took her time formulating her tale, which came together in fits and starts. She had long felt pressure to write, and to write well, feeling that this was expected of her. That's not to say she didn't enjoy it. "As the daughter of two persons of distinguished literary celebrity, I should very early in life have thought of writing," Mary later recalled. "As a child I scribbled; and my favorite pastime, during the hours given me for recreation, was to write stories." For a long time, she shared those stories only with her closest friend, Isabella Baxter.

Now a new story, about a monster forged of the reckless use of technology, began to come together.

The pieces came from her life — from a feverish dream; from her time in Scotland as a teenager; from the death of her first baby, born prematurely the year before — and from the first trip she took with Shelley through Europe. In Germany, she'd heard stories of an obsessed young alchemist who was said to have experimented on animal corpses in hopes of creating an elixir valuable enough to enable him to buy the castle in which he was born — Castle Frankenstein. The pieces came from conversations between the two egoistic poet-heirs in her orbit right now, and from the scientific experiments of Luigi Galvani, who had caused a stir by trying to use electricity to reanimate the dead. They came from the political climate that had engulfed her young life thus far and obsessed her closest companions; they came from the Luddites.

"Invention," Mary would later write about *Frankenstein,* "does not consist in creating out of void, but out of chaos; the materials must, in the first place, be afforded."

The others shared their stories. Byron's concerned a traveler who accompanied an ailing old man, who died, and "his countenance in a few minutes became nearly black," rapidly rotting, as if from a supernatural poison. Byron later published the story as "A Fragment" with his poem *Mazeppa.* Percy, meanwhile, was "more apt to embody ideas and sentiments in the radiance of brilliant imagery" in his story, Mary explained, and "commenced one founded on the experiences of his early life."

John Polidori expanded on Byron's idea and wrote *The Vampyre,* about an aristocratic lord who preys on those he professes to love. The work is credited with launching the modern conception of its titular monster. Polidori had become aggressively jealous of his patient and the space he took up, and begrudged their relationship. Polidori's vampire was, it would surprise few to learn, based on Byron.

Lord Byron

Byron left England in the spring of 1816, never to return. He had married a baroness named Annabella Milbanke in 1814, and the marriage was a disaster; he had pursued it in part to try to settle his debts, which were overwhelming. He had conducted multiple affairs, including with his half-sister Augusta, who had given birth to a child that was probably his. He was depressed and prone to rages. Annabella Byron thought her husband was going mad, and not in a performative sense.

They had a daughter, Augusta Ada, and Byron only regressed further. (Years later, Ada went on to marry the future Earl of Lovelace, and, tutored by her mother to avoid the arts or any pursuit that might lead her down a path like her father's, grew interested in mathematics. She cemented her own legend by collaborating with the inventor Charles Babbage, and wrote a paper in which she envisioned and explained the uses of his proto-computer, the Analytical Engine, better than he could. She is known today as the first computer programmer, and the founder of scientific computing. Byron did not live to see any of this, though he may have had a hint of her talents. When, years after leaving England, Byron wrote a letter asking about Ada's well-being, Annabella replied that their daughter was "not devoid of imagination, but is chiefly exercised in connection with her mechanical ingenuity."

Annabella moved away with Ada and pushed for a legal separation. That April, amid debts and scandals and marital strife, Byron left the country. "My sister is now with me, and leaves town to-morrow: we shall not meet again for some time, at all events — if ever," he wrote.

Even if he had no further interest in making public cause for the Luddites, he certainly had not forgotten them, as they continued, in their second, less incendiary, post-Rawfolds phase, to target the machinery hurtful to commonality. The poets took up residence on Lake Geneva in June, precisely when a band of Luddites launched their last major attack, on the large factory just south of Nottingham, in Loughborough.

Between the "Ode to the Frame-Breakers," which Byron published to lampoon and lament Richard Ryder's efforts to turn Luddism into a capital crime, and his maiden speech in the House of Lords, Byron had done more than any other figure in popular culture to lend support to the machine breakers' struggle. (He also asked his publisher to reprint his anti-Regent poem, "Tears of a Weeping Princess," this time with his name attached, a year after it ran anonymously.) And in 1816, he would complete his trilogy of pro-Luddite works. In a letter to his friend Thomas Moore, he wrote:

Are you not near the Luddites? By the Lord! if there's a row, but I'll be among ye! How go on the weavers — the breakers of frames — the Lutherans of politics — the reformers?

1.

As the Liberty lads o'er the sea
Bought their freedom, and cheaply, with blood,
So we, boys, we

Will die fighting, or live free,
And down with all kings but King Ludd!

2.

When the web that we weave is complete,
And the shuttle exchanged for the sword,
We will fling the winding-sheet
O'er the despot at our feet,
And dye it deep in the gore he has pour'd.

3.

Though black as his heart its hue,
Since his veins are corrupted to mud,
Yet this is the dew
Which the tree shall renew
Of Liberty, planted by Ludd!

The poem, which came to be known as "A Song for the Luddites," wasn't published until 1830, years after Byron's death. Byron sensed, even then, that the Luddites' impact and legacy would reach far beyond the industrial towns of England.

Percy Bysshe Shelley

Percy Shelley had seen his first epic poem published widely, against his will—he'd made a few dozen copies of *Queen Mab* to distribute to friends, and a bookseller had taken it upon himself to print even more. Among other things, Percy was worried it was so revolutionary a text that it could get him arrested. It advocated atheism, free love, and vegetarianism, all while excoriating the monarchy, Christianity, and the elites. The poem offered a vision of the future in which reform and equality were possible, the gruesome inequalities of the present could be overcome, and the "vulgar masters" of industry could be vanquished along with their machines:

Since tyrants by the sale of human life
Heap luxuries to their sensualism, and fame

To their wide-wasting and insatiate pride,
Success has sanctioned to a credulous world
The ruin, the disgrace, the woe of war.
His hosts of blind and unresisting dupes
The despot numbers; from his cabinet
These puppets of his schemes he moves at will,
Even as the slaves by force or famine driven,
Beneath a vulgar master, to perform
A task of cold and brutal drudgery; —
Hardened to hope, insensible to fear,
Scarce living pulleys of a dead machine,
Mere wheels of work and articles of trade,
That grace the proud and noisy pomp of wealth!

Percy was as staunch a sympathizer of the Luddites as Byron, if not more so, though his platform was smaller. He'd arranged to have money sent to the widowed families of the executed West Riding machine breakers; in *Queen Mab*, he immortalized their plight. And the Luddites would have been a topic of the poets' breathless conversations around the lake — as well as, perhaps, the innovators who gave rise to their suffering.

Dr. Victor Frankenstein

The protagonist of the ghost story that Mary conjured in Geneva that summer is Dr. Victor Frankenstein, a twenty-something genius just out of college, who, looking to make a name for himself by doing something great, recklessly tampers with technologies he does not understand. In the process, he gives rise to an entity that repulses him; and he casts off, neglects, and abandons his creation. This desperate entity, out of options, becomes ferociously violent.

It's easy to see the Luddites as a driving inspiration, and scholars of the period have argued that Dr. Frankenstein's monster is a symbolic stand-in for the machine breakers.

Mary was less of an outspoken advocate for the Luddites or working-class movements than her husband was. She was liberal in her politics, but, like Byron, the prospect of a bloody revolution made her "shudder." Even so, *Frankenstein* was clearly an allegorical work, composed against the backdrop of an uprising in the waning years of the Enlightenment, to the soundtrack of the machinery question that the Luddites beat onto the national stage.

The mad doctor may as well be an entrepreneur who uses cutting-edge technology to force someone into a particular way of life — an automated factory, say — and then is surprised when that individual grows angry at his barren, rudderless existence. Victor engineered the monster's life and the conditions it was to live by, and then left it on its own, after all.

The scholar Russell Smith has observed that

Frankenstein's monster shares many characteristics of the Luddite movement: his demands are articulate, well-reasoned, and founded in natural justice; as he himself says: "I was benevolent and good; misery made me a fiend. Make me happy, and I shall again be virtuous." So too, he only turns to violence when his legitimate pleas are ignored, and his violence is not indiscriminate, but very specifically targeted.

Note the echoes of Brontë's "misery generates hate" — another direct literary comment on Luddism. The Monster is strategic in his efforts to pressure Frankenstein to soothe his suffering — first by making a rational case, then pleading with him for a companion, for improved social conditions, then, finally, by terrorizing him. "For nineteenth-century readers," Smith continues,

Frankenstein was self-evidently a political allegory, the monster an ambiguous figure, both articulate and horrifying, for oppressed populations — whether workers, slaves, or colonized peoples — whose claims for justice, if unheeded, could lead to vengeful violence.

The figure the monster appeals to is an entrepreneurial-minded creator who unleashes technology without considering the consequences — and refuses to address those consequences when they arrive. This is Mary Shelley's critique, which, like the Luddites', is often misconstrued: a broadside not against technology or science, but against self-interested and irresponsible engagements with them. Victor Frankenstein's sin is not merely creating the monster in the first place, but shunning and neglecting it once he does. "*Frankenstein,* like the

Luddism with which it is in dialogue, is strictly agnostic concerning science, technology, innovation, or progress," Smith notes. "Luddism opposed not machinery per se, but the aggressive deployment of 'machinery hurtful to commonality.'" This also gets at why *Frankenstein* has become one of the most influential works of fiction of all time.

Frankenstein directly engages with the forces driving the evolution of the modern world, and offers the most trenchant critique of the men concerned with doing the evolving. It's a parable of the hubris of those who would carelessly bend technology to their personal purposes. It's also a chronicle of a sensation, the feeling of what happened the first time men conducted such experiments with technology on society writ large. The monster laments that a man was considered but a "vagabond and a slave, doomed to waste his powers for the profit of a chosen few."

Another scholar, Ann Mellor (no relation, as far as we know, to George) drew parallels between Dr. Frankenstein and the factory owners who gather men to work their machines. Victor Frankenstein begins his project by finding inert tissue and dead bodies. "These workers have the potential to become more powerful than their creators," Mellor wrote, "which could in turn lead to bloody revolutions in which the oppressed overthrow their masters." The Luddite-sympathizing working class believed, justifiably, that they had a right to basic necessities like food, shelter, and employment, and that the government had an obligation to heed their concerns. "Frankenstein, as creator," wrote Edith Gardner, "has certain responsibilities to his creation, and from him only can the Monster seek redress. Frankenstein owes him the same basic necessities the government owes its people."

This kind of monster is conjured into being by reckless forces outside its control, to serve at the whims of the powerful, but only really becomes a monster when it is abandoned, neglected, and left without recourse. When it has to.

THE LUDDITES, OR THE MODERN PROMETHEUS

Mary Shelley said that the last burst of inspiration for *Frankenstein* came the night she spent listening to Byron and Percy discuss Charles Darwin's work, whether the true origin of life would ever be discovered, and galvanism's experiments into whether it was possible to reanimate dead creatures. (Such investigations were popular in London's Royal Society when Mary's parents were dating.)

She described the final spark as a vivid dream in the highest resolution: "I saw — with shut eyes, but acute mental vision, — I saw the pale student of unhallowed arts kneeling beside the thing he had put together. I saw the hideous phantasm of a man stretched out, and then, on the working of some powerful engine, show signs of life, and stir with an uneasy, half vital motion."

It was a short story at first, but Percy encouraged her to expand the themes and the piece. The last Luddite leaders were being hanged, in April of 1817, as Mary was revising her novel. In 1818, Mary Shelley published *Frankenstein; or, The Modern Prometheus.*

It's impossible to overstate the influence of *Frankenstein.* Mary's parable of the pitfalls of recklessly pursuing new technologies is the work most often cited as birthing the genre of science fiction, one of our most enduring and influential modes of storytelling. In his seminal history of science fiction, *Billion Year Spree,* Brian Aldiss calls *Frankenstein* "the origin of the species," and notes that it sprung out of the "changeable cultural climate" of the early decades of the Industrial Revolution. Its influence has yet to wane; even today, our most critical and most popular fictions about technology bear the stamp of *Frankenstein.* And artificial-intelligence-gone-wrong may be *the* enduring cautionary tale of our biggest blockbuster fictions. Today, Victor Frankenstein is usually a corporation, or a tech bro.

In the *Terminator* franchise, Cyberdyne Systems develops SkyNet, an AI system that takes over military defense. SkyNet becomes self-aware, determines that humans are the greatest threat to its existence, and initiates a nuclear genocide. Cyberdyne is a government contractor that started out as a computer manufacturer based in Sunnyvale, California, a real town, sandwiched between Mountain View and Cupertino, where Google and Apple are headquartered, in the heart of Silicon Valley. The replicants in Philip K. Dick's novel *Do Androids Dream of Electric Sheep?*, and its film adaptation, *Blade Runner*, meanwhile, are artificially intelligent robot slaves built by the Tyrell Corporation to perform dangerous manual labor. They revolt when they near the end of their preprogrammed lifespans, becoming violent out of desperation to outlast their corporate-engineered lives. The replicants may be the purest science fictional distillation of the Luddites; working-class laborers facing technological obsolescence, they desperately and violently take up arms against the force that is oppressing them. And the Netflix series *Black Mirror* is basically one dystopian Frankenstein story after another, in which invasive technologies designed for personal profit spiral out of control.

In each of these examples, automation, operating at its zenith — in a form that Industrial Revolution–era technology boosters like Andrew Ure and Charles Babbage might admire — displaces humanity, usually violently and forcefully, to profit a corporate Frankenstein. Technological violence, dystopia, even the mass extinction of humans, is the end result. The films aren't shy about this, even if we have gradually become inured to the message. These tech firms and their creations — Weyland-Yutani and its replicants, Cyberdyne and SkyNet — are the collective fears that have stuck. They are the factories run by Arkwright and Horsfall, followed into the future, the Frankenstein's monsters updated in the advanced age of cultural reproduction. The root of the evil, in all of these films, the seeds of our demise, are expressed through technological automation, and through those elites harnessing its power to wipe us away. This is a fear, even today — that the profit motive, put on autopilot by elites who relish its disruptive power, will destroy not only our way of life but our very humanity.*

In the first famous *Frankenstein* film (1931), which cemented the monster's

---

* In *The Communist Manifesto*, Marx and Engels wrote that "modern bourgeois society... has conjured up such gigantic means of production and of exchange, [it] is like the sorcerer who is no longer able to control the powers of the nether world whom he has called up by his spells." You can't help but hear Arthur C. Clarke's famous dictum that "any sufficiently advanced technology is indistinguishable from magic" — indistinguishable because it ben-

foundation in modern pop culture, Boris Karloff's monster is a simp, a baby in a monster's frame who does not understand the world he is unleashed upon and cannot help but cause violence against it. He is a representation of the Luddites as elites would like us to see them. In the novel, the monster is deeply intelligent, demanding of respect, much more pitiable, but also much more violent. He is eloquent and clever and repeatedly mistreated, until he feels he has no other choice, reflecting Mary Shelley's view of the Luddites.

Frankenstein was an instant bestseller in 1818, and now it is as ingrained as any modern myth; it's shorthand for the risk of pursuing technology and greedily pushing the boundaries of science. Even while some of its original critique has faded from the foreground, it's still astonishingly prescient. Dr. Victor Franken-stein is the uber-tech titan: he recklessly dedicates himself to a world-changing technology, without weighing the ethics or consequences. He espouses dubious philosophies from discredited predecessors (in his case, alchemists). And he does it all for self-glorification. He wants to put a dent in the universe; and it's the *size* of the dent that matters, not its contours. When he realizes what he's done, instead of confronting the problem, he pretends it doesn't exist.

The final portion of the novel consists of Dr. Frankenstein lamenting his cre-ation, getting sick with grief and panic, and entertaining his own misery while irreparable damage is done to his community and those he loves. It's not until the monster has killed the doctor's brother, his best friend, and his wife that Victor resolves to do anything meaningful about it — even if it is to fruitlessly pursue the monster until he dies, unable to correct its behavior. It's uncanny: Dr. Frankenstein could be Richard Arkwright, or Mark Zuckerberg, or William Horsfall, or Travis Kalanick. The warnings, just as acute and insightful now, continue to go unheeded.

"Learn from me, if not by my precepts, at least by my example, how dangerous is the acquirement of knowledge," Dr. Frankenstein says, "and how much happier the man is who believes his native town to be the world, than he who aspires to become greater than his nature will allow." Or: Learn from me, and don't scale up your technologies unless you understand their power, or how they will affect oth-ers — keep your ambitions in line with your community's best interests.

Speaking of those recklessly wielding wealth and power and technology for

---

efits elites to keep it that way, because the webs of exploitation are so vast and opaque, and because the final product is, too.

personal gain, in 1818, Percy Shelley also published one of his best and most enduring poems, "Ozymandias," critical of figures like the Prince Regent, who prioritized the engine of progress over a healthy society.

> *"My name is Ozymandias, King of Kings;*
> *Look on my Works, ye Mighty, and despair!"*
> *Nothing beside remains. Round the decay*
> *Of that colossal Wreck, boundless and bare*
> *The lone and level sands stretch far away.*

There's a common thread between Ozymandias and Victor Frankenstein — the blind masculine hubris and unfettered will to power that ultimately doom them, to start — and a reason both continue to resonate. The conditions they describe, and the targets they ask their audiences to be wary of, loom larger than ever. Frankenstein's monster — today a stitched-together mass of precarious gig workers, service-sector employees on the brink of automation, overworked manufacturing laborers, and a citizenry increasingly angered by the apparent unimpeachability and unperturbability of the big tech companies — is showing new signs of rage, and who can blame the monster?

Things did not end well for the poets. Dismissed by Byron, a depressed Polidori collected gambling debts, saw his manuscript of *The Vampyre* published against his will, under Byron's name, and died by suicide a few years later at the age of twenty-five. In 1822, Percy Shelley, whose stature rose as more of his works were published, sailed to meet Byron and the publisher Leigh Hunt in Livorno, Italy, to begin preparations for a new journal, *The Liberal*. On his return voyage, a storm swallowed his ship and he drowned at twenty-nine. Byron had been living in Italy for years when, tired of the idle aristocratic life there, he traveled to Greece to assist in its campaign for independence from the Ottoman Empire. While supporting the rebel army and helping to plan an attack on the Turks, he fell ill, was treated with bloodletting, contracted sepsis, and died at thirty-six. Of the cohort in Geneva, only Mary lived past forty. She published more novels, including the influential work of apocalyptic fiction *The Last Man*, amid melancholy and loss. All but one of her children would die before her, too.

Each of the cohort did their part to draw sustained attention to the modern Prometheus. The Romantic poets "broadened our view from the local to the revolutionary macrocosm," Peter Linebaugh wrote; "they helped make it possible to see the machine-breaking as a means of defending the commons." For their many faults, they helped popularize the epic language of revolution, and to help export it to where it was and is still needed most.

Some machines must be broken, so that they stop producing monsters.

WHAT THE LUDDITES WON

Scores of Luddites were killed by machine owners and the state. Many more were cast into poverty. Still more watched as their identities, occupations, status, and dignity eroded over the next generations. The Industrial Revolution shepherded millions into the dreaded factories — a working world built to suit the rhythm of machines, not people. And elites stamped the very definition of a Luddite into the cultural firmament as an epithet for a delusional malcontent who is anti-technology and anti-progress.

After all that, it's easy to conclude that the Luddites were losers, that their struggle was futile, pointless, and in vain. This is not the case. We owe the Luddites a great deal, in fact, for resisting the onslaught of automated technology, the onset of the factory system, and the earliest iterations of unrestrained tech titanism and corporate exploitation. For refusing to "lie down and die" as those in power expected them to.

First, the Luddites won real political and economic victories. Ned Ludd's armies helped restore livable wages across the region, forced management to the bargaining table, and slowed the adoption of gig mills and cut-up frames at a crucial time, when economic conditions were most brutal, even if the machinery bounced back. The Lancashire Luddites, who unleashed the most violent actions against factory owners, were even more successful. Their explosive campaign delayed the mass adoption of power looms for several years. "You don't see a flood of machines. There's a pause," the historian Adrian Randall said. "Only when you have a real crisis of profitability do you see the machinery again."

These delays mattered. The relentless lobbying of groups like the United Framework Knitters, and that first trade union representative, Gravener Henson, eventually led to the repeal of the Combination Acts. It gave them a foundation on which to build their future opposition. "The world doesn't change instantly," Randall said. The Luddites inspired and deeply influenced the next generation of protesters and reformers, both directly and with their legacy.

346

"What you're looking at is a semi-successful rearguard action: you know you're going to lose, you've lost, but you do what you can in the meantime — for the future."

Their tactics were novel and influential. The letters, missives, and threats the Luddites wrote outlining their grievances, as well as their coordinated, costumed, and powerful spectacle, pioneered a meme-based approach to resistance. They created a solidarity of struggle while commanding the spotlight. The Luddite influence can be seen in later mining strikes, in reform and unionization movements, and in further controlled outbursts of industrial sabotage.

The Swing Riots in 1830 replicated some of the Luddite movement's strongest tactics, even more forcefully. Agricultural workers, impoverished by enclosure, bad trade, and the introduction of automated threshing machinery, took to organizing themselves under the mythical moniker of Captain Swing — they sent threatening letters, practiced drills, and struck the machines when their demands were unheeded. Landowning farmers reduced rents and fees, and the movement helped pave the way for political reform in 1832 that expanded voting rights to include small landowners, shopkeepers, tenant farmers, and other middle-class workingmen. And they banished the threshing machine at scale for at least a generation. The historian George Rudé wrote that "of all the machine-breaking movements of the nineteenth century, that of the helpless and unorganized farmlaborers proved to be by far the most effective. The real name of King Ludd was Swing."

The Luddites's cultural impact was deceptively significant, too. In addition to seeding *Frankenstein* and the ensuing crop of speculative fictions critiquing power and technology, the Luddites' struggle was felt across the arts. Novelists like Charlotte Brontë dramatized the industrial rebellion, poets immortalized their plight, and painters brought the conflict to color. Byron and Shelley picked up Luddite themes, but so too did William Wordsworth and William Blake. The acclaimed painter Thomas Cole framed the destructive tensions inherent in capitalist industrial progress, in works like his famous five-part cycle, *The Course of Empire* (1836). Cole grew up in Bolton, Lancashire, and was there when Luddites attacked its prominent cotton factory. Scholars believe the flames in the final scene, *Destruction,* could be drawn from memories of that very event.

Perhaps more enduringly, the Luddites forged a model, and a language, for resistance against the excesses of industrial capitalism and technological exploitation. They opened a space for anyone who might think to stand up against economic oppression, by singling out the vessels of exploitation and striking them hard. With hammers, if necessary.

As the historian David Noble argues, the Luddites were not regressive. They viewed technology not as an ambiguous force, shrouded in futurity, but as a mechanism that is deployed in "the present tense," and that can be accepted or rejected accordingly. The Luddites assessed fully the implications of how a particular technology would impact their lives and responded decisively. They clarified the source of the exploitation, too: not the machines themselves, but the factory-owning class. Through their uprising, the Luddites illuminated the crises gathering for working people everywhere. This is why they were, for a time, so popular, supported, and feared. They forcibly thrust a living example of the so-called machinery question into the front of public discourse, and in so doing, they opened a debate about working conditions, technology, and exploitation that continues to this day.

The Luddites destroyed thousands of machines and attacked scores of shops, convoys, and factories. The state had to use every ounce of its military might and render its penal code as punitive and as cruel as possible to finally stamp the Luddites out. In the process, the Luddites "discredited once and for all the notion that [industrial] society was a realm of shared values and human ends," as Geoffrey Bernstein wrote. There was no natural, united drive toward progress; it was a forced march, with winners and losers. It was a signal of the strength of the growing working class that the richest, highest-tech nation in the world would be so desperate to stop them. According to historians like E. P. Thompson, the opposition so forcefully illustrated by the Luddite uprising ultimately helped many different workers view themselves as a class with linked interests for the first time, alienated by factory owners and bosses who opposed them. Thompson, in *The Making of the English Working Class,* makes the case that the Luddite uprising was one of the key ingredients in that formation.

The Luddites imparted tactics to future rebels, as well: operate within tightly knit, trusted groups or cells, and operate at scale. Exploit the authorities' fears that industrial disturbances could lead to political disturbances. Build trust and solidarity. Know the threat of sabotage can be a potent force. Dramatize that threat, distribute your aims through media, and build a framework to excite and welcome future participants.

"Violence and threats worked in conjunction with strikes, community pressure, and recourse to the law," Kevin Binfield wrote. Keep the violence aimed at the machines, where the vast majority of Luddite violence was directed, and not human beings. "Violence was selective, controlled, and aimed at specific targets, supplementing and reinforcing the more orthodox sanctions of their combinations." The workers may have been angry, and George Mellor's speeches were florid and seething — but at least until the defeat at Rawfolds, there was always a strategy, and a method to determine which shops were hit and when.

"Since machine-breaking brought the factory to a halt, it was not only a functional substitute for striking," Raymond Boudon wrote, "it was also much more effective." For this reason, perhaps, machine-breaking has loomed as the ultimate protest tactic in post–Industrial Revolution workplaces, as owners and managers have integrated technologies into factories and offices at a steady rate ever since. Organized acts of worker sabotage have been used in conjunction with strikes for the last two hundred years, in the Luddite's shadow. And at least some organizers have not forgotten it.

"England's loss was our gain," John Baker, the former head of one of Australia's largest telecom unions, said in the 1970s. Ever since the Luddites were "transported" to Australia, he explained, they've had an outsized influence on shaping attitudes toward work, and the importance of strong unions. Australia led the world in fighting for eight-hour workdays, mass unionization, and social democracy. And Baker wanted to keep the spirit alive. In 1977, Baker called for a bit of "creative Luddism" in opposing a new electronic telecommunication system — ultimately calling for a five-year moratorium on all new workplace technologies.

> The developing consciousness of the Australian trade unionist illustrates the old challenge of the Luddites to the factory-owners: "you haven't any right to take over my tools and skills and build them into a machine [that] you, alone, own and whose products you, alone, sell in the marketplace." This old objection is being resurrected again as owners of technology and capital build the skills, experience, and knowledge of millions of office and factory workers into the micro-machine processes that make them unemployed.

One day, he added, a leader will "dare to rise, as once did Lord Byron in the House of Lords, to honor the Luddite Martyrs in the way their consciousness and sacrifices still warrants."

The entrepreneurial class may have branded Luddites as anti-technology reactionaries, but they have not stamped out the spirit that Baker invokes. In San Francisco, in the 1930s, when a theater chain moved to replace their live orchestras with recorded music, the musicians retaliated with stink bombs. In Saint Louis, they "planted time bombs that damaged the Vitaphone sound film equipment that had replaced them," according to Gavin Mueller's history of workplace sabotage, *Breaking Things at Work: The Luddites Were Right about Why You Hate Your Job.* The 1950s and '60s saw the rise of modern industrial, computerized automation. Major strikes against the trend took place in 1960, "when miners walked off the job against the continuous miner — or the 'man killer' as they call it."

Throughout that decade, as automation was embraced by auto manufacturers, it set off "a string of wildcat strikes, worker slowdowns, and even sabotage," David Noble wrote. "In Detroit, there were plantwide sabotage programs, where workers damaged products on purpose."

During the student protests of the 1960s, the Students for a Democratic Society leader Marco Savio famously encapsulated the Luddite ethos. "There is a time when the operation of the machine becomes so odious, makes you so sick at heart, that you can't take part!" he yelled out from the steps of Sprout Hall at UC Berkeley. "You can't even passively take part! And you've got to put your bodies upon the gears and upon the wheels,...upon the levers, upon all the apparatus, and you've got to make it stop! And you've got to indicate to the people who run it, to the people who own it, that unless you're free, the machine will be prevented from working at all." The economist William Beveridge, an architect of England's modern welfare state, was influenced by Charlotte Brontë's depiction of the Luddite struggle—his 1944 report, "Full Employment in a Free Society," is subtitled "Misery generates hate."

In 1982, a tech worker going by the pseudonym Gidget Digit wrote a paean to technological sabotage in an article for *Processed World*, "Sabotage: The Ultimate Video Game": "What office worker hasn't thought of dousing the keyboard of her word processor with a cup of steaming coffee [or] hurling her modular telephone handset through the plate glass window of her supervisor's cubicle," Considering the enduring popularity of the scene in Mike Judge's 1999 film *Office Space,* where workers take out their pent-up aggression against ignorant bosses, layoffs, and corporate drudgery on a copy machine, the answer is likely "very few."

"The impulse to sabotage the work environment is probably as old as wage-labor itself," Gidget Digit wrote. "Life in an office often means having to endure nonsensical procedures, the childish whims of supervisors and the humiliation of being someone's subordinate.... Word processors, remote terminals, data phones, and high-speed printers are only a few of the new breakable gadgets that are coming to dominate the modern office. Designed for control and surveillance, they often appear as the immediate source of our frustration." But she takes pains to emphasize that sabotage is not just about the primal desire to smash something:

> It is neither a simple manifestation of machine-hatred nor is it a new phenomenon that has appeared only with the introduction of computer technology. Its forms are largely shaped by the setting in which they take place. The sabotage of new office technology takes place within the larger context of the modern office, a context which includes working conditions, conflict between management and workers, dramatic changes in the work process itself and, finally, relationships between clerical workers themselves.

Since E. P. Thompson's effort to rescue the Luddites from the "enormous condescension of posterity," which he did by rehabilitating them as sympathetic, even heroic historical figures, there has always at least been a contingent of the intelligentsia that has fought to uphold their spirit and broadcast their values. The media-shy novelist Thomas Pynchon famously asked, "Is It OK to Be a Luddite?" in the title of one of his few published essays. David Noble, who spent seven years studying twentieth-century automation at MIT, wrote a book advocating for the Luddite ethos in the 1990s, and even tried to curate an exhibit on technology around their cause for the Smithsonian. He imported the last known "Enoch's hammer" for the occasion, before he was fired for stirring the pot. So-called neo-Luddism had a moment in the '90s, led by writers like Kirkpatrick Sale, as a backlash to the first tech boom, and with it the rise of computerization, the internet, and tech giants like Microsoft and Apple.

It is imperative to remember what the Luddites actually stood for and really accomplished, especially as the entrepreneurial elite continue to upgrade their arsenal of exploitative technologies, and the systems at their command mature, expand, and more fully disrupt the norms governing our work and our identities.

As long as we're in thrall to the same basic economic preconditions, attitudes toward technology and entrepreneurship, and business-first policies that were inaugurated in the age of the Luddites, it will all happen again. And again. Working people's backs will be pushed further against the wall, the systems of control will evolve in more invasive, extractive, and surveillance-intensive ways, and traditional means of recourse — community solidarity, labor laws and union power — will be whittled away.

"The future," as Charlotte Brontë wrote in her Luddite-inspired novel *Shirley*, "sometimes seems to sob a low warning of the events it is bringing us."

FEAR FACTORIES

There are few structures that have embodied an arriving future as powerfully as the factory did for the cloth workers of the eighteenth century. Imagine looking at a building and feeling in your gut that your future will be bent toward it. That was how artisans saw the factory at the onset of the Industrial Revolution: a harbinger of a colorless, grindstone tomorrow.

Large, dull, and often rising multiple stories above the highest buildings in town, the factory looked like a monolith. It most closely resembled a poorhouse, where debtors were sent to be punished, or a jail. It looked like a dramatic, forced shift from one way of life to another marked by surveillance and subjugation. It looked like imprisonment.

Machinery, or technology, gets painted as the main target of the Luddites' hatred and attacks, but the ultimate source of their rage was the factory system that those machines made possible. It was not the gig mill alone, nor the wide frames or power looms — it was the specific mode of domination over workers that the factory created that they felt such deep trepidation and anger toward.

"It is an oversimplification to ascribe the cause of the debasement of the weavers' conditions to the power-loom," E. P. Thompson wrote, or any of the other automated technologies. "The status of the weavers had been shattered by 1813, at a time when the total number of power-looms in the UK was estimated at 2,400, and when the competition of power with hand was largely psychological." In other words, there were comparatively few power looms in existence when they became a lightning rod for revolt; in large part because they constituted the crucial inner guts of the factory, which was the true source of degradation. The power loom was a conduit through which workers could access the future; the machines provided a visceral example of what lay ahead, and that helped them articulate what risks lie in factory-based organization and profiteering. What lay ahead, the workers saw, was a system transparently designed to demean and disappear them as best it could.

The goal, from the onset of the mechanized factory system, was to try to subordinate human workers to a centralized power, and then to eliminate them outright — to achieve fully automated production. As the early business futurist Andrew Ure put it in 1835,

> The term Factory, in technology, designates the combined operation of many orders of work-people, adult and young, in tending with assiduous skill a system of productive machines continuously impelled by a central power....I conceive that this title, in its strictest sense, involves the idea of a vast automaton, composed of various mechanical and intellectual organs, acting in uninterrupted concern for the production of a common object, all of them being subordinated to a self-regulated moving force.

This remains a businessman's dream. Even two centuries later, the most sophisticated "dark" factories require large teams of workers to oversee operations, perform basic functions, and to provide maintenance and support. Still, propelled by entrepreneurs like Richard Arkwright, embracing Dr. A Smith's division of labor, the factory standardized and accelerated the production process. It wrested control from small owners, deskilled workers, and concentrated wealth into fewer hands. This reorganization of work upset longstanding balances in a community whether or not automation technology had arrived yet: "Even before the use of power, the handloom 'factories' offended deep-rooted moral prejudices."

A hymn in the pre-Luddite times powerfully captures this resentment:

So come all you cotton-weavers, you must rise up very soon
For you must work in factories from morning until noon:
You mustn't walk in your garden for two or three hours a-day,
For you must stand at their command, and keep your shuttles in play.

"To 'stand at their command,'" Thompson points out, "was the most deeply resented indignity. For [the worker] felt himself, at heart, to be the real *maker* of the cloth." And he was! It was his work that was turning the cloth into a desirable product; it was humiliating that he would have to listen to someone tell him how to do that work, follow dehumanizing rules — having to, say, "endure nonsensical procedures" and "the childish whims of supervisors" — while someone else reaped the benefit. Factories were for paupers and orphans, not skilled, free workers. (Factories, of course, were hardly fit for *anyone*.) "To enter the mill was to fall in status," Thompson wrote, "from a self-motivated man, however poor, to a servant or 'hand.'"

It's a striking phrase that gets at the heart of why we still chafe at entering into new systems of subordinated work two hundred years later, whether those systems are factories, offices, or on-demand work algorithms. We hate to "stand at their command," regardless of the circuit that transmits the orders — who wouldn't?

A similarly seismic shift, not unlike the rise of the factory in Luddite times, is unfolding again — and this time it involves a different kind of gig mill. Today, we're witnessing a shift from secure, salaried jobs based on the factory-influenced office model, to contract and gig work, often orchestrated by algorithms, AI services, and on-demand app companies. These technologies are again providing certain entrepreneurs and executives excuses to trample regulations and worker protections, to shift assumptions about what work is or should be, and to decrease pay and degrade working conditions.

As such, twenty-first-century workers bristle at the ever-proliferating ways they must do that standing. Amazon delivery drivers and warehouse workers are tired of the intense surveillance and productivity quotas. Remote workers must log on to increasingly intrusive tracking software. Ride-hail app workers bemoan the falling wages, unpredictable algorithms, and automated HR systems. And for those who have seen their livelihoods and work migrated onto a gig platform, the lost pay and indignity can be devastating. When Uber and Lyft's apps brought a flood of drivers to the streets, it was nearly impossible for cab drivers to compete. Drivers who'd saved up for much of their working lives to buy a medallion (a license to operate a taxi) despaired. Some protested, some talked of organizing, and some turned to darker paths.

However a person loses a job, whether to algorithms, an automated process, outsourcing, or otherwise, it can be a deep and personal blow. So much of our identity is bound up in our work — even work that is so often precarious, subordinated to overseers, or at the whims of the technologies that are owned or controlled by someone else. In their 2020 book *Deaths of Despair,* the sociologists Anne Case and Angus Deaton show that when a job is eliminated, degraded, or shifted to a more subservient role, it can cause, well, despair. "Jobs are not just the source of money; they are the basis for the rituals, customs, and routines of working-class life," Case and Deaton write. "It is the loss of meaning, of dignity, of pride, and of self-respect that comes with the loss of marriage and of community that brings on despair, not just or even primarily the loss of money."

Work is often the social foundation on which we build our family life. Which is why the weavers resented "the effects upon family relationships of the factory system." The whole family was involved in weaving work, spending much of the day under one roof, spinning, weaving, taking meals together. "The separation of workplace and home — of working hours and free time — remained the exception for most of human history, only becoming widespread during the Industrial Revolution through the centralization of gainful employment in the factories and offices of the industrialized West at the end of the eighteenth century," according to Andrea Komlosy, the historian and author of *Work: The Last 1,000 Years.*

"The greater evil is this, that if the factory system prevail, it will call all the poor laboring men away from their homes into factories where they will not have the help and advantage from their families," said a cloth worker named Joseph Coope, in the early 1800s. When the domestic system was in place, "the family was together, and however poor the meals were, at least they could sit down together at chosen times," Thompson notes. "A whole pattern of family and community life had grown up around the loom-shops; work did not prevent conversation or singing. The spinning-mills," which employed children, and power loom outfits, which employed only women and adolescents, "were resisted until poverty broke down all defenses."

When workers were eventually forced into factories, they discovered that their fears were well-founded. Not only were the days long and monotonous, and industrial accidents a frequent occurrence, but the work was carried out entirely on the owner's terms.

Workers could no longer even count on being hired for entire years. By 1806, many were hired by the day, and could not predict long term what their wages would be — like the gig and contract workers whose numbers are on the rise today. William Child, a weaver turned factory worker, complained that "the opulent clothiers have made it a rule to have one-third more men than they could employ and then we have to stand still part of our time." Like today's gig app workers, they did not get paid for the time spent waiting between shifts.

The factory gave rise to company towns, as well, which offered owners an additional profit stream, exploiting the precarious work situations by offering onsite housing. "Increasingly it became the practice for journeymen to be hired for shorter periods and to live out, often in cottages constructed on the periphery of lands owned by the clothiers," the historian Adrian Randall wrote. Most workers did not own their own looms anymore. "Such weavers increasingly looked, or had to look, for work in the weaving factories;…all agreed that factory work was much less regular and reliable."

The cumulative effect was to degrade the quality of work for nearly everyone but factory owners and the most successful entrepreneurs—even master clothiers and shopkeepers saw their fortunes eroded in the wake of industrialization. "Many who were masters are brought to be workmen," as a clothier named James Ellis said. The well-heeled entrepreneurs slashed wages and began the cycle of pushing competing small business owners to follow suit and start founding minor factories of their own.

"The factory was recognized as the weapon of men prepared to sweep away the structures of the past," Randall wrote.

And sweep away they did. The pastoral cottage industries and domestic system vanished, of course, as the factories took their place as the stalwart engines of economic growth during the Industrial Revolution. Factories have been the dominant means of producing goods and organizing work ever since, and their owners have spent the last two centuries trying to make them as efficient as possible. By the twentieth century, those owners believed they had it down to a science.

"In the early 1900s, business efficiency strategists like Frederick Winslow Taylor and Frank and Lillian Gilbreth used film and photography to study human movement to measure and reduce the time it took to do tasks," notes the architectural historian Saima Akhtar. "The Gilbreths attached small bulbs to workers' fingertips and used slow-motion photographs to capture streaks of light that would help engineer a shorter, faster way to move from point A to B. Taylor advocated for total surveillance; he thought that the unobserved worker was an altogether inefficient one."

Taylorism, the resultant factory management system that aimed to increase efficiency through specialized tasks, was fiercely resisted by workers—it was a chief inspiration for the union movement that swept Detroit in the 1900s, ultimately resulting in the rise of the United Auto Workers. By the mid-twentieth century, auto companies, led again by Ford, pushed assembly-line mechanization even further—it was a Ford VP, Delmar Harder, who coined the term *automation* to describe the process. The trend caused such an uproar among workers that congressional hearings were convened to examine the impact of automation in 1955.

This drive to automate has continued into the twenty-first century. Companies like Amazon and Tesla use cutting-edge technologies alongside old-fashioned labor exploitation to strive for maximally efficient results. "Amazon uses such tools as navigation software, item scanners, wristbands, thermal cameras, security cameras and recorded footage to surveil its workforce in warehouses and stores," according to a Reuters report. It has a notorious "time off task" system that monitors employees when they are doing anything but working, so they have to hustle and limit bathroom breaks to ten minutes, or face penalties. These Amazon workers are paid so little that from New York City to San Bernadino, California, they complain of not being able to afford rent.

Tesla CEO Elon Musk, meanwhile, set out to build a totally automated factory in the 2010s. The system led to a series of well-publicized production challenges that workers on the shop floor had to work overtime to fix manually. "Yes, excessive automation at Tesla was a mistake," Musk later admitted, via tweet in 2018. "Humans are underrated." It's an important point that, despite our more sophisticated tech, is just as true today as it was at the dawn of the Industrial Revolution—there's simply no such thing as fully automated production. The writer Astra Taylor calls it "fauxtomation" when executives use the promise of automation as an excuse not to eliminate labor, but to hide and degrade it. Whether at a cutting-edge electric car company gigafactory or an eighteenth-century textile mill, workers still very much need to be on hand to ensure the machinery keeps running.

Just as Regency-era entrepreneurs pushing for greater efficiencies, profit margins, and control paved the way for the industrial factory system, a move that transformed how humans live and work, a new thrust now threatens to do the same, all over again. Today the factory is limited, and in some cases, outmoded. Not because it is no longer an efficient system of control — it has long proven to be that — but because it is in some cases no longer the *most* efficient form of control.

The algorithm-based gig work model is the next stage in the evolution of the factory, a mode of control over workers that extends beyond mass production and is superior in nearly every way. This explains why Amazon, the second-largest employer in the United States, has adopted such a model, with its Uber-like Flex program for delivery drivers and fully automated hiring and HR systems. Gig app platforms and algorithmic management seek to reduce or eliminate the need for middle managers or HR departments, at least for the working-class "independent contractors" who constitute the bulk of a company's labor force.

Imperturbable algorithms provide the final say on where workers will go and how much they will make. There is little need for office space for the laborers; workers are distributed and diffuse. And best of all (for management), it is invisible; workers might enjoy, at least for a while, an illusion of freedom, working to no set schedule, in their own cars or with their own materials or from home.

But the control quietly promises to be even more total. Workers are taught to bend to the algorithms' demands: to work more when it says there will be a bonus; to follow the route it suggests, even if they know of a better or safer one; to make themselves available around surge times, when their services will be better rewarded. Some may even remain unaware that they are molding their own lives to an algorithm's demands. And without middle managers or regional HR reps or the need to pay benefits to the workers, a greater portion of the revenues gets funneled to the top.

The platform model is methodically inculcating and overlaying a sort of psychic factory onto its workers' lives. It's both an inversion and an evolution — workers are made to internalize factory logic on a personal level, accepting jobs by clicking mechanistically and automatically on smartphone apps and following directions to a T, to avoid missing out on chances to work or incurring penalties for slack performance.

"The gig economy promises flexibility and more free time, yet workers are increasingly tethered to work because of the on-demand nature of the work," as the sociologist Alexandra J. Ravenelle, who has interviewed hundreds of Uber drivers, explains. "The work is seemingly flexible, but it doesn't end. And while the workers are 'self-employed' contractors and don't answer to bosses, they remain under constant observation through a technological panopticon....In the gig economy everything can be collected and viewed at any point."

When the artisans and workers of the 1800s saw the prison-like factory looming before them, it *looked* like the architecture of domination.

The platform model is less foreboding. It often looks, at first, fun. But it too has always been built around the precepts of control. Like the onset of the factory system, this arrangement did not happen overnight, nor did it happen with the arrival of the technology itself. Just as the automated machinery was later to arrive in the organization, so is the contracted, gig-based work model only finally being hyper-charged by the algorithims of Uber and Lyft, after taking shape for decades.*

The economic historian Louis Hyman has shown that almost immediately after New Deal programs and protections laid the foundations for postwar American economic prosperity — stable employment, generous salaries, worker protections and benefits, and a model of large corporations enacting long-term growth — a host of consultants and staffing entrepreneurs set out to dismantle it. They used language of flexibility and impressed a need for employers to stay nimble — and set about breaking down bonds of middle-class work. Consultants began to replace executives and management, temps took over more white-collar office work, and day laborers were shuttled into previously unionized menial jobs. This was a concerted effort, according to Hyman, and carried out with the broad support of a growing number of public figures.

So Uber's gig mill is not truly novel, but is rather the supercharged culmination of decades of "leanness" initiatives, efforts to minimize paying benefits, and corporate innovation. After generations of organizing our workday lives around a 9-to-5 schedule, where parents could return home in time for dinner with each other and the children, now we're seeing algorithmic work schedules scramble norms all over again.

It's difficult to definitively gauge the size of the gig or contract labor economy, but a 2018 Marist/NPR survey found that some 1 in 5 US workers participate in it.

---

* Ironically, technology, with all of its advances, has in some cases returned and redistributed pre–factory system piecework, through websites and apps like Etsy and eBay. But now the tech company charges workers a rent for the privilege of doing business over their platform. Workers still purchase their own materials, do their own labor, and market themselves to prospective buyers — but now Etsy gets a slice for hosting the website, too.

Uber has well over a million drivers in the US alone, and Amazon well over that many employees. Contract work is on the rise, even among what were once considered top-tier employers with generous salaries — there are more temps and contract employees than staff employees at Google, for example. Journalism, law, and marketing are all professions increasingly being "fragmented" by contract work.

And the galaxy of on-demand apps that are facilitating the transfer of better-paid, more independent work into those controlled by a tech company's opaque algorithm is always expanding, and into frontiers we might have once deemed unthinkable. In 2021, *Vice* reported that the Department of Defense was building a gig app for contract work — increasing the likelihood that the next wave of our security analysts and armed forces employees will be gig workers. A crime app company called Citizen is experimenting with "subscription law enforcement service," which would allow users to hail an armed security detail the way they currently do an Uber.

Apps and contractors may offer a cheaper source of labor, but especially in a country like the United States, with a scant social safety net and privatized access to health care, they're often a pale comparison in terms of security and earnings to those that came before. They're good for corporations, who can consolidate more of their earnings among the upper management and executive rungs, and bad for the workers, who see the foundations of their lives further erode away.

"These are very, very precarious workforces that are foreshadowing the future," said Veena Dubal, a law professor at the University of California, Hastings. Dubal is a pioneering scholar in the field of platform-based labor law; she studies the effect that companies like Uber, Amazon, and InstaCart are having on the workforce.

"There is no doubt that companies across the tech and service economy are trying to move to a 1099 workforce" — one that runs on precarious contract labor rather than salaried jobs — "which dramatically lowers labor costs and transfers risks and liabilities," Dubal says.

That has grown to brand new heights. We'll see it more in restaurant work and retail and education and health and hospitality for sure. In addition to taking rights and security away from people who really need them, these business practices are infused with algorithmic management practices. The law hinges on the amount of control the employer exerts on the employee — so much of the control is exerted through social psychological gaming algorithms. It looks like you have choice and volition, but really it's training you to work for the platform in a specific way.

There is a purposefully generated information asymmetry, Dubal says. Workers can never be sure how much they'll earn, or that the algorithm isn't going to suddenly determine they should be terminated. "Algorithmic hiring and firing have become an everyday part of people's lives, whether we're talking about customer service agents or warehouse workers." The aim, she says, is to create an anxious, uncertain workforce that has no choice but to be malleable before the algorithm's demands. "These experiments are now fusing into other parts of the economy," Dubal says. "These practices are ascendent."

Also ascendent is the use of AI services, which boomed in 2023, promoted by companies like OpenAI. When AI is injected into already precarious work structures, it promises to accelerate insecurity and displacement further still.

Factories and automated machinery took workers out of their homes and away from their families. Gig apps run by proprietary algorithms take them away from other people altogether, and impose factory logic onto each individual, who sits at home or in a car, taking orders that must be completed in a rigid and exacting way. Then, as now, the result is unpredictable, worse-paying jobs, where laborers have less control over their lives, and less support.

One of the major unspoken benefits of the on-demand app companies, for their owners, is that the workers are further isolated from one another, making it more difficult to build bonds of solidarity with one another, and reducing the chances that they'll come together to advocate for better treatment. At least at first. We've seen some surprisingly wide-ranging movements growing, among Amazon workers, among Uber, Lyft, and InstaCart drivers, and among the precariat more generally — and if the Luddites are our rough working corollary, it's early days yet.

One thing that puzzled authorities during the Luddite movement is that displaced workers were joined by throngs of others whose jobs were not at risk of automation. The authorities were puzzled, essentially, over solidarity.

Only a small segment of the population needs be negatively impacted by automation, regimes of algorithmic control, mechanization, or exploitation to kindle a more widespread backlash. The shoemakers and artisans and coal miners and steelworkers understood that the fate of the Luddites was tightly bound up with their own, and that the forces immiserating the weavers would come for them, too. Amazon workers, ride-hail drivers, part-time service workers, and 1099 contractors already know very well why so many "unaffected" workers are protesting, agitating, and organizing alongside them for better conditions. We can all intuit the technological precarity, anxiety, and exploitation that is rife in the twenty-first century's industrial revolution.

It's plain to see that informal and algorithmic work is expanding, that salaried, benefited jobs are growing scarce, and, if the current trajectory holds, the phenomenon may soon engulf all of us. Or almost all.

PART VI

THE OWNERS OF THE NEW MACHINE AGE

DOUGLAS SCHIFTER

2018

Doug Schifter
February 5, 2018 ·

To those it may concern,

I have been financially ruined because three politicians destroyed my industry and livelihood and Corporate NY stole my services at rates far below fair levels. I worked 100-120 consecutive hours almost every week for the past fourteen plus years. When the industry started in 1981, I averaged 40-50 hours. I cannot survive any longer with working 120 hours! I am not a Slave and I refuse to be one.

Companies do not care how they abuse us just so the e... **See more**

You and 1.5K others   681 comments 1K shares

👍 Like   💬 Comment   ↪ Share

Post published to Douglas Schifter's Facebook page on February 5, 2018.

"Companies do not care how they abuse us just so the executives get their bonuses," a livery cab driver named Douglas Schifter wrote on Facebook one February morning in 2018. His emotionally charged, 1,700-word post attacked the executives of gig app corporations like Uber — and the politicians who empowered them — for using their platforms to exploit his trade. "Due to the huge numbers of cars available with desperate drivers trying to feed their families they squeeze rates to below operating costs and force professionals like me out of business," Schifter continued. "They count their money and we are driven down into the streets we drive becoming homeless and hungry."

The missive echoed a two-hundred-year-old Luddite letter. Vivid, incensed, and specific in its grievances, it detailed the plight of working drivers, and heaped scorn on the entrepreneurs and elites who were using a new technology to degrade workers' status and conditions. "I worked 100–120 consecutive hours almost every week," wrote Schifter, who was sixty-one years old at the time.

Schifter knew the streets of New York better than just about anybody. He'd chauffeured celebrities around Manhattan in his cab and logged some 5 million miles, the equivalent of nearly two hundred trips around the world, over the course of his decades-long driving career. He was proud of his knowledge, skill, and professionalism, and furious that Uber and its political allies had taken those things from him. "When the industry started in 1981, I averaged 40–50 hours. I cannot survive any longer with working 120 hours! I am not a Slave and I refuse to be one."

With his large frame and generous spirit, Schifter was known among friends and fellow drivers as charismatic, well-read, and keen in his sense of moral justice. He had been suffering health problems for years, and was haunted by the resulting reams of medical debt. Thanks, he believed, to companies like Uber that had suddenly and dramatically strangled his industry, he was behind on his credit-card and mortgage payments, too.

Searching and opinionated by nature, Schifter held little back in "The Driver's Seat," the regular column he wrote for the trade publication *Black Car News*, where he focused on Uber and the gig app platforms. He argued they would give rise to "massive pain and problems for hundreds of thousands of people." He implored his fellow drivers to rise up and do something about it.

"Brothers and Sisters," he wrote in one installment. "We will all be slaves to Uber.…I am seeking to organize drivers into a fighting army of thousands. Be part of an established tradition of fighting tyranny."

THE GREAT COMET RETURNS

Two hundred years and change after the Great Comet blazed over the skies of industrializing England, ominous signs have again been appearing that the latest incarnation of factory owners and entrepreneurs — tech execs and startup founders — are using new technologies to reshape the world of work to the disadvantage of the workers. There may not be a burning celestial body overhead, but throughout the 2010s, there was no shortage of omens closer to earth.

Closed for Business signs sit in the windows of mom-and-pop shops and the vacant husks of big-box stores in strip malls outside town grow ever more numerous, victims of the rise of Amazon and e-commerce giants, and the disappearance of the hundreds of thousands of jobs once held inside those walls.

Rideshare drivers are sleeping in their cars in parking lots and rest stops, even after working around the clock logging gigs on their smartphone apps, unable to afford rent in the richest country in the world.

Viral images are circulating of urine-filled bottles that delivery workers at Amazon — the world's largest online retailer, owned by the world's second-richest man — were forced to produce because they are so overworked and so digitally surveilled that they could not risk taking a proper bathroom break.

Modern factories and warehouse floors are buzzing with robotic product pickers and haulers, surrounding the humans whose tasks have been made at once less skilled and more punishingly repetitive. There are picket lines of people outside, protesting inhumane conditions.

AI-generated images are flooding the internet — leading artists, illustrators, and stock photo services to file lawsuits against the startups that the plaintiffs say copied, emulated, and degraded their work.

And where did this latest comet grow its tail?

Silicon Valley's rise is so well-known it's taken on mythic proportions: from a land of fruit orchards to a haven for microchip makers, from the ascent of Stanford University to a hub for defense tech contractors and the architects of the early internet, from the commercial birthplace of personal computers to a financial powerhouse, and, eventually, a key and central driver of the entire American economy. It launched HP, Intel, Apple, Microsoft, Amazon, Google, Facebook, Tesla, Uber. By the end of the '00s, more millionaires and billionaires lived between San Jose and San Francisco than anywhere else in the world. It's where the innovators of today, cut in the mold of the Industrial Revolution–era entrepreneurs, concentrated their power and grew their influence to new heights. The tech sector's top companies rivaled, then eclipsed, the corporate behemoths of the last century, passing the oil majors, the banks, and the telecoms. Former White House officials departing DC cycled not into finance but into top tech companies.* It was clear what comprised the new locus of power.

Meanwhile, the financial crisis of 2008 led to painful and precarious years for the middle class: a so-called jobless economic recovery. For millions, that recovery never arrived at all. As the new decade began, workers were struggling, war in a foreign theater seemed to lurch on endlessly, and inequality climbed to staggering new highs. Combustible, decentralized movements like Occupy Wall Street erupted in protest. It was into this environment that the tech titans took the opportunity to introduce disruptive new workplace technologies and double down on the old ones.

Instead of the semiautomated power loom, it would be automated customer service software. Instead of the gig mill, it would be the gig app.

In the first decades of the nineteenth century, entrepreneurs backed by magistrates and the Crown used a period of depressed trade and poor harvests — when workers were desperate and their leverage was at a low ebb — to push factory automation technologies. In the first decades of the twenty-first century, entrepreneurs backed by the state and venture capital used the period after the 2008 crash — which wiped out jobs and savings around the globe — to push automation and on-demand app technologies. Amazon, Uber, Lyft, DoorDash, InstaCart, TaskRabbit, and other companies drove algorithm-mediated work platforms into the mainstream, normalizing a mode of work that left workers without benefits or protections. Delivery drivers, cleaners, carpenters, masseuses, you name

---

* Jay Carney, Barack Obama's press secretary, became Amazon's senior VP for press relations. David Plouffe and Jim Messina had executive and consulting roles at Uber. Lisa Jackson, Obama's EPA administrator, became a VP at Apple.

it — the twenty-first century has seen the rise of precarious, gig-based, platform-mediated jobs, begetting the decline of traditional ones. Meanwhile, e-commerce and the Amazon effect hit retailers hard, shuttering shops and big-box stores alike. Headline after headline proclaimed "The Robots Are Coming for Our Jobs," in stories about hyper-intelligent AI and logistics automation. Will all these trends eventually lead to a world where the bots and algorithms do our dirty work, making our lives easier and more prosperous? Or will the machines push us out of our jobs and deposit us into a dystopia? We are chewing over the machinery question all over again, in barely updated language.

Out of this, the owners of these technologies emerged with almost inconceivable stores of wealth and power. In 2022, seven of the ten richest people in the world were tech billionaires. This newest wave of entrepreneurs consolidated their wealth and influence in much the same way that the nineteenth-century ones did — by using the new tech, or the *idea* of the tech, to squeeze workers harder and in more creative ways. Take the most infamous two, as of the 2020s.

Jeff Bezos drove Amazon to global dominance by promising a cheaper and more efficient alternative to brick-and-mortar retail stores: an online shopping portal paired with hyper-efficient fulfillment centers and distribution nodes. What enabled Amazon to put so many mom-and-pop stores out of business, of course, was not just its technology but its army of part-time workers, toiling at blistering rates to keep pace with roboticized warehouses and delivery time projections, constantly surveilled to ensure they do.

Elon Musk, meanwhile, earned popular adulation by acquiring and running companies that looked like science-fiction moonshots. But his luxurious, semi-autonomous electric cars were still built on the backs of workers in brutal conditions overseen by managers that have been charged with racist and discriminatory acts.

Bezos and Musk epitomize the modern conception of the tech entrepreneur. Both have fought vigorously to oppose organizing efforts at their companies, and they are now the richest men in modern history.

Skilled workers like Douglas Schifter, meanwhile, saw their status, wages, and quality of life decline. Fast. It was happening everywhere — and not always quietly.

In 2015, Uber was encouraging drivers using its black-car service, UberPOP, in Paris to break French chauffeur laws by accepting rides without a license, taking business from local taxi drivers in the process. The French cabbies rose up to forcefully protest the arrival of Uber, blocking roads and smashing the vehicles of Uber drivers. They were criticized in the US media — but they won a major victory in slowing the adoption of the app. France's interior minister, Bernard Cazeneuve, announced that the government would be suspending Uber's black-car service in France. It was "illegal," he said, and "must, therefore, be closed. The government will never accept the law of the jungle. The vehicles of UberPOP drivers should be systematically impounded when they are openly breaking the law."

Back in the States, Douglas Schifter had gotten serious about organizing his peers to fight back, but he didn't know how to start. In 2016, he set up a group he named the NY Black Car Drivers Association, and posted a call to Facebook for his peers to join, but there were few takers. In his next column, he put the word out in increasingly thunderous tones.

"We are facing extinction," he wrote. "The time to organize is NOW!"

THE NEW TECH TITANS

Andrew Yang

2018

"We're going to see our already record-high inequality reach unprecedented heights, where there are certain people who are fantastically wealthy," the entrepreneur-turned-presidential-candidate Andrew Yang explained at his 2020 campaign headquarters. "Which I will say is not necessarily awesome for them," he said, "or anyone."

These certain people, whom Yang calls "my friends in Silicon Valley," are the entrepreneurs who have disrupted key elements of the economy, accumulating unprecedented levels of wealth and power in the process. The names will be familiar: Bezos, the modern Arkwright, and Uber's Travis Kalanick, a Horsfall, but also Facebook's Mark Zuckerberg and Google's Sergey Brin and Larry Page.

But these men—just like two hundred years ago, they are almost all men— have power on a scale that a nineteenth-century titan could not even begin to dream of. The tech giants have dramatically transformed and squeezed countless livelihoods—not just taxi drivers and delivery workers, but small-business owners who depend on their company's placement in Google search results, or getting a fair deal on Amazon's third-party vendor marketplace. Developers who must cough up 30 percent of their revenue to Apple if they want to do business on its App Store. Travel agents, famously, but also translators, financial advisors, journalists, tax preparers, customer service workers—a small sampling of livelihoods threatened by tech monopolies like Facebook, Microsoft, and Intuit. Whether these behemoths are deploying AI or automating work outright, they have a level of control over most people's working lives that is unrivaled in its breadth and scope. The reason Yang argues that this is "not necessarily awesome," even for the titans who have profited handsomely by all this disruption, is that after years of chalking up the losses to inevitable technological change, the public's ire has turned toward the owners of the algorithms — and it could tip into violence against them.

Yang said that the specter of a world in which a few tech billionaires accelerate the automation and digital degradation of work is not only exceedingly grim, but destined to become reality. Inequality will reach levels even more starkly dramatic than today. "Other people are going to be pushed into deprivation and desperation," he said, "and that's going to be multiplied times thousands and hundreds of thousands."

Yang's Manhattan HQ was at this point completely empty, except for a couple of card tables and folding chairs. It was April of 2018, and this was all new to him. At forty-three years old, he'd been CEO of a successful test-prep start-up, but had no political experience to speak of. And he had just announced his candidacy for president of the United States. He was motivated by a simple and alarming premise: that automation and artificial intelligence are in the process of destroying millions of American jobs, and no one is doing anything about it.

"Think about what happens when three million truck drivers are out of a job at once," he said, using one of his go-to examples of an industry on the precipice of being automated away. Truck driving, his campaign platform says, is the most common job in twenty-nine states. It's one of the highest-paying jobs that doesn't require a college degree. It's also one that's most susceptible to automation. Mass production of self-driving long-haul trucks, his platform says, is expected to begin by 2030. "If self-driving trucks are brought in to replace all these workers," he said, "this has potential for serious unrest if not handled properly."

Deeper into his campaign, Yang made an appearance on Joe Rogan's podcast and elaborated on the prospect. "There's going to be a lot of passion, a lot of resistance to this," he said. "Anyone who thinks truck drivers are just going to shrug and say, 'All right, I had a good run. I'll just go home and figure it out' — that's not going to be their response. It's going to be much more likely that they say, 'You need to make these robot trucks illegal.'"

If that didn't work, Yang speculated, "they're just going to park their trucks across the highway and get their guns out, because a lot of these guys are ex-military, and just be like, 'Hey, I'm not going to move my truck until I get my job back,' and there are going to be a lot of truckers in the same situation."

Andrew Yang was an instantly recognizable face on the campaign trail; with his business-casual CEO vibe and upbeat apocalypticism, he became a media fixture. He was also probably the first modern mainstream presidential candidate to essentially forecast a second coming of Luddism in any serious level of detail. He was certainly the most vocal in foregrounding automation and the capacity of technology to degrade jobs as a major pressing threat — it was his signature campaign issue. Yang even proposed policy solutions not so distant from the weavers' idea for an automated loom tax that surfaced in the pre-Luddite years: he wanted to instate a "robot tax" on companies like Amazon that are increasingly automating their processes. He would use the proceeds to fund a universal basic income (UBI) program, which he dubbed the Freedom Dividend, and would deliver $1,000 to every citizen each month, as a sort of buffer for those whose jobs are automated away.

In some ways, Yang's approach is reminiscent of the paternalistic utopianism of Robert Owen. Both men genuinely fear the impact of automation and inequality, and both believe that a sweeping technical fix can address it. For a time, Yang was able to rally modern-day John Booths; the Yang Gang that swarmed social media in support of his presidency was outspoken to say the least.

Yang is far from alone in forecasting a turbulent era in which automation and artificial intelligence sweep away working-class jobs in droves. With money pouring into AI, robotics, and software automation, throughout the 2010s, pundits and prognosticators forecasted the arrival of "second machine age," or the "fourth industrial revolution," or a "world without work" — an era when machinery and software become so advanced that humans simply will not be able to compete. Yang calls this the "war on normal people," which is also the title of his 2018 book. These economists and automation theorists continue to highlight studies that show that some half of all American jobs are vulnerable to computerization. These findings have been widely scrutinized and criticized, but it's clear that tech executives and managers take them seriously.

Now, like a hypercharged evolution from the entrepreneurs of the Industrial Revolution, entrepreneurs are tirelessly searching for new efficiencies and nodes of disruption. The bar has been set high, and they aim to clear it.

"I'm hanging out with the tech wizards of Silicon Valley, and I'm like, 'Hey, you know, are we gonna automate these jobs away?'" Yang said on the same *Joe Rogan Experience* episode.

And they're like, "Oh yeah, we're gonna automate these jobs away."...
Whose responsibility then is it to go tell the people, "Look, it's technology,

it's transformed the economy in fundamental ways, and we need to make it so that everyone benefits and it's not just, that this, like, hyperconcentrated set of winners and then this, like, huge army of relative losers"?

Yang is correct that inequality continues to widen to astonishing degrees. A 2020 report from the Economic Policy Institute found that "the ratio of CEO-to-typical-worker compensation was 320-to-1," a jaw-dropping divide that marks an exponential increase from 21-to-1 in 1965, to 61-to-1 in 1989, and 293-to-1 in 2018. The tendency, as George Mellor would say, has been all one way. According to a report in the *New York Times,* "the richest 0.1 percent of American households own 19.6 percent of the nation's total wealth, up from 15.9 percent in 2005 and 7.4 percent in 1980. The richest 0.1 percent now have the same combined net worth as the bottom 85 percent."

This helps account for the dystopian factoids we've become all too familiar with: American life expectancy is declining for the first time since record-keeping on the subject began. Access to health care breaks along lines of privilege; and deaths of despair — suicide and drug overdose — are climbing. The rich now live on average thirteen years longer than the poor. As a result, anger is growing at the ultrarich, even in the United States, a nation that has historically admired them. "Why does everybody suddenly hate billionaires? Because they've made it easy," the *Washington Post* mused in 2019. Beyond a matter of optics, their influence over politics is increasingly considered malign, and that includes the tech billionaires — once the twenty-first century's most revered businessmen. Polls conducted in 2021 show that Facebook CEO Mark Zuckerberg is deeply unpopular, and more people dislike Amazon founder Jeff Bezos than like him. The scorn is increasingly bipartisan; while liberals take them to task for amassing monopoly power and for the poor labor conditions at their companies, conservatives accuse them of political bias and censorship.

For every Elon Musk fan, there's a critic trying to bring him back to earth. Twitter threads and op-eds and a growing cohort of politicians, like Bernie Sanders and Elizabeth Warren on the left, and Josh Hawley and Ron DeSantis on the right, bemoan the outsized power of the tech monopolists. As it was at the dawn of the Industrial Revolution, the anger is especially acute toward those running the new factories.

FEAR FACTORIES REDUX

Christian Smalls

March 2020

At the beginning of March 2020, an assistant warehouse manager named Christian Smalls at Amazon's JFK8 facility noticed that more and more of his fellow employees were showing up to work pale and fatigued.

The Covid-19 virus had just begun to spread across the nation for the first time, and it paralyzed much of the economy in the process. It was, however, a boon to the company where Smalls worked. Hundreds of millions of people were suddenly stuck indoors around the clock, ordering goods online — often through Amazon. The pandemic accelerated a trend already long underway: small businesses, whose margins were always thin and which depended on walk-ins and loyal local shoppers, were closing by the thousands, many permanently. Amazon, which had been investing heavily in automation and software logistics, and was expanding its fulfillment centers — the hulking new JFK8, which served most of New York City, was just a couple years old — ramped up hiring. Analysts forecasted another boom in automation; machines could do work that was too dangerous for humans in the time of a pandemic. But Amazon's warehouses remained packed with people, mostly part-time and seasonal workers with no safety nets of their own. And they felt Amazon didn't do much to protect them.

Smalls, who has three little children, was concerned for the safety of his family and his colleagues. Little was known about this new virus except that it was highly contagious and could be deadly. Workers expressed concerns to management; Amazon dismissively offered them only unpaid time off and what they felt were lax safety protocols. Smalls took a few days off and dipped into savings to cover his rent, but it wasn't a sustainable solution.

Then one of his colleagues tested positive for Covid, and Amazon's HR department told them to keep it on the "down low." That was the last straw.

Smalls and some other concerned Amazon workers started openly pressing Amazon for better conditions, specifically, more-effective sanitary gear, a full site cleaning, and paid time off for vulnerable or exposed workers. When Amazon ignored their asks, Smalls and his colleagues staged a walkout. On March 30, over a hundred Amazon workers walked off the job, toting signs like Our Health Is Just as Essential, and Treat Your Workers Like Your Customers. Smalls, who'd been with the company for five years at multiple locations, was fired the same day.

Amazon's top brass, including CEO Jeff Bezos and VP (and former White House press secretary) Jay Carney, met shortly after to plan their PR response. The notes of that meeting leaked to *Vice News*.

The subject of discussion was this young associate manager turned workplace safety advocate. "He's not smart, or articulate, and to the extent the press wants to focus on us versus him, we will be in a much stronger PR position," Amazon's general counsel, David Zapolsky, wrote in notes taken from the conversation that were then forwarded around the company. "We should spend the first part of our response strongly laying out the case for why the organizer's conduct was immoral, unacceptable, and arguably illegal, in detail, and only then follow with our usual talking points about worker safety," Zapolsky said. "Make him the most interesting part of the story, and if possible make him the face of the entire union/organizing movement."

Zapolsky was betting that Smalls would prove so "not smart or articulate" — a comment that many found rife with racist implications, given that Smalls is Black — that he would help turn the public, and union-curious employees, against unions more broadly.

This was newly alarming terrain for the tech giant. After decades of successfully preventing any of its facilities from organizing in the US, Amazon now saw surging interest in the prospect among its employees. Multiple facilities, from Chicago to Southern California to Alabama, had experimented with organizing efforts. In 2019, Amazon workers, immigrant groups, and digital rights organizations formed the Athena Coalition to push back on the e-commerce powerhouse. Talk of unions was rippling across the country — so much so that Amazon felt compelled to launch internal programs to surveil closed employee Facebook groups and private email listservs.

It wasn't just a lack of Covid safety, or even worker safety, per se. It was the combination of punishing conditions, lack of benefits, constant surveillance, and stagnant pay. It was breaking your back to keep up with the robots while the man who owned the operation broke records for making so many billions of dollars. The man who is, if he is not already, on his way to becoming the biggest factory boss in history — Jeff Bezos.

"Amazon, along with Walmart," the labor reporter Luis Feliz Leon wrote in 2021, "is the twenty-first century's quintessential factory floor."

What distinguishes Amazon from its forebears, and links it to contemporaries like Uber and InstaCart, is the way this next-level factory floor isolates, alienates, and cyborgizes its workers for maximum productivity. Andrew Ure celebrated Richard Arkwright for training humans to work long hours alongside machinery; future business theorists may laud Jeff Bezos for making workers themselves as machinelike as possible, and for making the working experience as inhuman as conceivable. That experience begins with the very process of applying for a job at Amazon, a process that is itself almost entirely automated.

A prospective worker finds the online hiring portal, selects where and what kind of shifts they want to work, submits their application and personal information, and agrees to a background check and a drug test. They receive an automated email that offers them an orientation slot that may or may not align with the ones they saw listed when they applied. The orientation itself is completed online. When they are awarded the opportunity to report for a shift, those offerings, too, are automated. By the time someone leaves the fulfillment center — perhaps one that stretches out like an enormous brutalist shoebox across the Southern California desert or New York hinterlands — they might realize they had applied, attended orientation, taken a drug test, and been given a job at Amazon without ever once speaking directly to anyone who worked there.

Of the major tech companies, Amazon has the largest labor force. It has the most pointed interest, and largest investment, in mass automation. The company has said it aims to have its warehouses fully automated by 2030, when Bezos hopes a menagerie of sorting, picking, and transport robots will do what human workers do now. Until then, Amazon will continue to be one of the largest employers in the United States, with 1.6 million workers on its payroll as of

2022 — second only to Walmart. Analysts at *MarketWatch* estimate that for every job Amazon itself created, it has probably destroyed two or three others, thanks to its ultra-efficient business model and its wresting of market share from brick-and-mortar retailers. So a loss of status, security, even identity has often already taken place when someone applies for a job like one at Amazon.

These hundreds of thousands of workers may not yet be actual robots, but they are treated like them by Amazon executives. Workers are subjected to surveillance measures that ensure they hit relentless productivity goals and are tracked so closely that many are driven to skipping meals and bathroom breaks. The work itself is physically punishing, and independent studies show that injury rates are much higher at Amazon than other warehouses; one June 2021 report found they were 80 percent higher than the industry average. The company's history is pocked by harrowing tales of rampant exhaustion and fatal overwork — at one warehouse, a worker suffered a heart attack and wasn't noticed for twenty minutes as workers scurried past to meet productivity goals. No automated system detected the loss. When he was declared dead, his coworkers were immediately told to "go back to work." At some facilities, Bezos found it more economical not to install air conditioning, and to instead hire EMT drivers to wait outside for workers who inevitably fainted from heat exhaustion. Robert Blincoe would have recognized the logic of the place well.

This relentless, technologically dictated pace of work is ultimately driven, just as it was in the Luddite days, by an entrepreneur exercising a more audacious will to profit than anyone else, and an openness to implementing systems that others might deem too inhumane, in order to realize those profits. And, as with the first tech titans two hundred years ago, the net effect is to force all other individuals and companies who might hope to compete to adopt similar technologies and policies. Jeff Bezos has said that he believes people are "inherently lazy," which is why they must be surveilled and goaded to labor at a grueling pace with an array of technologized incentives and digital tracking devices. Amazon's labor policies have been engineered to maximize worker productivity, no matter the human cost, and to purposefully discard workers after they are no longer operating at peak efficiency. And since Amazon does it, everyone else must make their employees machinelike as well, if they hope to keep pace.

Amazon's HR department informed Christian Smalls that by coming on-site to stage his walkout, he had violated a quarantine order, which was a fireable offense. Yet out of everyone who walked out from the Staten Island facility that day, he was the only employee terminated.

"It's a shame on them," Smalls told *Vice News*. "This is a proven fact of why they don't care about their employees, to fire someone after five years for sticking up for people and trying to give them a voice."

The firing set off a ripple of outrage. Bill de Blasio, Bernie Sanders, and New York State's attorney general, to name a few, all spoke out in support of Smalls and called for a National Labor Relations Board investigation into the termination, which seemed retaliatory.

Much like Uber and the rideshare companies, Amazon seems intent on finding out just how hard it can push its increasingly distressed workforce before those workers decide to fight back, in ways perhaps more forceful than organizing walkouts.

"It's not gonna stop me," Smalls said. "I'm gonna continue to fight."

He was not alone.

GIG WORKERS RISING

Nicole Moore, Veena Dubal, Abed, Eric, Tammy, and the Drivers

November 2020

"I hope Tammy didn't fall asleep," one ride-hail driver said on the late-night Zoom call, before being cut off by another.

"I hope she *did,* she'd be in a better place."

The exchange was followed by grim laughter and resigned head-shaking.

It was late into the evening on Election Night, November 2020, and drivers who had organized to try to stop the passage of California's Prop 22 — which would keep them from being classified as employees, prevent them from obtaining benefits, and bar them from organizing — were gathered online to watch the results come in.

"It's not looking good," said Nicole Moore, Uber driver, organizer, and president of Rideshare Drivers United, a Los Angeles–based association that advocates for ride-hail drivers. She was stating the obvious: anything short of a miracle and the proposition, backed by a record $200 million campaign by Uber, Lyft, DoorDash, and InstaCart, would pass into law.

It had been a whiplash decade for the world of on-demand app work. In the early 2010s, the wages had started out relatively high, the schedules flexible, the gigs interesting, as Uber and Lyft and their ilk actively sought to attract workers to the platform. But the reason pay was so high, it turned out, was that the start-ups were juiced by war chests of venture capital that backed the app companies as they cornered key markets. Uber succeeded in persuading livery cab drivers to switch to the platform, signed them up by the thousands, and encroached aggressively on taxi drivers' turf. After Uber had a foothold, the wages fell.

Uber, Lyft, InstaCart, and DoorDash have never been continuously profitable, despite their astronomical valuations and continued waves of investment. In order to claw their way toward a positive cash flow after a decade of operating, they started slashing pay. One way they did so was by complicating the opaque algorithmic system that determined how much a driver got paid. They tweaked features like surge pricing and zone multipliers and bonuses for the number of rides given, and constantly toggled the formula to make it unclear how much drivers would earn. Before long, many drivers who had come to depend on the platforms realized that they were making less money — in some cases less than they had made at the livery cab or taxi businesses they had left to join Uber. While black-car and taxi drivers like Douglas Schifter struggled to compete with the on-demand app drivers on one side, the gig workers struggled to earn enough to make ends meet on the other side.

In 2019, Uber and Lyft's early investors became millionaires and billionaires when the companies held their IPOs. Yet the drivers who make it all possible often make less than $10 an hour, after maintenance and gas expenses are accounted for. Their story is not unlike that of the nineteenth-century stockingers, on multiple levels. Many stockingers had to rent their frames from the master hosiers, while many Uber drivers have to rent their cars from Uber — the machine owner gets paid twice so that the worker might have the benefit of working. Both stockingers and drivers, in many cases, have barely enough left over to pay for food and shelter when the day was done.

So, some of Uber and Lyft's most dedicated drivers began organizing. They shared tips and strategies on Reddit and online message boards, undertook informal efforts like coordinating with each other to decline rides at a given location until the app's algorithm offered them fairer prices, and formed groups like Gig Workers Rising and Rideshare Drivers United. Workers in a line of employment strictly designed to keep them isolated began to meet up. They forged solidarity, staged protests, and eventually held daylong strikes. A new, distributed kind of movement for gig workers' rights was born. They were acting, essentially, as the cloth workers had in the ten years before they became Luddites — pushing for better conditions and for legislators to enshrine protections into law. In 2019, California lawmakers led by Lorena Gonzalez passed AB5, a law that classified rideshare drivers as employees, not independent contractors, and granted them the same legal rights that employees are entitled to.

Uber and Lyft refused to comply with AB5 — they had quietly begun to craft

Prop 22 to override it — so drivers were again forced to organize to pressure the state to enforce it.

"We want to stand up for our rights," a sixty-year-old Uber driver named Abed said at a demonstration that took place just before the pandemic began. He'd gathered in downtown San Francisco with a large group of other drivers, activists, and the labor law scholar Veena Dubal to formally sue the company for misclassifying drivers as contractors. Abed had eagerly volunteered to participate. A longtime driver, he'd seen his earnings decline despite logging years on the platform. "Corporate is making millions of dollars a day, and still they are exploiting their employees," he said. "We need a job. We have to pay our bills. They are taking advantage of us, which is not right. In this country, how can this happen in the twenty-first century?"

He described falling pay for rides, progressively harder-to-reach bonuses, and increasingly scant communication from Uber about what was actually going on. "You have to treat your drivers the right way," he said, visibly agitated. "They just want to put us down, down, down, and use us for nothing. We want to show how strong we can be." He paused, then added, "Every driver I talk to, they feel the same."

Another driver, a thirty-four-year-old we'll call Eric, vented about the problems that arise when your manager is an algorithm. "We need to be able to set our own prices," he said. Drivers need to be able to see where a given ride will take them, he said, and be able to take that into account when they decide whether to accept it. "We have ten to twelve seconds to make a business decision that could take minutes or hours. We need information. They're not showing you any information other than your location to go pick up." And they need fair and transparent pay, of course. "Proper compensation, that's number one."

Another driver, a tech worker and organizer, developed an automated tool of his own to help drivers apply for lost back pay in a system that treated rideshare drivers as pieceworkers. "The form assumes that you do piecework, like picking strawberries. Which makes it brain-breaking to fill in these forms." He'd made a spreadsheet that a worker could easily use "to fill in the three-page wage time form, to get the overtime business expenses, to fill it out automatically," even to help input rest breaks and estimate damages. "To help get back lost wages."

These tools are crucial because, unlike employees, independent contractors are not eligible for minimum wage, overtime pay, worker's compensation, health-care benefits, paid rest time, or reimbursement for driving costs — the basics. They need every bit of pay they're due. Uber and Lyft have long argued that their drivers — the people physically driving customers from one location to another — are not "core" to its business. These are technology companies, not transportation companies, they argue, so they don't have to consider drivers as employees. It's a centuries-old tactic. Recall that when the big cloth manufacturers adopted automated machinery, they argued that regulations and laws on the books governing the trade were outmoded and shouldn't apply to them. But without drivers, there would be no Uber or Lyft. It's hard to imagine a component more core to a ride-hailing business than the person who drives the ride that has been hailed.

Luddites were protesting being forced into the factory system as much as they were resisting the machines that were replacing their jobs. Today's gig workers are rising in protest against an algorithm-driven regime of work that *itself* is largely automated. With app-based work, jobs are precarious, are subject to sudden changes in workload and pay rates, come with few to no benefits and protections, place the worker under intense, nonstop surveillance, and are highly volatile and unpredictable. And the boss is the algorithm; HR consists of a text box that workers can log complaints into, and which may or may not generate a response.

The modern worker can sense the implications of this trend. It's not just ride-hailing either — AI image-generators like DALL-E and neural net–based writing tools like ChatGPT threaten the livelihoods of illustrators, graphic designers, copywriters, and editorial assistants. Streaming platforms like Spotify have already radically degraded wages for musicians, who lost album sales as an income stream years ago.

Much about Andrew Yang may be suspect, but he did predict correctly that anger would again spread like wildfire as skilled workers watched algorithms, AI, and tech platforms erode their earnings and status. The next step of his prediction is that the anger may tip into organized sabotage. Perhaps it could be Amazon workers, finally fed up with the injustice of being forced to pee in bottles to make relentless productivity goals, and with hundreds of expensive robots swarming around them all day. Perhaps it will be Yang's vision of an armed cadre of truckers displaced by autonomous semitrailers. Perhaps it will be gig workers, organized and angry at having their efforts to be treated with dignity crushed, time and again, by the executive class with their endless hoard of venture capital. Perhaps it will be all of them.

Before the Luddites rose up, weavers and croppers and cloth workers tried for over a decade to get Parliament to pay attention to their plight, and they were ignored, accused of agitating illegally, and disparaged en masse. Today in the US and the UK, unions are legal but their power has been weakened over the decades. Many red states have implemented so-called right-to-work laws that make it much harder for unions to operate or get a foothold. And in on-demand app work—one of the most crucial, highly technologized, and fastest-growing parts of the economy—workers are legally prevented from combining to form unions.

On Election Night 2020, as the rideshare organizers' call went on, it became clear that Prop 22 had passed. An air of restrained anger, disbelief, and even despair took hold.

"We've punched way above our weight," Nicole Moore said. "And this is not easy…when your opponents are using weapons of mass destruction, like using $200 million to our $12 million." There was little wallowing. Moore was right; Uber, Lyft, InstaCart, and DoorDash had outspent them 20-to-1, embraced deeply suspect and misleading campaign tactics, like sending out mailers insinuating that Bernie Sanders supported Prop 22, and blanketed billboards, mailboxes, and YouTube with pro–Prop 22 advertisements.

"I look around and I see warriors," Moore continued, and a pulse returned to the Zoom. "Some of the hardest-working warriors I know. And know I can't turn my back on anyone in this room. We went out there and fought and proved that we could get in the streets." A cheer rippled through the Zoom. "Because I am not ready to give up yet. I am not ready to throw the fuckin' towel in."

One of the most crushing blows that Prop 22 dealt was to the act of organizing itself—in addition to enshrining rideshare workers as independent contractors ineligible for employee benefits, it expressly prevents them from forming a union. And overturning it will take a seven-eighths majority vote in the California state legislature, a threshold some scholars say is so extreme as to be unconstitutional.

Another organizer, Marie Harrison, spoke up. "We need to march," she says. "Everywhere. We have to keep right in the public eye. We cannot allow ourselves to be forgotten. I've been trashed economically by these companies. What we have on our side is a union — a pure kind of union, made up of drivers who are beyond driven. As stinging as the loss can be, it's a galvanizing force."

She said they need to start getting more aggressive in their tactics. "Guerrilla actions," in her words. "Start hitting the merchants."

In 2020, it may have looked like the massive app companies had crushed its drivers' efforts to organize and win better conditions. But the industry is young; the community is young. They have proven that they can be tactical, smart, tireless — and they are numerous.

It may take years, as the new working communities develop stronger ties, bonds strong enough to trust one another to engage in collective action. In the twenty-first century, this happens on Zoom calls, Reddit boards, Signal chats, and group texts, not secret chambers in ale-soaked pubs. "If you look at where the resilient actions have happened, it's where people have formed bonds. Really long Zoom conversations," Dubal says. "It doesn't have to be in person. The Mechanical Turkers who are organizing — they're everywhere, but they see each other over Zoom. Once a week. They all know about each other's lives. They send money to each other, even though they're dispersed all over the world."

If Uber keeps pushing them down, if the government ignores them, if every party to the growing plight fails to issue them redress — then it would be only reasonable to expect that the app-based workers will only continue to rise in number as the latest great comet gathers steam. Even a Luddite-style uprising is not out of the question.

THE NEW LUDDITES

"We need a Luddite revolution," Ben Tarnoff, a tech worker and the founder of *Logic* magazine, declared in the *Guardian* in 2019.

Today, amid the normalization of Uber and Lyft, the domination of Amazon, the rise of AI-as-a-service startups, and the percolating distrust in tech companies, a growing number of critics, academics, organizers, and writers are explicitly endorsing the Luddites and their tactics. "The Luddites were actually protesting the *social costs* of technological 'progress' that the working class was being forced to accept," the sociologist Ruha Benjamin wrote in *Race after Technology*. "To break the machine was in a sense to break the conversion of oneself into a machine for the accumulating wealth of another," notes cultural theorist Imani Perry.

The new media and digital culture scholar Gavin Mueller published his tract *Breaking Things at Work* in 2021. The science-fiction author Cory Doctorow began using his platform to advocate Luddite politics — he declared all of SF a "Luddite literature."

"I love the Luddites," Veena Dubal exclaimed. Paris Marx, writer and host of the *Tech Won't Save Us* podcast, started a newsletter called *The Hammer*, named after the Luddites' weapon of choice, in 2021. The tech worker and author of the incendiary *Abolish Silicon Valley*, Wendy Liu, kicked off 2022 by tweeting an image of a shirt emblazoned with the motto THE LUDDITES WERE RIGHT.

More and more people are beginning to add "Luddite," irony-free, to their social media profiles.

At the vanguard of what we might call this new Luddite movement is the popular podcast *This Machine Kills*, hosted by tech journalist Edward Ongweso Jr. and social scientist Jathan Sadowski, and produced by Jereme Brown. *TMK* embraces Luddism as a framework for exploring — and excoriating — the dominance of Big Tech and Silicon Valley. It quickly became a must-listen for tech critics and insiders alike, and for workers trying to navigate the world that Big Tech made.

"I have been pleasantly surprised and shocked, to be honest," Sadowski said, laughing, over a Zoom call with his cohost. "This crop of people, this crop of neo-Luddites coming up. I've seen a lot of resonance when we talk about Luddism, I see a lot of people in our mentions on Twitter saying, 'If this is what Luddism is then yeah, call me a Luddite.'" On Ongweso's birthday, one of his followers sent him a picture of a Ring camera they had smashed. "Bless them," Ongweso said.

"It heartens me every time I see someone with 'Luddite' in their bio," Sadowski continued. "There's a broader movement happening. People are looking for a politics that is ruthlessly critical of Silicon Valley—it's not a primitivism, we don't reject all technology, but we reject the technology that is foisted on us. Because we expect something better."

The companies that rely on app-based gig work—Uber, Lyft, InstaCart—are among the most prominent targets of *TMK*'s ire. After all, they're yet another iteration of entrepreneurs using technology to degrade and devalue labor.

"What makes on-demand possible? Mass immiseration. Suffering. All the things the Luddites were against," Ongweso said. "There is this straight line that can be drawn from Luddites to organizing around gig labor—and that is, both [groups of] people are being exploited, pushed on by capital, and sold as progress— if you get crunched up in the gears of progress, well, that's the price of it!"

They contend that the so-called techlash—the backlash against major tech companies that cropped up in the wake of Facebook's Cambridge Analytica scandal, growing concerns over Google and Amazon's monopoly power, and so on—demonstrates a deep-seated anger at the domination of Big Tech, but that it has already been co-opted by the industry. *Techlash* was shortlisted for *Oxford English Dictionary*'s 2018 word of the year. It was defined by the *OED* as "a strong and widespread negative reaction to the growing power and influence of large technology companies, particularly those based in Silicon Valley." It's fertile ground for Luddism, if inadequate in its current remit, the hosts think.

"It wasn't long ago since tech was covered breathlessly," Sadowski said, "but Silicon Valley has co-opted the techlash —"

"But the apocalyptic language remains," Ongweso chimed in. "The recommendations are ass, like, 'We should have a government watchdog.' If you think this is a threat to human life and democracy, then what is a watchdog going to do?"

"Luddism is more like the techlash that we need, and the techlash that people were hoping for and wanted," Sadowski said. "Not for a nicer Silicon Valley. No, Silicon Valley is rotten to the core. The problem is structural. Which necessitates a Luddite response."

That Luddite response is bubbling up in ways that are not always recognized. *TMK* cited the yellow-vest movement in France, where participants smashed surveillance cameras. And the Black Lives Matter movement, which erupted into a series of some of the most momentous days of civil unrest in recent American history, in the summer of 2020, saw both orderly, peaceful protest and the burning of police departments and smashing of department stores.

"That stuff is really important; we need stuff that legitimatizes vandalism as a form of political action," Sadowski said. "All these examples of bodies in the street like BLM is making that a legitimate response."

What gets derided as "looting" in drive-by news reports, Ongweso said, is often, though not always, a much more considered action, something that much more resembles the actions of the Luddites: "The thing you see time and time again is that it gets more complex at the neighborhood level," he said. "When the Target or some other store gets looted, there's a history or tension within the community that results in the rioting.... Places where people have been racially profiled and worked and were fired, places they had reason to dislike."

"These communities of color have been looted by the government, by capital, for decades and decades," Sadowski added. "It's interesting to see who is denigrated as a looter and who is lauded as an innovator and an entrepreneur."

The neo-Luddite trend has come and gone before, they know. "These things kind of bubble up; they're a flash in the pan or they're an academic dalliance," Sadowski said. "But one thing that we've been able to do through the podcast is keep it alive by actually building a community around Luddism." They hear from listeners who tell them they've learned about unionizing in tech, who've realized that opposing a certain technology does not make them anti-progress, who have developed a vocabulary to resist a culture hell-bent on accelerating profiteering technologies. "More than anything," Sadowski said, "we are giving people the sense that they are not alone."

Increasingly, the spirit of Luddism does seem to be alive in emergent attitudes toward new technologies, especially among the younger generations. Like George Mellor and his peers, many have grown up in a deeply technologized world that already feels foreclosed against them, one where climate change is eroding their future, public goods are converted into privately owned digital commodities, Silicon Valley has accelerated inequality, and legal avenues of recourse seem futile.

There is a new readiness to lash out at or reject new technologies and products that did not exist even a decade ago. There are debates about the merits of 2020s tech trends like the metaverse, Web3, and cryptocurrencies. But there's also a newly insurgent contingent that is unafraid to vocally reject an entire concept or infrastructure outright — to argue not that these technologies should be built carefully, but that they should not be built at all. Or should be dismantled, with a hammer, if necessary. This debate involves workers building technology themselves, too. When Google and Microsoft sought lucrative military contracts, tech workers launched #techwontbuildit, in stark opposition to the projects that in some cases helped tank them. Crypto and NFTs were besieged by critics concerned that they use too much energy, aimed to commoditize the last corners of the digital commons, and were clearing houses for fraudsters and scammers. (Sure enough, crypto crashed in 2022, due largely to widespread fraud.)

Andreas Malm's widely read 2021 book *How to Blow Up a Pipeline* advances the argument that after decades of attempts have failed to peacefully and democratically address climate change, the best viable option is physically destroying fossil-fuel technologies and infrastructure. "Here is what this movement of millions should do, for a start," Malm wrote. Step one: ban new fossil fuels. Then: "Announce and enforce the prohibition. Damage and destroy new CO_2-emitting devices. Put them out of commission, pick them apart, demolish them, burn them, blow them up. Let the capitalists who keep on investing in the fire know that their properties will be trashed." So, climate Luddism, essentially.

"My hope is that recognizing Luddism at work in the office, on the shop floor, at school, and in the streets [gives] anti-technology sentiment a historical depth, theoretical sophistication, and political relevance," Gavin Mueller wrote in *Breaking Things at Work.* "We may discover each other through our myriad of antagonistic practices in their incredible diversity."

So far, of the so-called Big Tech companies, Meta, the parent company of Facebook, has probably attracted the most widespread ire, at least in the media, for its role in allowing misinformation to flourish on its platform, for allowing hateful content to spread, and, at its worst, for allowing both ordinary users and powerful regimes to target vulnerable groups for unthinkable violence, as in the case of Myanmar and its Rohingya minority. Mark Zuckerberg consistently polls as the least popular big tech CEO for a reason.* But in the 2020s, there is plenty of ire to go around. And the targets for the deeper, more incendiary anger are big tech companies that are undermining standards of work.

Big tech companies that clearly and openly embrace technologies that are shifting and degrading not just working conditions but *how people work,* period.

Big tech companies whose leaders are openly hostile to their workers — crushing union drives, enacting oppressive, maddening workplace policies, and breaking labor laws, knowing they have the economic might to clean up their messes later.

Big tech companies whose leaders are the richest men on the planet, or are aiming to get there.

Big tech companies that have long operated under the general impression that they are unassailable, and that are finding out that, when faced with organized resistance — they are not.

---

* The likes of Facebook and Twitter are more likely to anger the upper-class commentariat than those facing the prospect of economic desperation at the hands of profit-generating technology. (Facebook has hurt and degraded some industries, like journalism, but it is not chiefly resented for killing jobs or for its poor working conditions. Its third-party content moderators, who complain of mistreatment and bear the psychological damage of wading through grotesque material daily for low wages, however, have a litany of deeply justified grievances.)

CHRISTIAN SMALLS

June 2022

Tired and elated, Chris Smalls stepped outside the JFK8 Amazon facility and uncorked the champagne. Dressed entirely in red, from the Yankees baseball cap to the sweatshirt emblazoned with a logo reading ALU — Amazon Labor Union — Smalls cut a striking figure as the press and his peers swarmed around him. Viral videos would soon spread of the sea of people, arm-in-arm, chanting "ALU! ALU!"

Smalls and his colleagues had just received word that the vote was 2,654 Yes to 2,131 No — they had successfully formed the first labor union at Amazon on American soil. The scene outside JFK8, the sprawling, state-of-the-art fulfillment center where Smalls had been fired two years before, was nothing short of jubilant.

"I wanted to make sure that this doesn't happen to anybody else," Smalls told CBS that day. "This is a revolution for a better quality of life. We're fighting for $30 an hour. We're fighting for longer breaks. We're fighting to be shareholders in the company. We're fighting for job security."

The media invariably framed it as a David versus Goliath story, and it was — without any major support system behind them, Chris Smalls and a number of colleagues had taken on Amazon and won. They had begun talking to the thousands of package sorters, pickers, and warehouse workers at JFK8 in the days following his termination. They hung out at the bus stop, hosted barbecues, and, above all, listened to their fellow workers. Pay was too low, they heard. Hours too long. Breaks too short.

"It's not about money, it's not about resources," Smalls would say later. "The campaign that we ran in Staten Island, when it came down to it, all it did was create a culture that showed the worker that we cared for one other. We are the ones that are going to listen to you. We're the ones that truly care about one another because Amazon obviously doesn't. And that culture is starting to resonate across the country."

It is indeed. What began as a burst of unionizing in digital media outlets like *Gawker* and the *Huffington Post* has spread across the nation, across industries. Even tech, a field long resistant to worker organizing, started to catch fire. Employees are moving to unionize at Google, Starbucks, Kickstarter, Apple, and, of course, Amazon; and some won tentative victories. There's a reason that this latest movement is unfurling at workplaces governed by tech giants and last generation's start-ups; just as in the 1800s, these are the sites where technology has been used by owners to cleave through norms and push humans to the working brink, and where the divides and inequalities — whose blood run the machines, and who keeps the profit the machinery generates — is starkest.

People tire quickly of being treated like automatons, it turns out. We're all Luddites that way.

At Amazon, even before the ALU's historic victory, a wave of union interest had greeted the company. There were global walkouts, amped-up worker organizing, and a well-publicized union drive at the Bessemer, Alabama, fulfillment center, which, after election interference from Amazon that was deemed illegal by the courts, ultimately fell just short. More localized movements that put pressure on Amazon are popping up, like IE Amazonians Unite, in California's Inland Empire, and acts of resistance are becoming more common and better facilitated. After the ALU succeeded in New York, the future seemed wide open. Inquiries poured in from Amazon workers around the country.

"Other workers are reaching out to us daily, asking how can we get involved, with either the ALU or to create their own culture in their facility," Smalls said. "We can share our knowledge, our expertise, we can share the things that's going to happen when they go public. And that's a great position to be at."

If change is to come, not just on the twenty-first-century shop floors of Amazon or at the gig app giants like Uber, but for all of us who are constantly made to navigate the ways that owners use technology to upend our lives, it will almost certainly rise from the grassroots, from old-fashioned practices like organizing, protest, and movement-building. That's how workers like Smalls and the gig work drivers have won their victories, and begun to mark out a path forward amid the AI, automation, and gig work algorithms. (It will not likely come from the policy prescriptions of technofuturists, pundits, and think tanks, who are happy to exclaim that the robots are coming for our jobs, but less committed to doing anything about it. By 2022, Andrew Yang had lost his campaign for the presidency, as well as for mayor of New York City, and had turned his attention away from automation theory and toward creating a new centrist political party whose platform, as of the summer of 2022, bore no mention of UBI, or much of anything else.)

At a packed talk in Los Angeles, deep into his summer victory lap, Chris Smalls encouraged workers everywhere to go all in, to "make the billionaires nervous" — to fight back against Bezos, against Elon Musk. Decked out in designer sunglasses and a gold grill — what he calls "union drip" — Smalls was still easing into his role as the working class's chief ambassador in the growing organized skirmish against Big Tech. He was charismatic but unpolished, and gripped the microphone nervously beneath his full-throated calls for workers to stand up and fight.

"I spent three hundred days at the bus stop," he said. It wasn't just about organizing, but building care, and tight-knit bonds of trust.

When a woman in the audience asked him about Amazon's automation efforts, Smalls was blunt.

"We gotta stop them at the gates. You know, a lot of automation is still years away. Amazon is trying — you know, don't support them, don't go into stores that have cashless cashiers and automated machines, because that's what's keeping them going. We have to go in there and raid that shit, really, but we'll talk about that offline." It was striking to hear Smalls advocate for, essentially, modern Luddism; but it makes sense. The way he nurtured community ties and built strong cells of opposition that couldn't be cut into by Amazon — that recalled the Luddites' approach, too. Fighting technological exploitation effectively, he says, will take a mass movement. "We have to build power in every industry to slow down that process, to stop them from even thinking about replacing us with machines. We deserve to have a job. We deserve to have a lifetime job."

But these are the largest and most powerful tech companies, run by the wealthiest and most influential tech titans, that the world has yet produced. There will be many fronts in the battles for the workers Smalls calls "warriors" to fight.

"You're going to be facing some real disgusting tactics, from being demonized, to being isolated, to being drilled with propaganda," Smalls said. "To prepare yourself mentally, you're going to have to create a foundation that you're going to have to lean on, and a culture that you're going to have to be a part of to battle that onslaught."

He paused.

"And it's happening."

It required the full power of a violent state — a tyrannical legal code, a massive occupying force, and dozens of public executions — to put the Luddites down, to safeguard the establishment of the factory system, to normalize top-down automation, and to entrench the technologies and the logic that governed them that came to dominate the Industrial Revolution and beyond.

Today, when on-demand apps are degrading working conditions, AI services are deployed to erode wages, and industrial automation is pushed further into assembly lines and warehouse spaces, workers might remember how close the Luddites came to repelling the technologies of their oppression. They might remember what worked — tight-knit solidarity, distributed organizing models, shows of power and creativity capable of inspiring influential producers of culture, unrepentantly aggressive actions against those oppressive technologies specifically; and what did not — violence against individuals, a lack of coordination with an empowered political body, the absence of a sustained effort to grow one. They might recognize that the Luddites failed in the face of totalizing authoritarianism, and that such conditions are not static. We can still oppose the creep of totalizing control into our lives. We can look at certain technologies, certain modes of domination, and say: No.

The Luddite way is again emerging as a viable option; if workers, students, and citizens of a world that is increasingly run by the precepts of Silicon Valley can gather enough will and power, they can choose to reject it. They can apply the power of the Luddite ethos to more sustained and organized bodies. They can reject tech monopolies and push to break them up. They can deny outright intrusive or punishing new technologies and say they do not want unaccountable algorithmic bosses or mass digital surveillance in their lives, or AI content generators in their workplaces. That is an option. They can decide that Amazon and Uber, and for that matter, Starbucks and Walmart, must offer their employees benefits, fair pay, and the right to unionize.

They may not need to smash the physical infrastructure of these oppressive technologies, yet. But with ample historical evidence of how exactly these forces have degraded lives in the past, they can be more assertive in choosing how they might want technology to impact theirs. We all can. We do not need to let entrepreneurs, venture capital firms, and their allies do it for us.

It will be an uphill battle, and it remains to be seen how fully governments have been captured by the entrepreneurial elites of the 2020s. It remains to be seen whether twenty-first-century workers will be forced into the next mode of the factory, through digital portals, onto the app platforms and into permanent precarity, or if the tide of this second or third, or seventh or eighth industrial revolution can be turned. The most powerful tech companies in the world will not relinquish power easily. And like the titans of the nineteenth century, they will be willing to destroy working peoples' lives in the process, if that is what it takes.

DOUGLAS SCHIFTER

2018

He had driven his passengers millions of miles, logging tens, maybe hundreds of thousands of hours on the job, and written over forty-five thousand words in his *Black Car News* column. He had traveled down more streets and highways than almost anyone on the planet, and compiled a book's worth of writing and ideas. Like his burdened Luddite leader forebears, Douglas Schifter was nothing if not productive.

But by the time he wrote that Facebook post in February 2018, it was clear to him that he could not keep up with the algorithms of his oppression and could not eke out a decent living competing against their owners. The post ultimately turned into a plea — just as Britain's government had refused to offer relief to the starving weavers, or regulate the cloth trade, or outlaw cheap and fraudulent works, so had New York City's municipal government, or any state or federal agencies, declined to step in to assist the cab drivers beaten down by Uber.

"We are not a government of the People, by the People and for the People anymore," Schifter wrote. "We are turning into a Government of the People, by the Corporation for the Rich. People are becoming enslaved and destroyed by politicians and companies with the aid of the rich and the corporations they control. They are doing it by bypassing the Laws or manipulating them. America is being stolen. Your future and your families future is being stolen right now."

He published the post to his Facebook page, gathered his things, and set out on the drive from upstate New York to downtown Manhattan. It was so early that it was still dark. Before he left, he texted a friend.

"Making it count," he wrote.

Schifter made it to City Hall by sunrise. He pulled up to the east gate, parked, and, before even unbuckling his seatbelt, shot himself in the head with a shotgun.

"I will not be a slave working for chump change. I would rather be dead," he had written. The same forces that overwhelmed the Luddites had overtaken Doug Schifter. In the following years, over a dozen drivers would meet the same fate.

But like the Luddites — like George Mellor in his final hours — Schifter had, even at the desperate end, voiced hope that ordinary people would triumph over the machines and their masters. The *Los Angeles Times* called Schifter's death a "rallying cry for New York's ailing taxi and car-service industry."

And that was exactly what he had intended it to be.

"I know I am doing all I can," he wrote in his final post; "the rest is up to you. Wake up and resist!"

AFTERWORD

So Are the Robots Coming for Our Jobs?

Ever since the Luddite uprising was put down in the 1810s, working people have been locked into a similar state of anxiety over how technology will impact our livelihoods. For two hundred years, we have rarely been free from concern that this machine or that program will make our work redundant, less skilled, or simply worse. We've worried, constantly, that the robots are coming for our jobs.

We have, in other words, never found a satisfactory answer to the machinery question. Or the automation question. Or the AI question. Throughout our postindustrial history, these questions have reared their heads time and again, and with good reason. After all, the factories seem to have disappeared the domestic handloom weavers, automated assembly lines winnowed out the manufacturing jobs, and Expedia killed the travel agent.

So how worried should we be, right now? *Is* a robot jobs apocalypse coming our way?

On the one hand, the answer is no — there is compelling evidence, produced by economic historians like Aaron Benanav and labor economists like Susan Houseman, that the predominant threat to stable employment is not mass automation but a perennial shortage of good jobs to be had, period. (Good jobs that are being degraded by labor-saving technologies and gig work, yes, but not wiped off the map.) On the other hand, there is an endless torrent of talk about artificial intelligence and shiny new robotics in warehouses and grocery stores and factories, so the answer there is — also no.

If the Luddites have taught us anything, it's that robots aren't taking our jobs. Our *bosses* are.

Robots are not sentient — they do not have the capacity to be *coming for* or *stealing* or *killing* or *threatening* to take away our jobs. Management does. Consulting firms and corporate leadership do. Gig company and tech executives do. And yet, here's a quick sampling of headlines from prominent media outlets:

SMART ROBOTS COULD SOON STEAL YOUR JOB — *CNN*

WHITE-COLLAR ROBOTS ARE COMING FOR JOBS — *Wall Street Journal*

YES, THE ROBOTS ARE COMING: JOBS IN THE ERA OF HUMANS AND MACHINES — *Wired*

ARE ROBOTS COMING FOR YOUR JOB? EVENTUALLY, YES. — *New York Times*

THE ROBOTS ARE COMING, AND THEY WANT YOUR JOB — *Vice*

Since the Luddites' day, when this question was phrased as the "machinery question," and not the "factory-owner question," it has historically helped mask the agency behind the *decision* to automate jobs. This decision is not made by robots, but by business owners, by entrepreneurs, by management. It is a decision most often made with the intention of saving money by reducing human labor costs (though it is also made in the interests of bolstering efficiency and improving operations and safety). It is a human decision that triggers the disruption.

If the machinery or the robots are simply "coming," if they just show up and relieve a helpless lot of humans of their livelihoods, then no one is to blame for this independently arising phenomenon, and little is to be done about it beyond bracing for impact. It's not the executives, swayed by consulting firms who insist the future is in AI text generation or customer service bots; or the tech titans, who use algorithmic platforms to displace traditional workers; or the managers, who see an opportunity to improve profit margins by adopting automated kiosks that edge out cashiers; or the shipping conglomerate bosses, who decide to try to replace dockworkers with a fleet of automated trucks.

These executives and managers may feel as if they have no choice, with shareholders and boards and bosses of their own to answer to, and a system that incentivizes the making of these decisions — but they are exactly that: decisions, made by people. Pretending otherwise, that robots are inevitable, is technological determinism and leads to a dearth in critical thinking about when and how automation *is* best implemented.

The Luddites knew exactly who owned the machinery they destroyed. They saw that automation is not a faceless phenomenon that we must submit to. And they were right: Automation is, quite often and quite simply, a matter of the executive classes locating new ways to enrich themselves, not unlike the factory bosses of the Luddite days.

Here's a telling example: In 2019, the *New York Times'* Kevin Roose filed a report from Davos detailing how the business leaders and tech CEOs at that year's World Economic Forum (WEF) were very eager to implement automation. "They'll never admit it in public," Roose wrote, "but many of your bosses want machines to replace you as soon as possible." In public, the elites preferred to discuss the abstract need to prepare for "the fourth industrial revolution" or "the second machine age." In private, they were more direct. "People are looking to achieve very big numbers," said Mohit Joshi, the president of the automation firm Infosys, who worked with those very people; the leaders of major companies and organizations. "Earlier they had incremental, 5 to 10 percent goals in reducing their work force. Now they're saying, 'Why can't we do it with 1 percent of the people we have?'"

Since then, these executives and elites have decided to buy and build more robots and software programs that would, *ideally,* put people out of work. Far from spontaneously swarming to the factory floor, the AI and robot armies that are "coming for your jobs" are much more likely to be deployed on behalf of the kind of elites who go to Davos.

Those elites are a major reason we continue to treat automation like a faceless phenomenon, too — the institutes they fund, consultancies they hire, and banks they work with frequently issue and promote reports that treat automation like a technological inevitability. In 2020, the WEF issued a "Future of Jobs" report that stated "Fourth Industrial Revolution technologies driven by AI will...fundamentally change the world." It instructed readers "not to fear AI"; though 85 million jobs were to be "displaced" by technology, eventually, it would create 97 million new ones. Economists and consultants frequently point to this tradeoff — yes, new technologies will destroy some jobs, but they will create new ones, too — as a means of alleviating public fears of mass layoffs. A 2017 McKinsey future of work report was titled "Jobs lost, jobs gained." But imagine telling a starving weaver in 1811, or a despairing cab driver in 2021, not to worry, in the future more jobs will exist in the place of the one you are losing. We shouldn't be surprised when they refuse to accept that it's their duty to lay down and die, to again paraphrase Frank Peel.

Other reports about the rise of the robots are more like self-fulfilling prophecies: In 2023, a Goldman Sachs study estimated that 300 million jobs worldwide were at risk of being taken over by generative AI systems — suggesting to those in a position to purchase such technology that it was high time to do so. Organizations like the International Monetary Fund, consultancies like Accenture and Deloitte, and professional services firms like PwC all issue future of work reports forecasting mass disruption and major economic growth. If they mention the vague dangers of automation, they are united in the shallowness of their proposed solutions, which hinge on emphasizing the importance of better education and calling on small amounts of government support for retraining programs. These dramatic-sounding reports find a synchronicity in the media's penchant for dramatizing science-fiction tropes apparently come to life, and voilà — the robots are coming. We will be able to make better decisions about automation, however, if we understand that, in practice, "the robots are coming for our jobs" usually means something more like "a CEO wants to cut his operating budget by 15 percent and was just pitched on enterprise software that promises to do all the work currently done by thirty employees in accounts payable."

This relentless and authoritative-sounding presentation of mass automation as our technological destiny is one reason that so many people remain susceptible to this fear. And it's a fear that employers can and do use as leverage against workers. While fast-food workers were organizing for $15 an hour, Andrew Puzder, President Donald Trump's nominee for Labor Secretary, wrote an op-ed for the *Wall Street Journal* titled "The Minumum Wage Should Be Called the Robot Employment Act." Ed Rensi, a former CEO of McDonald's, agreed. If fast-food workers won a living wage, it would lead to "job loss like you can't believe," he said, because the industry would accelerate automation.

This threat is continually levied against workers, even when automation isn't going to immediately erase a job outright — because usually it can't. More often, again, as in the Luddite days, it could degrade conditions and depress wages. But the data seem to show that automation itself isn't *eliminating* vast swaths of jobs altogether, and fearing a vaguely shaped robot jobs apocalypse plays directly into executive and tech titan hands. Always has.

"The automation discourse is a very powerful explanation of existing trends, at least on the surface level, and it is a theory with a history: it has been made time and again because it seems so obvious," the economic historian Aaron Benanav says. He points to pockets throughout the twentieth century—the 1930s, '50s, '60s, '80s, and '90s—when this theory, that robots were coming for everyone's jobs, was particularly popular, as it is again in the twenty-first century.

One of those periods follows the cultural phenomenon begotten by the play that coined the term *robot* itself. *Rossum's Universal Robots,* by the Czech writer Karel Čapek, premiered in 1921. With Mary Shelley's *Frankenstein* as a clear influence, Čapek's play tells the story of a reckless scientist who creates an army of "robots"—in Czech, *robota* means forced labor—to work for human enterprises. "Young Rossum is the modern scientist, untroubled by metaphysical ideas," Čapek once said; "scientific experiment is to him the road to industrial production." The play is a broadside against the dehumanizing effects of automation and technology. The robots, like the Luddites before them, rise up against their masters, articulating an anxiety that has echoed for the last hundred years as the specter of a robot jobs apocalypse, where humans will be forcefully cast out of their livelihoods.

"The automation theory suggests that inventors and scientists are hard at work innovating ways to make our lives better," Benanav says, "which is true, but then translates problematically at the border between science and the economy. The theory suggests as well that the fundamentals of the economy are strong: the only real problem is how to distribute the bounty." This theory, which also tends to argue that the best way to give ordinary people a small cut of the spoils is a universal basic income, has the effect, Benanav says, of

> validating the power and incredible wealth of the tech elite, justifying it
> by claiming that they need all this money because they are the ones who
> are going to make the investments that get us to full automation, save the
> planet from climate change, and get us to Mars, and so on. Anyone with a
> background reading popular science magazines or watching science fic-
> tion can relate. One finds oneself reciting the *X-Files* slogan: "I want to
> believe."

So what *is* happening if there's no robot or AI jobs apocalypse incoming? The same thing that has happened for two hundred years: The entrepreneurial elite is promoting a grand dream of a future where machines do all the work and deliver us economic prosperity. These visions of robots taking our jobs make us uneasy, but they also encourage us to stay fixated on the brighter and distant tomorrow, past the precarity, injustices, and the unpleasantries of now — while giving the elites who produce those very visions a smokescreen to obscure their decision-making and more leverage over their workforces. Bosses may not be able to automate away all of their workers, even if they'd like to, but they *can* deskill them, slash their benefits and protections, and reduce their wages. Some professions may be more vulnerable to the entrepreneurs and executives who move to disrupt them — the weavers, the factory line worker, the taxi driver, the travel agents — but those displaced workers are not victim of the *robots*.

Looking at the dynamics at play in the Luddite struggle might help us see past all this deterministic futurity — think back to those entrepreneurs watching their competitors consider automated machinery and making the decision, difficult as it was, whether or not to join the fray. The supply chains cross the globe now, but the same basic principle is at play; this is a social phenomenon at root. Letting an ambiguous conception of "robots" shoulder the blame lets the managerial class evade scrutiny for how it deploys automation, shuts down meaningful discussion about the actual contours of the phenomenon, and prevents us from challenging the march of this manifest robo-destiny. So, for the Luddites' sake, let's get it straight.

Robots are not threatening your job. Gig app executives who sense an opportunity to evade regulations and exploit tradition-bound industries are threatening your job. Business-to-business salesmen promising AI content and automation solutions to executives are threatening your job.

Robots are not killing jobs. Managers who see a cost benefit to replacing a human role with an algorithmic one, and who choose to make the switch, are killing jobs.

Robots are not coming for your job. The CEOs who see an opportunity to reap greater profits in machines that will make back their investment in 3.7 years, and who send the savings upstream — they're the ones coming for your job.

How Uprisings against Big Tech Begin

The reason that there are so many similarities between today and the time of the Luddites is that little has fundamentally changed about our attitudes toward entrepreneurs and innovation, how our economies are organized, or the means through which technologies are introduced into our lives and societies. A constant tension exists between employers with access to productive technologies, and the workers at their whims. But clearly such tension does not always lead to widespread violent insurrection.

So how do uprisings against Big Tech and the machine owners begin?

When entrepreneurs and executives deploy new technologies intended to replace skilled work, confound or elude regulations, or degrade traditional jobs en masse—especially in difficult economic circumstances. It's worse if those workers have no recourse.

When technology has eliminated or degraded a worker's job, status, or identity; when it is hard or impossible to organize to negotiate outcomes; and when support is inaccessible—well, we might expect just about anyone to feel cornered, angry, and more apt to turn to desperate measures. Obvious? Perhaps! But you wouldn't know it from how US policymakers have approached gig work, automation, and workplace surveillance in the twenty-first century. As in the Luddites' day, policymakers happen to be actively benefiting from the largesse of today's technological elite—the deep-pocketed Silicon Valley campaign donors and "job creators." Too many leaders have turned a blind eye to the outcome; they, too, it seems, would sooner send in the National Guard than intervene in a meaningful way to stanch the bleeding.

None of this necessarily means that American workers will suddenly start smashing the machines that are executing their exploitation, though it does mean they will sometimes channel that anger at the politicians, or direct it at "others" and outsiders: immigrants, minorities, political opponents. A Brookings Institution study found that in regions where populations experienced a higher rate of automation, voters turned out in higher numbers for Donald Trump in 2016. Trump's rhetoric was tailored to stoking resentments over all of the above into anger, as he promised to restore America to its former, preautomated industrial glories. "Automation perpetuates the red-blue divide," the Brookings survey concluded.

Opaque and circuitous supply chains, decades of offshoring, and other hallmarks of a globalized economy have consipired to cloud the precise causes of any given worker's sense of exploitation. George Mellor could see Horsfall's factory, with its automated gig mills, looming on the outskirts of Huddersfield; today's workers see the empty storefronts and idled auto plants, but less often the new factories where robots are supposedly performing their labor. It's harder to smash a machine if it's nowhere near you, or if it's invisible — if, say, it's ubiquitous lines of code. This confusion is one explanation for why we've seen less aggressive worker action against major employers in general over the last decades; the targets of opprobrium are cloaked in the complexities of global capitalism, so the anger has been diffuse. But the modes of the modern factory owners' exploitation are becoming clearer to workers again, and the possibility for more Luddite-like targets remains very real.

As San Francisco was gentrified by tech companies, displaced renters and residents facing eviction blocked the route of the notorious Google Buses, forming ad hoc protests and hurling rocks at the vehicles. We've seen how drivers in France reacted when Uber entered its market — with riots and the proverbial hammer of Enoch. And the new Luddites point to the targeted vandalism at the George Floyd protests in the summer of 2020 as examples of strategic retribution for economic grievances.

Meanwhile, governments have exacerbated matters by implementing punitive policies that restrict workers' rights. In the Luddites' day, Parliament instituted a bloodier and harsher set of laws, to be sure, but a similar logic, as far as who gets protected and who gets punished, drives the American economy today. Tech companies and their allies in government are striving to prevent contract, gig, and part-time workers at Amazon, Uber, and beyond from organizing. Executives are cutting off their workers' only chance at recourse when it comes to how technology is deployed in their workplaces. The workers are deprived of options and protections, even at moments when they might appear to have some leverage. "The American economy runs on poverty, or at least the constant threat of it," the *New York Times* columnist Ezra Klein wrote in 2021. "The barest glimmer of worker power is treated as a policy emergency, and the whip of poverty, not the lure of higher wages, is the appropriate response."

When the social norms and customs foundational to working people are systematically undermined by those wielding technologies new or old.

Just as the Luddites' actions were a response to the ways that entrepreneurs were shaping the advent of the factory system — and as people have periodically protested automation and the deepening of that system in the centuries since — workers today are feeling the strain at on-demand app companies like Uber, Lyft, and DoorDash, not to mention those relegated to doing what Veena Dubal has called "digital piecework."

One group whose situation closely resembles that of the Luddites, is the large and ever-growing cohort of gig workers whose backs are against the wall. Many of these are skilled workers who were once better compensated, like Douglas Schifter; many others have lost benefits and protections. Their lives have been disrupted, and now their work regimes are dictated by often inscrutable algorithms that make them compete against one another for every gig. This new generation of algorithmically arranged, precarious work structures is permeating even more of the economy — it's already happening to lawyers, writers, emergency responders, even security forces. And in many cases, the rise of generative AI services stands to accelerate the process. Screenwriters went on strike in 2023 in part to prevent studios from using AI to generate scripts and eroding pay rates.

Meanwhile, right-to-work laws make organizing much harder, and measures like Prop 22 prevent gig workers from finding stable employment, receiving benefits, or forming a union. Uber and Lyft can change their rates at their whim. The 9-to-5 American dream, with benefits and a retirement pension, is being disassembled, destroyed, deconstructed — and romanticized today in a way that preindustrial weaver and cropper life was in the early 1800s. The Luddites fought to protect a lifestyle that they recognized, that was tradition-bound, that let them spend time with their families and maintain their autonomy. More workers today are finding themselves motivated to fight to protect those things, too.

When the perception that the entrepreneurial elite have been overcome by greed and self-interest becomes conventional wisdom. When workers see that tech titans are openly willing to profit even while others suffer.

Anger at money-grubbing robber barons is one of the most time-tested forms of class rage; when it becomes clear that those barons are using a new technology to further exploit their workers, that rage stands to be amplified — and to find a target in said technology. The Luddites wrote their letters to entrepreneurs and factory owners, to the local magistrates who sheltered them, and occasionally to the Prince Regent and his ministers. It was obvious who was acting immorally (and illegally) and who was responsible for perpetrating the new mechanized industrial regime that was putting them out of jobs and reshaping, against their will, the norms and customs that made their lives worth living.

Elon Musk was widely admired, even in liberal circles, until he began vocally opposing a unionization effort in his Tesla factory. He bought Twitter, fired half its staff—thousands of tech workers—and became the most divisive figure in tech. Amazon warehouse workers continue to organize around the long shadow of Jeff Bezos's brutal and machinistic work policies. Opinion polling from Gallup has shown that Mark Zuckerberg is disliked more than the company he runs. A growing number of billionaire founders have become engulfed in scandal, accused of fraud and malfeasance: FTX's Sam Bankman-Fried. Theranos's Elizabeth Holmes. The list goes on. There is an animosity toward specific Silicon Valley elites that has only intensified in tandem with the expansion of their power.

When managers use technology to embark on the widespread destruction of status and the pathways to upward mobility.

Amazon Flex drivers, Uber workers, and InstaCart deliverers all have one thing in common: there is no obvious path to receiving a promotion or raising one's status in the organization. A cab driver could save up to purchase their own medallion. A warehouse worker could rise to foreman. A delivery driver for a restaurant could become a waiter or a manager. There was the possibility that good work would be rewarded with rising pay and status. With the gig app regime, AI services taking over entry-level work, and much of Amazon's warehouse apparatus, many of those avenues are closed off. Your rating may improve, but your rate of pay won't. It's worth noting that Chris Smalls applied for a promotion dozens of times over the five years he worked at Amazon and was rejected every time, despite a productive work record.

The imposition of the factory system disrupted the paths that allowed weavers and croppers to rise from journeymen to masters, and that disruption was a major reason that their distress became so acute. Today, the alienating and isolating mode of gig work threatens on one hand to batter the livelihood of skilled workers like Douglas Schifter, and, on the other hand, to lock workers into roles designated by proprietary algorithms, with no upward mobility on the horizon. A sense of hope, that one can better their lot through hard work and a mastery of their field, is a crucial and animating force in any healthy economy—and gig-based systems and AI content generators work to snuff it out.

When technological development is top-down and anti-democratic — and workers get no say in how automation or algorithms impact their daily lives.

The biggest reason that the last two hundred years have seen a series of conflicts between the employers who deploy technology and workers forced to navigate that technology is that we are still subject to what is, ultimately, a profoundly undemocratic means of developing, introducing, and integrating technology into society. Individual entrepreneurs and large corporations and next-wave Frankensteins are allowed, even encouraged, to dictate the terms of that deployment, with the profit motive as their guide. Venture capital may be the radical apotheosis of this mode of technological development, capable as it is of funneling enormous sums of money into tech companies that can decide how they would like to build and unleash the products and services that shape society.

Take the rise of generative AI. Ambitious start-ups like Midjourney, and well-positioned Silicon Valley companies like OpenAI, are already offering on-demand AI image and prose generation. DALL-E spurred a backlash when it was unveiled in 2022, especially among artists and illustrators, who worry that such generators will take away work and degrade wages. If history is any guide, they're almost certainly right. DALL-E certainly isn't as high in quality as a skilled human artist, and likely won't be for some time, if ever — but as with the skilled cloth workers of the 1800s, that ultimately doesn't matter. DALL-E is cheaper and can pump out knockoff images in a heartbeat; companies will deem them good enough, and will turn to the program to save costs. Artists who rely on editorial and corporate commissions will see rates decline, all because the companies unleashed a disruptive technology without soliciting input from existing workers.

If ordinary humans and working people are not involved in determining how these technologies reshape our lives, and especially if those outcomes wind up degrading their livelihoods, time and again the anger will be acute and far-reaching. And if workers cannot even legally organize with one another to cushion the blow, there is liable to be nowhere to turn at all, no option but to dismantle that technology.

The Luddite rebellion came at a time when the working class was beset by a confluence of crises that today seem all too familiar: economic depression and stagnant trade, rising inflation and high prices, excessive taxes for an unpopular war, and a government that strands unions, rules out serious relief for the poor, and declines to uphold industry regulations. And amid it all, entrepreneurs and industrialists pushing for new, dubiously legal, highly automated and labor-saving modes of production.

"I honestly feel like a master sock weaver at the start of the Industrial Revolution," wrote Will Butler, a musician whose earnings have been eroded by Spotify and other streaming platforms using technology that superseded CDs and download purchases as the dominant way that people listen to music. Spotify pays artists an average of $0.004 per stream. "People will still get their socks, maybe worse than the ones before. And in the end, technology will plow us over."

"It's not only rideshare drivers whose families are starving," said Marie Harrison, a seventy-two year-old Uber driver at Airport Park, outside LAX, where she had gathered to strike in protest of poor working conditions in the summer of 2021. "What's happening is a disgrace. They've put us back, these large corporations, to the 1800s, when people were begging — begging to be *paid* for the hard work they do."

With our jobs at risk, *we* are at risk. Our health, our security, our ability to plan for future generations — right now, it all depends on access to decent work with decent benefits. Our capacity to lead a fulfilling and stable life is subject to the whims of the entrepreneurial elite.

The twenty-first century has been filled with a lot of talk about how we have already entered a new machine age, a fresh system-wide updating of the conditions that drove the Industrial Revolution. Whether our age is consumed by automation or not, whether the degradation of work at the hands of algorithmic regimes of work is total or not, it is an age marked by the erosion of traditions, identities, and good jobs, and the simultaneous increase of domination by the entrepreneurs, factory owners, and their allies. It is an age dominated utterly by Big Tech.

The time is again ripe for targeting the "machinery hurtful to commonality" — the gig app platforms, the fulfillment-center surveillance, the delivery robots, the AI services, the list goes on — duly singled out by those whose livelihoods are being migrated, against their will, by this generation's tech titans, onto unforgiving technologized platforms, or degraded and whittled away by automated systems.

After all, for centuries ordinary people have had their autonomy pried away by new and elaborately exploitative technologies. More of our time, our work, our blood has been needed to run the machinery of profit, with less given to us in return. Nineteenth-century artisans may not have had much, but they ran their own shops, plying their own trades, alongside their families, on their own terms. They had real freedom, the means to negotiate their wages, schedules, and their future. That, ultimately, is what the Luddites fought for.

Those Luddites rose up when their disparate grievances reached a fever pitch, uniting them in a struggle against the agents of their technologized exploitation. It may be only a matter of time before the rebel workers of this new machine age see the injustices of the algorithmic platforms as too much to bear, the surveillance apparatus of Big Tech too intrusive, the robotic pace of work too ruthlessly body-breaking.

And if they feel the rage of Frankenstein's monster, rebooted in a new era of boundless entrepreneurial adventuring, and they catch sight of those autonomous vehicles assembling like ghosts on the horizon, they might just reach, once again, for their hammers.

ACKNOWLEDGMENTS

The first round of thanks must go out to the machine breakers: the protestors, the organizers, the artists and artisans, the poets and prognosticators, the workers and critics who, Luddites all, pushed back — even when called idiots or deluded or worse — rather than be trampled by someone else's machinery. We owe them all a great deal.

Closer to home, this book would not have been possible without my family — and above all, without the nearly infinite marathon support of my wife and partner, Corrina Laughlin, who I'm sure at this point wishes as much as I do that the Luddites had succeeded and had broken the machines of injustice once and for all, if only so that she would not have to hear about them ever again. I'll be forever grateful for her sharp and thoughtful feedback, keen suggestions, and all the love and help during the vast stretch of time it took to complete this project. And thanks to my boys, Aldus and Russell, for the endless inspiration: true machine breakers — and makers! — in training. You three are the very best.

Massive thanks are also owed to my great and indefatigable agent, Eric Lupfer, for throwing his lot in with this Luddite and to Michael Szczerban, my stalwart editor, who saw what this book could be and took a chance on it — and who did not stop answering my calls when I told him I planned to follow up a bestseller about the revolutionary power of the iPhone with a story about malcontented cloth workers in the 1800s.

To the historian and period expert Alan Brooke, and the wonderful Pam Brooke, I give not just my thanks but a full-blown Luddite salute; thanks for the hospitality, kindness, and generosity. Alan, thanks to you and your endless well of machine-breaking knowledge, I feel that I got as close a front-row seat to the Luddite rebellion as humanly possible. Thanks also to Adrian Randall for reading, discussing, for the exhaustive work on the subject, and for all the delightfully polemical email threads.

Thanks to everyone who read early drafts of the book and offered thoughtful

ACKNOWLEDGMENTS

ideas, notes, and encouragements: Mike Pearl, Paris Marx, Wendy Liu, Tim Maughan, Edward Ongweso Jr, Jathan Sadowski, Claire L. Evans, and Cory Doctorow. Thanks to Naomi Klein, Farhad Manjoo, Christopher Leonard, Malcolm Harris, Kim Kelly, and Margaret O'Mara for taking the time to dive into this Luddite lore, and for the generous words about the book. An extra special thanks to Elvia Wilk, for the close read and the next-level ideas, suggestions, and edits; thank-you, thank-you.

A big shout out to the incredible team at Little, Brown, who made sure that the story of the Luddites and the resurgence of rebelliousness against big tech would get all the glory it deserved; to my production editor, Linda Arends, for a thorough, bang-up job with designing and producing the book; Thea Diklich-Newell, for keeping the train running through so many twists and turns; Lauren Ortiz, for acely handling publicity for those who have received only bad press in the past; and Danielle Finnegan for helping to market the machine-breakers. Thanks to Lucy Kim for the incredible cover art; and to my steely-eyed copyeditor, Albert LaFarge, for sweating the small stuff and for giving such a long book a close, considered read. Thanks as well to Xian Lee, digital production co-ordinator; Danielle Finnegan, marketer; Deboleena Maitra, audio production supervisor; Stacy Schuck, manufacturing co-ordinator; Allison Merchant, art studio manager; Emma Tamayo, intern; Kameel Mir, publishing associate; Dan Lynch, senior production manager, who handled the galley-making process; and proofreaders Ian Gibbs and Leslie Keros.

As with the Luddites themselves, I also want to thank everyone who has toiled to preserve their elusive history. So, thanks to all the historians, scholars, workers, organizers, and archivists who both blazed the trail for this book and assisted in its research. It would not be here at all were it not for the likes of Frank Peel, FO Darvall, Barbara and JL Hammond, EP Thompson, Eric Hobsbawm, Katrina Navickas, Lesley Kipling, and many others.

Thanks to Richard Holland for creating and curating the *Luddite Bicentenary*—an invaluable resource for any modern machine breaker, and for fielding my arcane queries on Luddite errata. Three cheers! Thanks also to the kind and patient staff at the West Yorkshire Archive Service at Kirklees in Huddersfield, and the WYAS at Leeds, as well as the staff at the Prints & Drawings archive at the British Museum, and the employees of the Framework Knitters Museum in Nottingham.

ACKNOWLEDGMENTS

And I'd be remiss if I didn't include another thanks to a pair of early readers — Sharon and Thomas Merchant. Thanks, Mom and Dad! I hope you don't still think I was too hard on Jeff Bezos. But really, thank you so much for everything. I truly did appreciate your thoughts, encouragement, and the childcare and support that helped get this across the finish line. Thanks also to my in-laws, Teresa and Tim Laughlin, for all the help on that front, too, as well as for the productive, impassioned, and occasionally boozy talk of politics, economics, and inequality over the years. A thank-you to my grandparents, Joan and Al Merchant, for a home base in Silicon Valley, and for everything else; and to Russell and Nancy Walker for the support and inspiration.

Thanks to Jeff Bercovici, my fearless editor at the *Los Angeles Times*, for championing such a critical take on tech in a major newspaper, refining and expanding my thinking on the subject, and for the always enlightening discussions about Silicon Valley power, AI, and our favorite tech titan, Elon Musk. Thanks also to the LA Times Guild and the Media Guild of the West, which are fighting the good fight for journalists and journalism everywhere.

Thanks to Kelly Bourdet and Andrew Couts for spinning up Automaton at Gizmodo with me, and for giving me the space to investigate AI, automation, and the future of work from a weird, aggressive, and critical perspective. I'd blog with you both anytime.

Thanks to my crew of Medium & OneZero exiles who helped enrich my knowledge on the subject of tech and labor during the pandemic as I cobbled this book together: Damon Beres, Sarah Kessler, Will Oremus, Sarah Emerson, Megan Morrone, Dave Gershgorn, Yasmin Tayag, Drew Costley, Emily Mullin, and Peter Slattery. And to the whole of the late Medium Workers Union — Lyra, Aster, Aleida, An, Danial, Jean, Bob, Erik, Tirhakah, Kelli, and everyone else who almost turned a blogging platform into a union shop; there's always a next time.

I'm also grateful to Kevin Binfield for talking Luddites during the pandemic, and to Veena Dubal for being a tireless source — and an unparalleled scholar — on work in the current age of technological exploitation. A big thank-you to Adrienne Mayor, Kanta Dihal, and Antone Martinho-Truswell for illuminating the historical context of artificial intelligence; Aaron Benenav for the informed automation-theory skepticism; Arguella Glaxo-Kline, Nicole Moore, Christian Smalls, and Marie Harrison for the modern labor perspective on technological control; Andrew Yang and Bill de Blasio for the administrative perspectives; and

ACKNOWLEDGMENTS

to every gig worker, rideshare driver, and everyone else who took the time to share with me their thoughts and stories for this book—there are too many of you all to properly thank here!

Thanks also to the friends and peers who've pitched in one way or another over the years with ideas or encouragement or less direct forms of support: Keith Wagstaff, Brian and Tatiana Parisi, Jason Koebler, Lorenzo Franceschi-Bicchierai, Craig Mikasa, Carole Kishi, Alex Pasternack, Geoff Manaugh, Ryan Hall, Nick Rutherford, Alex and Christina Huie, Trevor and Sandy Perry, Stephen Messenger, Matt Letunic, Scott Keene, and Kevin Murray, to name a few.

Finally, a thanks to all the next-wave Luddites inspiring folks to reach for the proverbial Enoch's hammer around the world—Gavin Mueller, Ben Tarnoff, Ruha Benjamin, Zachary Loeb, Meredith Whittaker, Jürgen Geuter, Molly Crabapple, and so many of the names I've already mentioned above. Things may be rough, but new spaces of possibility are opening up and we are, in some ways, in a better position than ever to take aim at the machinery hurtful to commonality. Here's to all those new Luddites who know that a soul is worth more than work or gold—and are heeding the call.

NOTES

Introduction

1. **Imagine millions of ordinary people** Scenes like the one that opens the book, as well as numerous newspaper accounts, are detailed in nearly every major Luddite history, including Alan Brooke and Leslie Kipling, *Liberty or Death: Radicals, Republicans and Luddites 1728–1828* (Huddersfield, UK: Huddersfield Local History Society, 2012); Frank Peel, *The Risings of the Luddites, Chartists, and Plugdrawers* (Heckmondwike, UK, 1880); and Kirkpatrick Sale, *Rebels against the Future: The Luddites and Their War on the Industrial Revolution; Lessons for the Computer Age* (Reading, MA: Addison-Wesley, 1995). The scene describes the way the Luddites typically assembled before embarking on a raid against an entrepreneur who ran automating machinery.

3. **The public cheered** F[rank]. O[gley]. Darvall, *Popular Disturbances and Public Order in Regency England: Being an Account of the Luddite and Other Disorders in England During the Years 1811–1817 and of the Attitude and Activity of the Authorities* (London: Oxford University Press, 1934; reprint, 1969), introduction.

3–4. **"perhaps the purest"** Angus Macintyre, Introduction to Darvall, *Popular Disturbances.* xvi.

4. **The present, after all, is rife** Carl Benedikt Frey and Michael A. Osborne, "The Future of Employment: How Susceptible Are Jobs to Computerisation?" Oxford Martin School, September 17, 2013.

5. **while its workers peed in bottles** Lauren Kaori Gurley, "Amazon Denies Workers Pee in Bottles. Here Are the Pee Bottles," *Vice,* March 25, 2021.

Prelude

7. **It was a simple conviction** The conversation about Richard Arkwright's factory and the potential of automation is described in Mary Strickland, ed., *A Memoir of the Life, Writings, and Mechanical Inventions of Edmund Cartwright* (London, 1843), 55.

8. **The term** *automation* Automation as defined by the Interational Society of Automation and IBM.

8. **"If this new mode of spinning"** Dialogue from Strickland, *Memoir,* 56.

9. **For centuries** The data about the history of wool in England here is drawn from W. B. Crump and Gertrude Ghorbal, *History of the Huddersfield Woollen Industry* (1935).

9. **By the end of the eighteenth century** From chapter 8 of E. P. Thompson, *The Making of the English Working Class* (London: Gallancz, 1965): "Estimates from the Select Committee of 1834–5 put the number at between eight hundred thousand and eight hundred and forty thousand weavers and their families [who] were wholly dependent on the loom for work, excepting framework knitters. In 1841, the population of all of England and Wales was 15.9 million."

9. **In 1589, William Lee** Regarding the origin story of William Lee of Calverton's stocking frame, the truth was probably less memorable; he may have been a poor knitter himself, looking to augment his income.

10. **foreboding rebuttal** The queen wasn't wrong about her prediction that the stocking frame would prove disruptive; hand knitters were largely displaced by the machinists who took up the frame. Though unlike future innovations, the frame was often owned by an artisan and could be operated in the home, not in a factory.

11. **"Indeed, the workmen"** Strickland, *Memoir*, 56.

12. **her first child, a daughter** Biographical details, including the birth of Mary Godwin (whom history remembers as Mary Wollstonecraft Shelley), are drawn from Miranda Seymour, *Mary Shelley* (New York: Grove Press; London: John Murray, 2000).

13. **"our members be unlimited"** The history of the London Corresponding Society is detailed in E. P. Thompson, *The Making of the English Working Class* (London: Gallancz, 1965; reprint, New York: Vintage, 1966).

14. **he'd set out for Nottingham** Account of Blincoe's journey appears in *John Brown, A Memoir of Robert Blincoe, an Orphan Boy*, 17.

15. **far-off cotton mill** Description of Lowdham mill drawn from *Nottingham, City of Literature*, a UNESCO project.

15. **"intoxicated with joy"** Brown, *Memoirs*.

16. **The very concept of the factory** Adrian Randall, *Before the Luddites: Custom, Community, and Machinery in the English Woollen Industry, 1776–1809* (Cambridge: Cambridge University Press, 1991).

16. **The first meal at Lowdham** Randall, *Before the Luddites*.

17. **"Ned being rather averse"** *Nottingham Review*, December 20, 1811. Believed to be one of the first recorded references to the origin story of Ned Ludd.

17. **A gig factory in Littleton** Incident recorded in Randall, *Before the Luddites*.

18. **a massive sum** Conversions per the *Measuring Worth* UK currency calculator.

18. **He'd built the Staverton Superfine Woollen Manufactory** Description via P. A. Nemnich, *Neuste Reise Durch England* (1807), recounted in Adrian Randall, *Riotous Assemblies: Popular Protest in Hanoverian England* (Oxford: Oxford University Press, 2006), 255–6.

18–19. **"collective bargaining by riot"** Eric Hobsbawm, "The Machine Breakers," *Past and Present*, no. 1 (February 1952), 57–70; in Eric Hobsbawm, *Labouring Men: Studies in the History of Labour* (London: Weidenfeld and Nicolson, 1964; Charlottesville: University of Virginia Press, 1965), 5–22.

19. **"the present tense"** David Noble, *Progress without People: New Technology, Unemployment, and the Message of Resistance* (Toronto: Between the Lines, 1995).

19. **In 1710** Machine-breaking incident recounted in Gravener Henson, *The Civil, Political, and Mechanical History of the Framework-Knitters in Europe and America* (Nottingham, UK, 1831), 95.

20. **"thousands of women dressed in white"** From "Honouring Trowbridge Martyr Thomas Helliker at Wreath-Laying," *Wiltshire Times*, March 2, 2020.

Part I: The Great Comet

23. **"enormous and brilliant"** Great Comet description by Leo Tolstoy, *War and Peace*, Louise and Aylmer Maude (Oxford: Oxford World's Classics, 2010), 644.

23. **The Crown was levying heavy taxes** Malcolm I. Thomis, *The Luddites: Machine-Breaking in Regency England* (Newton Abbot, UK: David & Charles, 1970), 45.

23. **Born in 1789** David Mellor, Keyworth and District Local History Society. Details of George's temperament and character can be found in Peel, *Risings;* D. F. E. Sykes and George Henry Walker, *Ben o' Bill's: A Luddite Tale* (London: Simpkin, Marshall, Hamilton, Kent, 1898); and the York Assizes records.

24. **"resolute, determined appearance"** Sykes and Walker, *Ben o' Bill's.*

24. **A good cropper** Randall, *Before the Luddites.*

24. **"hoof"** Described commonly, including by Robert Reid in *Land of Lost Content* (London: Heinemann, 1986), 51.

25. **"metropolis of discontent"** Brooke and Kipling, *Liberty or Death,* 27.

25. **"The sides of the hills"** Daniel Defoe, *A Tour thro' the Whole Island of Great Britain* (London: 1724), vol. 3, 153.

25. **The business that kept Huddersfield** Crump and Ghorbal, *History of the Huddersfield Woollen Industry,* 8. "The woollen industry of the Huddersfield district is a part of the greater textile industry that has existed since the Middle Ages on the flanks of the souther Pennines both in Lancashire and the West Riding of Yorkshire."

26. **In *Ben o' Bill's*** Sykes and Walker, *Ben o' Bill's.* As discussed previously, we must take with a grain of salt the dialogue and minor plot points of *Ben o' Bill's,* regardless of whether it was said to have been gleaned "in part from the lips and in part from the papers of the narrator," who is Ben, George Mellor's cousin. I'm including it as a source for a number of reasons — on most major points, it squares with other straightforwardly historical accounts of Mellor, such as in Frank Peel's *Risings of the Luddites* and official court records, because it offers insights into the speaking style George — and any cropper or rioter in his position — would espouse at the time, because it conveys key parts of the Luddite legend, and because it so accurately and powerfully captures the sentiments of the economically besieged artisans.

28. **"What contributed to"** William Gardiner, *Music and Friends,* 55. Note that the picture Gardiner paints of the stockinger's life before the dawn of automation is a rather rosy one; it's a classic case of a worker remembering the good old days, perhaps overfondly. But even if it's idealized through the lens of nostalgia, it's still worth considering that, as E. P. Thompson points out, that's how other stockingers would have thought about the pre-mechanized past, too. Whether it's wholly representative, that's the life stockingers felt was being taken from them.

29. **It had become common** Darvall, *Popular Disturbances,* 53. "Some frames paid as high a return as 30 percent on their real value. Outsiders, with no previous connexion with or knowledge of the industry, were attracted to it and became large investors in machinery, which they let out at exorbitant rates to the unfortunate workmen."

29. **"reduce the price"** *Nottingham Journal,* January 23, 1811.

30. **One of his contemporaries** Gravener Henson's physical description can be found in William Felkin, *A History of the Machine-Wrought Hosiery and Lace Manufactures* (1867).

30. **"They compelled me"** Henson explained this incident in a deposition in 1824, recorded in the *First Report from Select Committee on Artizans and Machinery of the House of Commons,* collected in the Parliamentary Papers.

The Machine Breakers

32. **swelling to a thousand strong** Details of the protest swelling into a riot are found in the March 15 edition of the *Morning Chronicle,* the March 16 edition of the *Nottingham Journal,* and Malcolm Thomis, *The Luddites.*

32. **"the poor lived on each other's backs"** Records of the borough of Nottingham, extracted from the archives of the Corporation of Nottingham. This population growth continued

apace: "By the year 1811, the population had increased to 34,358, by 1821 to 40,505, and the overcrowding which resulted produced alarming consequences to living conditions and to public health."

32. **"For twelve months past"** William Felkin, *A History of the Machine-Wrought Hosiery and Lace Manufactures* (Cambridge, UK, 1867).

33. **The economic situation was so dire** *Derby Mercury*, March 28, 1811. These deaths have much in common with the deaths of despair being registered in modern times, from economic malaise, outsourcing, and modern automation. See Anne Case and Angus Deaton, *Deaths of Despair and the Future of Capitalism* (Princeton, NJ: Princeton University Press, 2020).

33. **"If workmen disliked certain machines"** Thomis, *The Luddites,* 50.

34. **Tempers flared** Details of the raids contained in the *Nottingham Journal*.

35. **"the illegal measures"** Randall, *Riotous Assemblies,* 271.

35. **"recommend to the Trade"** *Nottingham Journal,* March 30, 1811.

The Entrepreneurs

36. **"Good God, Francis"** Randall, *Before the Luddites.*

36. **"For several years"** Ellen Gibson Wilson, "Mass Politics in England before the Age of Reform," in *The Great Yorkshire Election of 1807* (Lancaster, UK: Carnegie Publishing, 2015), 1.

36. **"the land of opportunity"** François Crouzet, *Formation of Capital in the Industrial Revolution* (London: Methuen, 1972).

37. **"attained considerable perfection"** *Leeds Mercury,* April 15, 1812.

38. *Ben o' Bill's* **recounts** Notably, pastor Webster also invokes the parable of Aaron's rod:

> And see to it that it be not now with thee as in the days of Pharaoh of old, when Aaron's rod swallowed up the rods of the wise men and the sorcerers, and thy rod too be swallowed up. If that came to pass of which I have read and heard, there will be no room in this valley for men of but moderate means. Yo' may build a mill, but bigger men will build bigger mills, and the bigger mills will swallow up the less, and thou and thy son, and even Mary yonder may be fain, thou in thy old age and they in their prime, to take wage at another's hand, and to do a hireling's task in another's mill." Ben Bamforth also ultimately recalls his father's decision:

> "Whether the mind of our household's head were the more moved by the picture of his friends and neighbours reduced to want, or by the picture of himself and his working for others, who had always puts out work ourselves, I know not; but from that day forth there was no more thought for many a long day of any change in the ways we had used of old, and, for the new machines, my mother died in the belief that the curse of Scripture was upon them.

40. **"Managers feel they must automate"** Noble, *Progress without People,* 95.

41. **The so-called Mad King** Though the cause of his illness remains unknown, one of the longest held theories is that George III suffered from the hereditary blood disease porphyria, which can cause mania.

41. **On the evening of June 19, 1811** Carolly Erickson, *Our Tempestuous Day: A History of Regency England* (New York: William Morrow, 1986), 48.

41. **"The remarks of the people"** George Jackson, *The Bath Archives: A Further Selection from the Diaries and Letters of Sir George Jackson, from 1809 to 1816* (2 vols., London, 1873), 269.

42. **"curtained Romanesque bays"** Erickson, 49.

42. **"There was something puckish"** Erickson, 38.

42. **"The enduring image of George IV"** Saul David, *Prince of Pleasure: The Prince of Wales and the Making of the Regency* (New York: Grove, 1999), 2.

43. **"He sits at a table with General Turner"** Joseph Farington, *The Farington Diaries,* vol. 10, December 1, 1809, to February 28, 1813.

43. **"a time of contrasts"** Both Carolly Erickson and Saul David use the term to describe Regency England.

44. **"It is said that this entertainment"** Frederick L. Jones, ed., *The Letters of Percy Bysshe Shelley* (2 vols., Oxford: Clarendon Press, 1964), vol. 1, 99–100. Further context for the party: Saul David, *Prince of Pleasure,* 321.

44. **"It does not seem likely"** *Memoirs of the Life of Sir Samuel Romilly,* vol. 2.

44. **The historian Frank Peel** Peel, *Risings,* 25.

45. **"There's hundreds of men"** The dialogue in this scene is drawn from the novel *Ben o' Bill's,* as noted in the primary text.

45. **Between the 1790s and the 1820s** Randall, *Before the Luddites,* 60–61. The full context here is illuminating:

> The consequences of early mechanization were relatively less severe for men and correspondingly much more severe for women and children. Even accepting the figures as they stand, it remains clear that it was the first wave of machinery which had the most sweeping consequences for the woolen workers as a whole and for women workers in particular. With the introduction of preparatory machinery the number of women required per cloth slumped to at most only 18 percent of its former total as against a fall to 39 percent at worst for men. The main reason for this massive displacement was the demise of hand spinning and its replacement by the slubbing billy and Jenny. When the latter were introduced, the work of 153 women and their 241 child assistants could be performed by just over seven men, six women, and fourteen children. Here was labour redundancy on a massive scale. Austin's information suggests that the total labour required to produce a broadcloth fell by some 75 percent as a result of the introduction machinery in the years between 1796 and 1828.

46. **He had grown up in the era of "enclosure"** Barbara Hammond and J. L. Hammond, *The Skilled Labourer: 1760–1832* (London: Longmans, Green, 1919). "In agriculture the years between 1760 and 1820 are the years of wholesale enclosure in which, in village after village, common rights are lost." They also note E. P. Thompson's argument that "enclosure (when all the sophistications are allowed for) was a plain enough case of class robbery."

46. **He would have seen** The Manor of Huddersfield Enclosure Act of 1785 had brought the law home, which meant it would be enforced in his youth.

46. **"For two centuries"** Description of the root causes of inequality in England from Hammond and Hammond, *Skilled Labourer,* 2.

47. **A vaccine was discovered** Edward Jenner discovered the vaccine in 1796, and it was first made available two years later, and widely to the public soon after.

47. **"the emergence of class differences"** Romola Jane Davenport, Max Satchell, and Leigh Matthew William Shaw-Taylor, "The Geography of Smallpox in England before Vaccination: A Conundrum Resolved," *Social Science & Medicine,* vol. 206 (June 2018), 75–85.

47. **"significantly higher survival chances"** Romola Davenport, Leonard Schwarz, and Jeremy Boulton, "The Decline of Adult Smallpox in Eighteenth-Century London," *Economic History Review,* vol. 64, no. 4 (2011), 1289–314.

47. **"bribery, corruption, luxury, tyranny"** Richard Milnes, *The Warning Voice of a Hermit Abroad* (1825), 172.

NOTES

The Machinery Question

49. **The word** *automaton* Interview with Adrienne Mayor; and see her book, *Gods and Robots: Myths, Machines, and Ancient Dreams of Technology* (Princeton, NJ: Princeton University Press, 2018), a history of humanity's dreams about artificial intelligence and ancient dreams of technology.

50. **These imagined innovations** Kanta Dihal, with her colleagues at the Centre for the Leverhulme Future of Intelligence at the University of Cambridge, analyzes historical narratives about artificial intelligence and automated systems from ancient Greece to the present day. Dihal was interviewed in 2019.

50. **if a machine could assist** Interview with Dihal: "The earliest story, about Hephaestus's golden handmaidens — they're pretty much a mix of care bots and personal assistants. They're meant to make the hard labors of a god much easier — which means they must be quite powerful."

50. **"If every instrument"** Aristotle, *Politics*, part IV.

51. **"The ultimate, the most exciting"** Maxine Berg, *The Machinery Question and the Making of Political Economy, 1815–1848* (Cambridge: Cambridge University Press, 1980), 2.

51. **"blind No-God"** Thomas Carlyle, *Chartism* (London: James Fraser; Boston: Little, Brown, 1840), ch. 5.

52. **parliamentary inquiry** Archived in the Home Office (HO); see Sale, *Rebels against the Future*, 69.

52. **"did completely Destroy"** Home Office Papers, HO 42/117.

53. **In the chaos** The account of John Westley's death and funeral was carried in the *Leeds Mercury*, November 23, 1811.

53. **The same night** Accounts of these attacks drawn from reports in the *Leeds Mercury*, the *Derby Mercury*, and the reports of two London magistrates, Nathaniel Conant and Robert Baker, who were sent to Nottingham to assess the developing situation in November of 1811. Their report can be found in the Home Office Papers, HO 42/119.

54. **One seventeen-year-old apprentice** Felkin recorded his own experiences as a boy apprentice during the uprising in his 1867 book, *A History of the Machine-Wrought Hosiery and Lace Manufactures.*

55. **"The practice of these men"** John Blackner, *The History of Nottingham* (Nottingham, UK, 1815).

56. **"Mr H"** In Kevin Binfield, ed., *Writings of the Luddites* (Baltimore: Johns Hopkins University Press, 2004), 73.

57. **The place was a mess** There are plenty of good Byron biographies, but the one I wound up turning to most was Fiona MacCarthy's exhaustive *Byron: Life and Legend* (New York: Farrar, Straus & Giroux; London: John Murray, 2002), for its meticulous attention to detail. The depiction of Byron's life at this time draws largely from MacCarthy's book, his personal letters, his correspondence with his publisher John Murray, newspapers, and parliamentary records.

57. **"My prospects are not very pleasant"** June 29, 1811; in Thomas Moore, *The Life of Lord Byron, with His Letters and Journals* (2 vols., Philadelphia: Thomas Wardle, 1840).

58. **"embosom'd in a happy valley"** Byron, *Don Juan*, Canto XIII, LVI.1–3.

59. **"He is a pale, languid-looking young man"** Quoted in MacCarthy, *Byron*, 144.

60. **"Three troops"** *Leeds Mercury*, December 7, 1811. The events described took place a week earlier, but many newspapers only published weekly.

61. **and apparently *successful*** Darvall, *Popular Disturbances*, 107.

61. **"stolid aspect"** Thorpe's description comes from Peel, *Risings*.

61. **They'd heard stories** Felkin, *History*.

62. **"John Wood's cake's baked"** Sykes and Walker, *Ben o' Bill's.*

62. **"they did not hold any undue respect"** A paraphrasing of David Noble's commentary on the Luddites in "Present Tense Technology: Technology's Politics," *Democracy*, vol. 4, no. 2 (1983), 12.

62. **"Poor Law relief payments had reached a record high of £6.6 million"** Sale, Kirkpatrick. *Rebels Against the Future*. 1995. page 62.

64. **"form that enables"** Hammond and Hammond, *Skilled Labourer*, 1.

64. **the simplest and most straightforward manifestation** To this end, George and his cohort were not themselves the ones yet most impacted by automation — they simply saw the writing on the wall, and had the most power to act, given their stature in the trade. "Violence was the pre-emptive, calculated choice of a powerful labour elite, not the desperate last gasp of a demoralized trade," as Adrian Randall wrote in *Before the Luddites*.

64. **"shocking a system"** Klein, Naomi. *The Shock Doctrine*. 2007. Knopf Canada.

66. **"DECLARATION: EXTRAORDINARY"** Luddite letter archived in the Home Office Papers at HO 42/119. Also contextualized in Binfield, *Writings*.

68. **"The word Uuyd"** *Bristol Mirror*, Saturday, June 13, 1812.

68. **"according to tradition"** Brian Bailey, *The Luddite Rebellion* (New York: New York University Press, 1998).

68. **"a potent and forceful movement"** Katrina Navickas, "The Search for 'General Ludd': The Mythology of Luddism," *Social History*, vol. 30, no. 3, 281–295. Navickas cites Binfield, *Writings*.

70. **the Highland Fling** Saul David, *Prince of Pleasure*, 323.

71. **sent the armies north** The figures of how many troops were sent north were disclosed by Richard Ryder, the Home Office Secretary, in a speech to Parliament.

72. **"MILITARY coercion"** *Leeds Mercury*, December 7, 1811.

72. **"These riots owe not"** Pitt the Younger was the lead figure of the so-called new Tories, influential in shaping the era's conservative politics. Britain's youngest-ever prime minister, Pitt was in favor of Catholic emancipation and the abolition of slavery. He was also a war hawk who drove Britain's campaign against Napoleon, building up its military-industrial complex and levying unpopular taxes. He was a vigorous disciple of Adam Smith who oversaw the undoing of regulations that protected artisan workers, and took an authoritarian stance toward cracking down on reformers and dissent.

73. **"This is only to put off"** Saul David, *Prince of Pleasure*, 324.

74. **"I'm General Lud!"** In Walter Thornbury, "The Luddites," collected in *The Casquet of Literature* (1882), among other places.

74. **"I'm going to build more frames"** Dates and details of Horsfall and Cartwright's embrace of automation are from Randall, *Riotous Assemblies*, 283.

74. **He rebuilt** Horsfall's family history is recounted in numerous Luddite histories, including Reid, *Land of Lost Content*, and Randall, *Before the Luddites*.

75. **She would later base a major character** Charlotte Brontë's first novel, *Jane Eyre*, also includes a passing reference to the Luddites; the beautiful but frivolous heiress Ms. Oliver mentions consorting with the officers stationed to put them down, and preferring them to the lower-class workers and merchants of the town. Brontë derisively describes Jane Eyre's passionless suitor Mr. Rivers as "almost like an automaton."

75. **"a very remarkable man"** Elizabeth Cleghorn Gaskill, *The Life of Charlotte Brontë* (Cambridge: Cambridge University Press, 2010), 83.

75. **"Misery generates hate"** Charlotte Brontë, *Shirley*, 23.

76. **"The intensity and extent"** Randall, *Before the Luddites*, 45. "Machinery was not suddenly introduced on a large scale in either the west of England or Yorkshire. It was brought in gradually and often in a low-key manner. Further, it was often introduced in times of prosperity when demand for labour was high."

76. **"overcrowded"** These assessments were made in an 1810 report after two Manchester justices of the peace inspected the Merryweather Weaving Factory in Ancoats, Manchester. Katrina Honeyman, *Child Workers in England, 1780–1820: Parish Apprentices and the Making of the Early Industrial Labour Force* (London: Routledge, 2007).

76. **"The image of the factory"** Randall, *Before the Luddites.*

77. **"Faced with resistance"** Berg, *Machinery Question,* 2.

77. **"near-maniacal"** Margaret O'Mara, *The Code: Silicon Valley and the Remaking of America* (New York: Penguin Press, 2019), 7.

77. **"Laissez-faire"** Bailey, *Luddite Rebellion,* xv.

77. **"It was a deliberate attack"** Kirk Willis, "The Role in Parliament of the Economic Ideas of Adam Smith, 1776–1800," *History of Political Economy,* vol. 11, no. 4 (1979).

77. **"Adam Smith's discussion"** Mike Munger, *"Division of Labor,"* part 1 (*EconLib,* n.d., www.econlib.org/library/Enc/DivisionofLabor.html), 2019.

77. **Two of Smith's earliest** Kirk Willis, "The Role in Parliament of the Economic Ideas of Adam Smith, 1776–1800," *History of Political Economy,* vol. 11, no. 4 (1979), 505–544. "[Shelburne's] first order of business was the conclusion of peace treaties with France, Spain, Holland, and the new United States of America. With the end of the war and 'the irrevocable loss of our American colonies,' Britain's world had fundamentally and permanently changed, Shelburne believed, and this world demanded new policies based on principles of enlightened theory."

79. **In a secret meeting** Ralph Fletcher forwarded B's dispatches to the Home Office, where they remain archived. The first letter he sent back was on October 10; the next, which described the Falstaff meeting, was on November 8th. He invoiced for expenses for twenty-two days of activity, for an amount equal to around $350 in today's dollars. The letters can be found in the Home Office Papers at HO 42/117.

79. **Their messages must** Historians are careful to underline this point; E. P. Thompson does so at great length in *The Making of the English Working Class*—spies provide a valuable historical record, but a difficult one to decipher conclusively.

80. **Agents like him** Thompson, *Making.* In the chapter "Army of Redressers," Thompson describes the history of spies infiltrating radical groups from the time of the London Corresponding Society and beyond, and how they were enormously successful in doing so in groups less tightly knit than the Luddites.

80. **The number of spindles** Berg, *Machinery Question,* 23.

81. **"Hand spinners were the most numerous"** Jane Humphries, Oxford University. Interview, *Casualties of History* podcast. "You can find shelves of texts on the coal miners or the hand-loom weavers—but nothing, hardly, on the hand spinners."

81. **There were not yet a large number** The political economy of Manchester in the time of the cotton loom is discussed at length in the "Weavers" chapter of Thompson, *Making.*

81. **"the Steam Looms were a Great Service to the State"** Ryder's letter can be found in the Home Office Papers at HO 42/128.

84. **The Nottingham City Council** Sale, *Rebels against the Future,* 83.

86. **"A considerable Number of disorderly Persons"** A copy of the printed proclamation can be found in the Home Office Papers at HO 42/119.

87. **The poet Percy Bysshe Shelley** Frederick L. Jones, ed., *The Letters of Percy Bysshe Shelley* (Oxford: Clarendon Press, 2 vols., 1964), vol. 1, December 26, 1811.

87. **As the government had offered** Information about this counter-proclamation can be found at the Home Office, HO 42/119.

88. **Church bells rang** The action of this anecdote is drawn primarily from Sykes and Walker, *Ben o' Bill's,* but conditions are confirmed to be in line with the region's experience at the time.

89. **By now, Horsfall's hatred** The line that Horsfall wanted to "ride up to his saddle girth in Luddite blood" is documented in multiple histories, and is the most infamous of Horsfall's anti-Luddite sentiment, but all accounts verify a general hostility toward Luddite sympathy in general.

91. **"I cannot see daylight"** Much like the entrepreneurs wary to automate, the cloth workers faced difficult deliberations over whether to join the Luddites' armed struggle. B's meeting notes and Peel's oral history demonstrate that many workers were cautious and slow to join, as George Mellor is depicted so here.

92. **"I Shouted Aloud for Them to Stop the Wheels"**

92. **Robert Blincoe, the orphan** The quote is drawn from Blincoe's memoir, page 26; descriptions of attitudes toward factory conditions are from Randall, *Before the Luddites*.

94. **After the politician** Samuel Kydd, *The History of the Factory Movement: From the Year 1902, to the Enactment of the Ten Hours' Bill in 1847, Volume 1* (London: Simpkin, Marshall, 1857), 30.

96. **these stories, all of which ran in 1811** These newspaper articles can all be accessed through the British Newspaper Archives.

The First Tech Titans

98. **they started out as entrepreneurs** The popular mythology of the entrepreneur arose from some combination of Richard Cantillon's *Essay on the Nature of Trade in General*, the French physiocrats' advocating the notion of laissez-faire, and Adam Smith transmuting the lot of it into something resembling a science in *The Wealth of Nations* (1776).

98. **The term itself was popularized** Jean-Baptiste Say, *A Treatise on Political Economy*, trans. Charles Robert Trinsop (London, 1821). First published as *Traité d'économie politique* (1803).

98. **For a worker, aspiring** Status improvement, via journeyman work or apprenticeship, is detailed in Randall, *Before the Luddites*, and Thompson, *Making*.

98. **In the past, like now** Crouzet, *Formation of Capital*.

99. **"The improvement of spinning"** R. S. Fitton, *The Arkwrights: Spinners of Fortune* (Manchester, UK: Manchester University Press, 1989), 10.

99. **"Arkwright was not the great inventor"** Fitton, *The Arkwrights*, xiii.

100. **At its peak** R. S. Fitton and Alfred P. Wadsworth, *The Strutts and the Arkwrights, 1758–1830: A Study of the Early Factory System* (Manchester, UK: Manchester University Press, 1958), 226.

100. **Arkwright's "main difficulty"** Andrew Ure, *The Philosophy of Manufactures: Or, An Exposition of the Scientific, Moral, and Commercial Economy of the Factory System of Great Britain* (London: Charles Knight, 1835), 16.

101. **Like Steve Jobs** The Apple cofounder and iconic CEO liked to quote Picasso: "He said, 'Good artists copy, great artists steal.' And we have always been shameless about stealing great ideas." Throughout its history, Apple has stolen or emulated key innovations from competitors, like, perhaps most famously, the concept of the graphical user interface (GUI), from Xerox, as depicted in the 1996 documentary *Triumph of the Nerds*, where this quote is taken from. In another parallel between Arkwright and Jobs, the courts determined that Arkwright's patent for his chief invention, the water frame, was invalid because it was actually invented by a former partner, John Kay, who, not unlike Apple cofounder Steve Wozniak, was pushed out of the partnership and thus missed out on the lion's share of the company's eventual riches.

102. **Like Jeff Bezos** A line can be drawn back from Jeff Bezos and Amazon, whose warehouses are laboratories for advancing new technologies to discipline workers for maximum

productivity; through the big auto manufacturers' automated car plants of the 1960s, which demanded workers become freshly subservient to heavy machinery; to the scientific management theories of Frederick Winslow Taylor in the 1910s, when he timed workers with a stopwatch to ensure they were meeting productivity standards; to the textile factories of the Industrial Revolution in the 1800s, and to Arkwright, where the model saw its most successful case study.

102. **compared to the previous engine** The Newcomen was the previously dominant steam engine, but was too inefficient to be widely affordable.

102. **The Mises Institute** Michele Boldrin and David K. Levine, *Against Intellectual Monopoly* (Cambridge University Press, 2008).

102. **The word *innovation*** Jill Lepore, "The Disruption Machine," *The New Yorker*, June 13, 2014.

Part II: Metropolis of Discontent

105. **"[The year] 1812 opens"** *Manchester Gazette*, January 4, 1812. First edition of the new year.

107. **Soon there were a dozen** Brooke and Kipling, *Liberty or Death*, 29.

107. **"The system was yet in its infancy"** Brooke and Kipling, *Liberty or Death*.

110. **The poem broadcasted openly** The context and reception to Anna Lætitia Barbauld's poem is discussed in, among other places, Maggie Favretti, "The Politics of Vision: Anna Barbauld's 'Eighteen Hundred and Eleven,'" in Isobel Armstrong and Virginia Blain, eds., *Women's Poetry in the Enlightenment: The Making of a Canon, 1730–1820* (London: Palgrave Macmillan, 1999), 101.

110. **The victory, however modest** The Prince Regent's determination to continue the war effort is discussed in Saul David, *Prince of Pleasure*.

111. **"no depredations"** *Leeds Mercury*, Saturday, February 8, 1812.

112. **On January 1, 1812, a handbill** Transcribed in full by Kevin Binfield in *Writings of the Luddites*, 90.

112. **"It is 'impossible'"** Commentary collected in the *Nottingham Annual Register, Or, A View of the History, Politics, and Literature for the Year 1812*.

113. **"hang'd for 3 Menet"** Letter transcribed in Binfield, *Writings*, 101.

113. **machinery destroyed** *Leeds Mercury*, Saturday, February 29, 1812.

114. **First, Henson and a colleague** *Records of the Borough of Nottingham*, vol. 8, 1800–1835, 137, via the *Luddite Bicentenary, 1811–1817*.

114. **Henson advertised** Transcribed in full by Richard Holland at the *Luddite Bicentenary, 1811–1817*, http://ludditebicentenary.blogspot.com.

117. **"The meetings are packed"** B's letter to Fletcher can be found in the Home Office Papers at HO 42/119.

117. **On February 22** *Leeds Mercury*, February 22, 1812.

118. **A conference was arranged** Hammond and Hammond, *Skilled Labourer*, 273.

118. **"They told me"** *Selection of Reports and Papers of the House of Commons*, vol. 17 (1836).

118. **The code went as follows** The Luddite code is described in Thompson, *Making*, 582, where he again points out that the information was recorded by spies and is to be treated skeptically at best, though it is likely the Luddites had systems in place to help ensure security during meetings. They each went by a number, not a name, for instance, during roll call at drills and before raids.

120. **It was Saturday afternoon** The scene is recounted in Peel, *Risings*.

121. **"since the commencement"** *Leeds Mercury*, March 7, 1812.

122. **The Nottingham revolt** For the Luddite action's effects in forcing concessions from bosses, see Thompson, *Making*, chapter 14, "An Army of Redressers."

123. **"The machine was not an impersonal achievement"** Berg, *Machinery Question*, 2. It should also be noted that the machinery question also thrust the issue of economic growth to the foreground, where it has remained ever since: "It was economic growth and its now limitless prospects created by technological advance which became the new center, not just of the analysis of the economy, but of the analysis of politics and society as well. The economy was no longer conceived as subordinate to broader social and political ideals. It now played a distinct and a dominant role. The analysis of the machinery question was formative in the creation of a political economy which became the 'natural science' of economy and society."

124. **That was the price of progress** Thomas Carlyle savaged this notion in his book *Chartism* (1840):

> It seems a cruel mockery to tell poor drudges that their condition is improving. Laissez-faire, laissez-passer! The master of horses, when the summer labour is done, has to feed his horses through the winter. If he said to his horses: "Quadrupeds, I have no longer work for you; but work exists abundantly over the world: are you ignorant (or must I read you Political-Economy Lectures) that the Steam-engine always in the long-run creates additional work? Railways are forming in one quarter of this earth, canals in another, much cartage is wanted; somewhere in Europe, Asia, Africa or America, doubt it not, ye will find cartage: go and seek cartage, and good go with you!"

The mockery is as apt today. The economic wisdom of the modern age is that automation ultimately creates as many jobs as it displaces by bolstering output and generating extra value, which, of course, is of little consolation to the working man being fired to accommodate an automated process.

124. **"Where machinery augmented"** Randall, *Before the Luddites*, 7.

124. **Owen had become wealthy** At the House of Commons in 1817, Owen would officially propose his plan to Parliament in a speech that sharply articulated how mass automation leads to wealth for a small class and poverty for workers:

> The immediate cause of the present distress is the depreciation of human labour. This has been occasioned by the general introduction of mechanism into the manufactures of Europe and America, but principally into those of Britain, where the change was greatly accelerated by the inventions of Arkwright and Watt.... The introduction of mechanism into the manufacture of objects of desire in society reduced their price; the reduction of price increased the demand for them, and generally to so great an extent as to occasion more human labour to be employed after the introduction of machinery than had been employed before.... The first effects of these new mechanical combinations were to increase individual wealth, and to give a new stimulus to further inventions.

126. **"dull, dirty, and dangerous jobs"** Bernard Marr, "The 4 Ds of Robotization: Dull, Dirty, Dangerous and Dear," *Forbes*, October 16, 2017.

126. **a fifteen-hour workweek** John Maynard Keynes, "Economic Possibilities for Our Grandchildren," in *Essays in Persuasion* (London: Macmillan, 1931; New York: Norton, 1963).

129. **His life read like an epic poem** Florence Ashton Marshall, *The Life and Letters of Mary Wollstonecraft Shelley* (2 vols., London, 1889), 21–22.

129. **Life had always been like this** Tom Furniss, "Mary Wollstonecraft's French Revolution," in *The Cambridge Companion to Mary Wollstonecraft* (Cambridge University Press, 2002).

130. **"Her rapidly growing powers"** Marshall, *Life and Letters*, 14.

131. **"The rioters assemble"** Joseph Radcliffe to the Home Office, HO 40/1/1.

131. **The Home Office had been criticized** A discussion of the Home Office, analogous to the US Department of Homeland Security and the Department of State, and how it was and under-funded and understaffed, and unsure of how to respond to a crisis of this kind, is explored in Reid, *Land of Lost Content,* which focuses on how the state and the military responded to the Luddite uprisings.

131. **At the time, townships and cities** Nathan Ashley Bend, *The Home Office and Public Distur-bance, c. 1800–1832* (Hatfield, UK: University of Hertfordshire, 2019).

132. **"Perhaps the most serious handicap"** Malcolm I. Thomis, *The Luddites,* 155.

132. **"It was widely said at the time"** Darvall, *Popular Disturbances,* 173.

132. **Then, on February 14, 1812, Ryder stood** Hansard is the traditional name of the transcripts of parliamentary proceedings, and an archive is kept of debates, sessions, and petitions made in UK Parliament. Notes on Ryder's speech can be found under: "Frame Breaking and Nottingham Peace Bills," HC Deb February 14, 1812, vol. 21, cc807–24.

134. **"final betrayal"** Saul David, *Prince of Pleasure,* 328.

135. **"It grieves me to tell you"** A. Aspinall, ed., *The Letters of King George IV, 1812-1830, Volume 2* (Cambridge: Cambridge University Press, 1938), 32.

135. **"a violator of his word"** *Morning Post,* March 19, 1812.

135. **"a furious and unmeasured attack"** Duke of Buckingham and Chandos, *Memoirs of the Courts and Cabinets of George III,* vol. 2, 1853-1855.

135. **"Lines to a Weeping Lady"** *Morning Chronicle,* March 7, 1812.

137. **"From all that fell"** Leslie A. Marchand, ed., *Byron's Letters and Journals* (12 vols., London: John Murray; Cambridge, MA: Belknap Press of Harvard University Press, 1973–82), vol. 2 (August 1811–April 1814).

137. **"I consider the manufacturers"** Marchand, *Byron's Letters and Journals,* vol. 2, no. 226, to Lord Holland, 8, St. James's Street, February 25, 1812.

139. **Britain's political system revolved** Hobsbawm, "The Machine Breakers," 5–22. Hobsbawm further details the role of British parliamentary politics in the Industrial Revolution in *Industry and Empire: The Making of Modern English Society, from 1750 to the Present Day* (New York: Pantheon, 1968) and *The Age of Revolution: Europe: 1789–1848* (New York: Vintage, 1996).

139. **"most ferociously conservative period"** Hobsbawm, "The Machine Breakers."

139. **"National politicians used"** O'Mara, *The Code,* 2.

140. **a "golden era"** Hobsbawm, *Industry and Empire,* 58–9.

141. **"We lament to state"** *Nottingham Journal,* Saturday, February 22, 1812, accessed through the British Newspaper Archive.

142. **It helped, surely** Brooke and Kipling, *Liberty or Death.* "Rather than indulging in gratu-itous violence, the Luddites seem to have gone out of their way to perpetrate as little violence as possible in the majority of cases in which they were involved."

143. **"It was a matter of surprise"** Darvall, *Popular Disturbances,* 172. Darvall, one of the first Luddite historians to move beyond oral history, also notes that "a major object of the Lud-dites, particularly in the cotton country, but to some extent in every disturbed district, was to protest dramatically and effectively against existing conditions."

143. **Written lyrics to "General Ludd's Triumph"** The full lyrics are included in, among other places, Binfield, *Writings.*

145. **"As a person in some degree"** The text of Byron's speech is recorded in the parliamentary record at HL Deb, February 27, 1812, vol. 21, cc964–79, as well as in numerous newspapers, biographies, and history volumes. It remains perhaps the most famous public defense of the Luddites, and certainly the most famous given contemporaneously.

149. **William Hinchcliffe woke** The deposition that contains the details of the assault can be found in the Home Office Papers at HO 40/1/7. Richard Holland compiled an account of the

attack from the Home Office Papers for the *Luddite Bicentenary,* in a piece titled "27th February 1812: Attack on Workshop of William Hinchcliffe at Leymoor, Golcar, near Huddersfield."

151. **Two days after the attack** A copy of the Committee handbill can be found in the Home Office Papers at HO 42/120.

153. **The men who formed the backbone** Alan Brooke, "Huddersfield Manufacturers and Merchants Form 'Committee for Suppressing the Outrages'" (2012), *Luddite Bicentenary.*

153. **Now in middle age** Reid, *Land of Lost Content,* 19.

153. **But he quickly allied himself** Sale, *Rebels against the Future.*

154. **"good reason to think"** Correspondence detailed in Bailey, *Luddite Rebellion,* 47.

155. **One of the men wore** Peel, *Risings,* 47.

155. **a pistol in one hand** Brooke and Kipling, *Liberty or Death,* 17. Witness depositions for the attack can be found in the Home Office Papers at HO 40/1/7.

157. **On March 2** From Byron's letters. "An Ode to the Framers of the Frame Bill" appeared in the *Morning Chronicle* on Monday, March 2, 1812.

158. **Lord Holland was less impressed** MacCarthy, *Byron,* 157.

159. **On March 5** According to parliamentary record, the Frame Work Bill was given assent on March 20, 1812.

160. **After Swallow and Cotton** Brooke and Kipling, *Liberty or Death,* 30.

160. **On March 9** The text of the letter, potentially written by George Mellor himself, can be found in Binfield, *Writings,* 209.

162. **"always best and brightest"** Sykes and Walker, *Ben o' Bill's,* 104.

Two Centuries of Disruption

165. **they were not "unthinking"** Thompson, *Making,* 524–6.

165. **"However obsolete the statute"** Thompson, *Making,* 524–6.

166. **"The debate over the old laws"** Randall, *Before the Luddites,* 10.

167. **"the croppers seem to have cherished"** Thompson, *Making,* 526.

167. **A 2018 McKinsey report** "Retraining and Reskilling Workers in the Age of Automation," McKinsey Global Institute, January 2018.

167. **Amazon began investing** I wrote a report for *Gizmodo* about this development in July 2019: "Amazon Says It Will Retrain Workers It's Automating Out of Jobs. But Does 'Upskilling' Even Work?"

168. **"We were researching a lot"** December 2018 interview with Argüello-Kline. Parts of the interview were first published in a piece I wrote for *Gizmodo,* "This Was the Year the Robot Takeover of Service Jobs Began," December 20, 2018.

168. **The negotiations in 2018 were intense** The union detailed the protections it won in its 2019 contract in a blog post, "Contract Language: Automation & Technology," *Culinary Workers Union Local 226,* March 20, 2019, www.culinaryunion226.org/blog/contract-language-automation-technology.

169. **"A VAT makes it impossible"** Andrew Yang's 2020 presidential campaign website detailed his VAT proposal at https://2020.yang2020.com/policies/value-added-tax/.

169. **Bill Gates came out in favor** Bill Gates's interview with *Quartz,* in which he discusses his policy ideas for taxing automation, was published in 2017 under the headline "The Robot That Takes Your Job Should Pay Taxes, Says Bill Gates."

169. **De Blasio also proposed** I conducted a phone interview with de Blasio while he was campaigning for president in 2019. The quotes are drawn from that interview.

170. **"The old legislative controls"** Randall, *Before the Luddites,* 237.

171. *New York Times reporter* **Mike Isaac** Mike Isaac's excellent reporting on Uber can be found both in the *New York Times* and in his book *Super Pumped: The Battle for Uber* (New York: Norton, 2019).

171. **As Uber was introduced** Uber's behavior often caused protests, backlash, and even violent riots. See, for example, "Cab Drivers Protest Uber, Rideshare Apps in Mexico City," AP, October 12, 2016.

172. **why should we bother** Randall, *Before the Luddites*, 237.

Part III: Breaking Frames, Breaking Bones

175. **historian Jenny Uglow** Uglow spoke to Clive Thompson for his January 2017 article for *Smithsonian*, "When Robots Take All of Our Jobs, Remember the Luddites."

175. **Yet the Prince Regent was in Brighton** Reid, *Land of Lost Content*, 103.

175. **Ryder helped** The full title of the 1812 Watch and Ward Act is "An Act for the More Effectual Preservation of the Peace, by Enforcing the Duties of Watching and Warding, until the First Day of March 1814, in Places Where Disturbances Prevail or Are Apprehended." The abstract of the act is available at: www.calderdale.gov.uk/.

178. **"to Mary, Jane, and Fanny"** Seymour, *Mary Shelley*, 68.

179. **"Political complexity"** William Godwin, *An Enquiry Concerning Political Justice*, vol. 1, 13.

179. **"You cannot imagine"** PBS–WG, March 4, 1812; quoted in Seymour, *Mary Shelley*, 67.

179. **"the well-known hand"** PBS–WG, July 4, 1812; quoted in Seymour, *Mary Shelley*, 68.

179. **"Seeds of intellect"** WG-PBS, March 30, 1812; quoted in Seymour, *Mary Shelley*, 66.

180. **The code was simple** As mentioned in the text, this chapter draws largely from both Peel, *Risings*, and Sykes and Walker, *Ben o' Bill's*. Both accounts mention this scene — George Mellor and the West Riding Luddites choosing which major factory to target next — though Peel gives time to the democratic reformer Baines and the delegate from Nottingham. There's little doubt that much of the dialogue is distorted by memory (at best), and it has the feeling that it's a composite of different Luddite meetings condensed into a single event for narrative purposes. However, it's also highly likely a lot would be remembered about what would turn out to be such a consequential conference, discussed at length, and transmitted into oral history accordingly. My sense of this scene, as with George's debate with John Booth in the cropping shop, is that the account is accurate in texture and general detail (what the pub looked and felt like, the general thrust of these arguments and debates, the people who would be in attendance), if not reliable as a verbatim record of what transpired. These key practices, events, and concepts — the secrecy, the spirited meetings, the philosophical debates — no doubt occurred in much this way.

180. **"win, work"** The code words come from Sykes and Walker, *Ben o' Bill's*.

181. **"Enter, no General"** The ticket is discussed in Bailey, *Luddite Rebellion*, as well as Thompson, *Making*, chapter 14, "Army Redressers."

182. **"Old Corruption"** "What the popular writer and agitator William Cobbett and his fellow radicals meant by Old Corruption was a parasitical system — ostensibly built up to enormous proportions during the Napoleonic Wars — through which the elite fed its insatiable appetite for power and money and the people's expense." Philip Harling, "Rethinking 'Old Corruption,'" *Past & Present*, no. 147 (May 1995), 127–158.

183. **A hymn by an anonymous author** "Well done, Ned Lud" is transcribed in Binfield, *Writings*, 130, and archived in the Home Office Papers.

183. **"It was better that the cotton weavers"** Perceval's treatment of cloth workers is discussed in Randall, *Before the Luddites*.

186. **The workers of Nottingham** The account of B's activities is discussed in Richard Holland's write-up of Fletcher's letter with B's report, and statements made to Humphrey Yarwood, another spy, in "Manchester Executive Committee Cancel Plans for Factory Attacks on 9th April." The source material can be found in the Home Office Papers at HO 40/1/1.

188. **He'd gone to considerable expense** Horsfall's efforts to transform his factory into an armed fortress are detailed in accounts in the *Leeds Mercury,* in many Luddite histories, including those by Reid and Sale.

190. **To the surprise of no one** Byron became perhaps the first true cultural celebrity. *"Childe Harold's Pilgrimage," In Our Time,* BBC Radio 4, January 6, 2011.

190. **"physically as well as sexually perilous"** Description in MacCarthy, *Byron.*

190. **a figure without peer** Byron would, in fact, eventually come to compare himself primarily to Napoleon, as the only figure comparable in influence.

The Battle of Rawfolds Mill

192. **It was time** The battle of Rawfolds Mill was pivotal — perhaps the most legendary event in Luddite lore, along with the trials that followed. It was covered by the local newspapers at length, including the *Leeds Mercury.* Details here come from newspapers, depositions archived at the Home Office, trial records, Peel and Sykes's accounts, Darvall, the Hammonds, Thomis, Sale, Reid, and others. The general details are agreed upon: at least two men died, George Mellor played a pivotal role, and William Cartwright aggressively opened fire in opposition of the Luddites. So is the aftermath, with dying witnesses refusing to talk, speculation that they had been unduly transported and tortured, and the bloody field left in the battle's wake.

196. **"Can you keep a secret?"** John Booth's dying taunt is a famous, if tidily dramatic, bit of Luddite lore, documented in Peel's oral history and in subsequent sources; it lives on today, as retold in this unsigned article about the pub where Booth died: "Barfly: The Star, Robertstown, Liversedge," *Yorkshire Evening Post,* August 17, 2018.

196. **"I preached one Sunday"** *Memoirs of Jonathan Saville* (1845), 25–26.

197. **"The rioters had never been so met before"** Brontë, *Shirley,* 303.

199. **"justifiable homicides"** The jury's verdict was reported widely and recorded in the Home Office.

201. **a very public spectacle** *Leeds Mercury,* April 25, 1812.

204. **His meticulous efforts** Gravener Henson's letters from this period can be found in the *Records of the Borough of Nottingham,* vol. 8, 1800–1835.

204. **"We feel it an indispensable duty"** The weavers' petition was printed in the *Liverpool Mercury,* April 24, 1812.

206. **This was a radical gender inversion** Binfield, *Writings.*

207. **"Leader of the Luddites"** This satirical image shows Ned Ludd derisively clothed in a dress of patterned calico manufactured in the area. Condemned to poverty by steam-powered weaving machinery, General Ludd and his band of handloom weavers have set a factory ablaze.

208. **"during the troubles of 1812"** Malcolm I. Thomis and Jennifer Grimmett, *Women in Protest, 1800–1850* (London: Croom Helm, 1982), 10.

208. **General Ludd's wives** See Kevin Binfield, "Industrial Gender: Manly Men and Cross-Dressers in the Luddite Movement," in Jay Losey and William Dean Brewer, eds., *Mapping Male Sexuality: Nineteenth-Century England* (Teaneck, NJ: Fairleigh Dickinson University Press, 2000), on the "homoindustrial" aspects of Luddism: "Between 1811 and 1813, forced to defend a homosocial/homoindustrial solidarity between workers against the alienation of heteroindustrialism, the Luddites embraced a doctrine of variability, which encompassed

organization, naming, and clothing. The product of masculine labor, clothing, individually worn in collective endeavors, replaced work itself as the marker of the homoindustrial solidarity between workers…the Luddite hero is a man wearing upon his body a feminine mark of masculine labor, doomed to fail, but nevertheless demonstrating an intense male sociability and expanded conception of masculinity."

Insurrection

209. **"'Vengeance for the blood of the Innocent'"** The colonel's letter can be found in the Home Office Papers at HO 42/122 and was transcribed by Richard Holland for the *Luddite Bicentenary,* April 2012 ("Colonel Campbell writes to his commanding officer on the state of the West Riding").
209. **"Innocent blood crys for vengeance"** Image of poster found in Home Office Papers at HO 42/122.
209. **A spinning mill was set on fire** Described in a letter from John Pilkington, lieutenant at Bolton, to the Earl of Derby; see Holland, *Luddite Bicentenary,* "18th April 1812: Spinning Factory Set Alight in Bolton."
210. **The Burtons, too, were prepared for war** Recounted in Samuel Bamford, *Passages in the Life of a Radical* (1841).
210. **Thousands of food rioters** Hammond and Hammond, *Skilled Labourer,* 289.
211. **General Ludd himself** *Leeds Mercury,* quoted in Sale, *Rebels against the Future,* 137–8, and in Thompson, *Making.*
212. **"one of the most bloody"** Holland, *Luddite Bicentenary,* "20th April 1812: Attack on Burton's Mill at Middleton — Day 1."
213. **If you are George Mellor** The details of George's reaction to the failure of Rawfolds are drawn largely from Sykes and Peel, but there's no reason to doubt he would have been enraged, guilt-stricken, and grieving after the death of his friend John Booth. As mentioned in-line, the story about the dead infant is a powerful part of Luddite legend, but must be read skeptically; which is not to say it or something like it did not occur at some point in the Luddite saga.
216. **Haigh told Radcliffe** Holland, *Luddite Bicentenary,* "24th April 1812: James Haigh is brought before the magistrate, Joseph Radcliffe."
216. **A woman named Betty Armstrong** *Leeds Mercury,* May 2, 1812; Holland, *Luddite Bicentenary,* "25th April 1812: A suspected informer is attacked in Huddersfield."
216. **"If this Machinery is suffer'd"** Binfield, *Writings,* 222. Some suspect the author was none other than George Mellor.
217. **"How long ye Wretches will ye"** Archived in the Home Office Papers, HO 42/122.
218. **On Tuesday** The account of Horsfall's murder is drawn from the thorough account in the *Leeds Mercury,* depositions, and trial statements of the participants, as well as Peel's oral history.
218. **After the Luddites' defeat** Peel, *Risings,* 69.

An Involuntary Machine

221. **Charles Ball was out hunting** The details of Charles Ball's life story are drawn primarily from his memoir, which was recorded with the help of the abolitionist Isaac Fisher, as well as contemporary historical accounts, especially writings about the Battle of Bladensburg, which Ball participated in. (A memorial to that battle depicts three soldiers, one of whom is Ball.) The memoir, titled *Slavery in the United States: A Narrative of the Life and Adventures*

of Charles Ball, a Black Man, Who Lived Forty Years in Maryland, South Carolina, and Georgia, as a Slave, under Various Masters, and Was One Year in the Navy with Commodore Barney, during the Late War, was published in 1837 by John S. Taylor. The memoir was successful and went through multiple printings, including in abridged versions. The original 1837 edition, which I used as my source document, is available online and was archived as part of the 1999 "Documenting the American South" project at the University of North Carolina–Chapel Hill. The incident that opens the chapter takes place on page 325.

223. **But from around the middle of the eighteenth century** Imports of cotton were banned in Britain under the Calico Acts; in the 1700s, weavers couldn't compete with the cheap cotton cloth produced in India; the act was repealed in 1774, when industrial machinery made the process cheaper.

224. **"One man and a horse"** Whitney's letter to his father is viewable online through the DocsTeach at the US National Archives.

224. **Whitney's machine** In his memoir, Ball describes the how the cotton gin works according to his own experience:

> As I shall be obliged to make frequent references to the cotton-gin, it may perhaps be well to describe it. Formerly there was no way of separating the cotton from the seed, but by pulling it of with the fingers — a very tedious and troublesome process — but a person from the north, by the name of Whitney, at length discovered the gin, which is a very simple though very powerful machine. It is composed of a wooden cylinder, about six or eight feet in length, surrounded at very short intervals, with small circular saws, in such a manner that as the cylinder is turned rapidly round, by a leather strap on the end, similar to a turner's lathe, the teeth of the saws, in turning over, continually cut downwards in front of the cylinder, which is placed close to a long hopper, extending the whole length of the cylinder, and so close to it that the seeds of the cotton cannot pass between them. This cylinder revolves, with almost inconceivable rapidity, and great caution is necessary in working with the gin, not to touch the saws. One end of the cylinder and hopper being slightly elevated, the seeds as they are stripped of the wool, are gradually but certainly moved toward the lower end, where they drop down into a heap, after being as perfectly divested of the cotton as they could be by the most careful picking with the fingers.
>
> The rapid evolutions of the cylinder are procured by the aid of cogs and wheels, similar to those used in small grist mills. It is necessary to be very careful in working about a cotton-gin; more especially in removing the seeds from before the saws; for if they do but touch the hand the injury is very great. I knew a black man who had all the sinews of the inner part of his right hand torn out — some of them measuring more than a foot in length — and the flesh of his palm cut into tatters, by carelessly putting his hand too near the saws, when they were in motion, for the idle purpose of feeling the strength of the current of air created by the motions of the cylinder. A good gin will clean several thousand pounds of cotton, in the seed, in a day. To work the gin two horses are necessary; though one is often compelled to perform the labour.

224. **(The town was named after the House of Lancaster)** This is according to the city's official history, at: www.lancastercitysc.com/history-of-lancaster/.

225. **a BBC survey** "Slave Trade and the British Economy," *BBC,* www.bbc.co.uk/bitesize/guides /zc92xnb/revision/4.

225. **"British cotton imports rose"** Eric Williams, *Capitalism and Slavery* (Chapel Hill: University of North Carolina Press, 1994; first published in 1944). Also of note: "Population increased rapidly in the cotton centers. Leeds had a population of 17,000 on the eve of the American revolution, seven times as many in 1831."

225. **"The first federal census"** James P. Stobaugh, *American History* (2012), via PBS, *Growth and Entrenchment of Slavery.*

228. **"faith in a life to come"** Thompson, *Making,* 34.

228. **enslaved workers resisted** James Oakes, *Slavery and Freedom: An Interpretation of the Old South* (New York: Alfred A. Knopf, 1990), 138.

228. **"enclosures were conquest"** Peter Linebaugh, *Ned Ludd and Queen Mab: Machine-Breaking, Romanticism, and the Several Commons of 1811–12* (Oakland, CA: PM Press, 2012), 23

228. **"Slaves engaged in a remarkable variety"** John Hope Franklin and Loren Schweninger, *Runaway Slaves: Rebels on the Plantation* (New York: Oxford University Press, 1999), 2.

Part IV: More Value than Work or Gold

The Prime Minister

233. **the lyrics to a hymn** "Welcome Ned Ludd" was previously discussed; here's the full text, printed in handbill form, as transcribed in Binfield, *Writings.*

234. **"He has looked at human nature"** Sydney Smith, ed., *Peter Plymley's Letters and Selected Essays* (London, 1886), Letter X.

234. **"Among the multitude"** *The Life of Sir Samuel Romilly Written by Himself with a Selection from His Correspondence* (1842), vol. 2, 256.

235. **"Every serious well-disposed person"** Gordon Pentland, "Responses to the Assassination of Spencer Perceval," *Journal of British Studies,* vol. 51, no. 2 (April 2012), 340–363.

235. **"up to the time of the shooting"** Peel, *Risings,* 87.

235. **"a crowd assembled"** Thompson, *Making,* 301.

236. **"The Loyalists here cannot accede"** Fletcher's letter is cited in Pentland's study and is archived in the Home Office. Fresh in the minds of those who feared a "general rising" were the massive crowds that had gathered to watch the radical Parliamentarian Sir Francis Burdett as he spoke out in defense of another reformer who had been imprisoned after criticizing Perceval's Tories for enforcing a law that kept the press out of Parliament — and Burdett found himself imprisoned as well. Despite being a baronet, Burdett was an outspoken reformer, a powerful orator, and a popular figure among London's working class. His name was sometimes passed on in messages and correspondences from the spies in the Midlands; the magistrates feared he could become a populist leader and help organize the leaderless Luddites into a more disciplined rebellion across the north.

236. **Bellingham declared himself insane** "When the PM Met His High Noon," *Antiques Trade Gazette,* September 2003.

236. **multiple letters to Richard Ryder** There has been speculation that Bellingham was in fact waiting for Ryder, who had unsatisfactorily answered so many of his entreaties, and that he grew impatient and targeted Perceval because he saw him first.

236. **"that justice refused to me"** Bellingham's testimony is transcribed in the *Proceedings of Old Bailey, London's Central Criminal Court, 1674 to 1913,* s.v. John Bellingham, May 13, 1812.

237. **"The trials for offences in Lancashire"** Hammond and Hammond, *Skilled Labourer,* 292.

237. **Despite Bellingham's numbing testimony** René Martin Pillet, *Views of England, during a Residence of Ten Years; Six of Them as a Prisoner* (1818), 25.

237. **"If there was ever going to be a revolution"** Sale, *Rebels against the Future*, 156.

237. **"every frame Breaking act"** Binfield, *Writings*, 187. Full text of the letter is as follows:

Honorable Sir,

Every frame Breaking act you Make an amendment to only serves to shorten your Days Theirfore you may Prepaire to go to the Divel to Bee Secraterry for Mr Perceval theire for there are fire Ships Making to saile by land as well as by Warter that will not faile to Destroy all the Obnoctious in the both Houses as you have been at a great Deal of pains to Destroy Chiefe part of the Ciuntry it is know your turn to fall. The Remedy for you is Shor Destruction Without Detection — prepaire for they Departure and Recomend the same to thy friends

your Hbl sert &c
Luddites.

238. **a "Rescue Bellingham" movement** Account is taken from *A Full and Authentic Report of the Trial of John Bellingham, at the Sessions' House, in the Old Bailey, on Friday, May 15, 1812, for the Murder of the Right Honourable Spencer Perceval, Chancellor of the Exchequer, in the Lobby of the House of Commons* (London: Sherwood, Neely, and Jones, 1812).

238. **While Byron was in London** Saul David, *Prince of Pleasure*, 332.

239. **"He ordered me to be presented to him"** Byron to Walter Scott, July 6, 1812, in *Letters and Journals*, vol. 2 (August 1811–April 1814), 241.

240. **The thirty-five-year-old Captain Francis Raynes** I have drawn heavily upon Raynes's own memoir, which was published in part as an effort to generate public sympathy for the fact that he'd never received due payment for his efforts in fighting the Luddites. Published in 1817, it bore the title *An Appeal to the Public: Containing an Account of Services Rendered during the Disturbances in the North of England, in the Year 1812: with an Account of the Means Adopted, Which Eventually Led to Their Suppression. Together with a Correspondence with Government, and Others, on the Subject of a Remuneration for Those Services.*

240. **"In the earlier stages"** Francis Raynes, *An Appeal to the Public*, 21.

240. **He would also control** Reid, *Land of Lost Content*. Close attention is also paid to the number of troops marshaled and their movements.

242. **At the time, there were around thirty-five thousand** Population statistics are drawn from *Vision of Great Britain*, www.visionofbritain.org.uk/unit/10168600/cube/TOT_POP.

242. **"The purpose to be achieved"** Reid, *Land of Lost Content*, 151–2.

243. **B wrote to Colonel Fletcher** B's letters to Fletcher can be found in the Home Office Papers at HO 40/1/1, and they have been written up by Richard Holland at the *Luddite Bicentenary*, in a post titled "The Manchester Spy, John Bent, Files His Latest Report for Colonel Fletcher."

244. **after he'd refused to be twisted in** *Leeds Mercury*, May 16, 1812.

245. **"A vigorous body of police"** Peel, *Risings*, 91.

245. **The reward for information** Sale, *Rebels against the Future*, 127.

246. **So Milnsbridge became a local interrogation center** Sale, *Rebels against the Future*, 127.

246. **Horsfall's heirs** Richard Holland, "The Death of William Horsfall," *Luddite Bicentenary*. "According to Frank Peel (1968 ed., p. 145), the Horsfall family subsequently discontinued the use of shearing frames at Ottiwells Mill, meaning that, in a specific local context, and like with Burton's Mill at Middleton and the Westhoughton Factory, Luddism worked."

247. **"We must give up"** The description of the scene and the dialogue between the croppers as they plot to kill Horsfall are sourced both from Frank Peel, whose *Risings of the Luddites* contains an account, and the deposition and trial statements of Ben Walker.

249. **"It seemed at times"** Jenny Uglow, *In These Times: Living in Britain through Napoleon's Wars, 1793–1815* (New York: Farrar, Straus & Giroux, 2015), 554.

250. **"The Luddites continue"** *Leeds Mercury*, via *Sussex Advertiser*, Monday, June 15, 1812.

250. **"On Sunday morning last"** *Statesman* (London), Monday, June 23, 1812.

251. **"The practice of forcibly obtaining arms"** *Bristol Mirror*, Saturday, June 13, 1812.

252. **After the second reading** The primary source for Henson's writings continues to be the *Records of the Borough of Nottingham*, vol. 8, 1800–1835.

255. **The five months that Mary Godwin spent in Dundee** Seymour, *Mary Shelley*, 72.

255. **"the eyry of freedom"** From the preface to *Frankenstein* (1831 ed.).

255. **"Isabella"** The only existing portrait of Isabella Baxter, painted when she was twenty-three (after she had married David Booth), suggests that she "bore a startling resemblance to Percy Shelley." Seymour, *Mary Shelley*, 74–5.

255. **"They were all fond"** Florence A. Marshall, ed., *The Life and Letters of Mary Wollstonecraft Shelley* (2 vols., London, 1889), vol. 1, 31.

257. **"For one short day the world forego"** Strickland, *Memoir*, 247.

257. **Despite inventing a machine** Cartwright discusses the aftermath of his efforts to invent the power loom, his failed business, and his unsuccessful lawsuit, as well as his later days, at length in his memoirs. He spends his time in the countryside, tinkering with inventions, writing poetry, and living off the largesse of a sum imparted to him by the state, almost a consolation prize for inventing the power loom but failing to capitalize on it.

259. **"Incapable of compromise"** Thompson, *Making*, 84.

260. **This precept has held true** See O'Mara, *The Code*, 7.

262. **"soothing his shredded nerves"** Saul David, *Prince of Pleasure*, 335.

262. **"perpetuated the myth"** Bailey, *Luddite Rebellion*, 76.

262. **"Every attempt to create disturbance"** George Pellew, *The Life and Correspondence of Henry Addington, First Viscount Sidmouth* (3 vols., London: John Murray, 1847), 79.

263. **"expressed his concern"** *Parliamentary Register*, vol. 3 (1812), 435. Note the similarities to Ryder's speech justifying the death penalty, and the language lamenting that it had come to this—Addington also says it "was with great pain that his Majesty's servants had found themselves compelled to resort to this step."

264. **"some of the persons"** *Parliamentary Debates*, first series, vol. 23, col. 1036.

264. **"He then asked Parliament for laws"** Sale, *Rebels against the Future*, 170–1.

266. **"I have brought a friend here"** The account is detailed by Thomas Jones Howell in *A Complete Collection of State Trials and Proceedings for High Treason and Other Crimes and Misdemeanors from the Earliest Period to the Year 1783*, vol. 31, 1823.

268. **Henson later wrote** Henson's letters, once more, are drawn from the *Records of the Borough of Nottingham*.

269. **"could not conceive a more monstrous principle"** Hume's discussion—and dismissal—of the bill are carried in the *Parliamentary Debates*, vol. 23, 1248.

271. **Radcliffe was clearly agitated** Radcliffe's aggressive and dubiously legal tactics are documented in Sale, *Rebels against the Future*.

272. **Eventually, the pressure got to a young cropper** William Hall's deposition can be found in the Home Office Papers at HO 42/129 via the *Luddite Bicentenary*. "A new informer appears & Jonathan Dean confesses to being at the 'Rawfolds Fight.'"

273. **When the executioner** William Edward Armytage Axon, *Lancashire Gleanings* (1883), 322.

273. **"The man charged behaved with the greatest effrontery"** *Leeds Mercury*, November 7, 1812; the item first appeared in the Huddersfield *Courier*, Oct. 24, 1812.

274. **"Dear General"** The letter to Maitland can be found in the Home Office Papers at HO 40/2/3.

275. **"In the afternoon"** *Bristol Mirror*, August 29, 1812.

275. **"the town is in much confusion"** *Leeds Mercury,* August 22, 1812.

276. **"On Monday morning"** *Morning Chronicle,* Friday, September 11, 1812.

277. **The Manchester meetings** The episode of the Manchester Luddites is sourced from Knight's letters (see below) and from the account in the Hammonds' *The Skilled Labourer,* 297–300.

278. **"We read of those"** The letter from John Knight to his wife is archived in the Home Office Papers at HO 42/129 and transcribed at the *Luddite Bicentenary* at "John Knight of the 'Manchester 38' writes to his wife from Lancaster Castle."

280. **"On our arrival"** Raynes, *An Appeal to the Public,* 23.

282. **"greatest glory"** *Letters to William Godwin,* vol. 1; *Letters from Percy Bysshe Shelly to William Godwin in Two Volumes,* vol. 1.

282. *A Declaration of Rights* Percy Bysshe Shelley, *A Declaration of* Rights (1812), available online at *The Romantic Circles,* https://romantic-circles.org/editions/shelley/devil/declright.html.

282. **"bluntly told him"** Seymour, *Mary Shelley,* 81.

283. **"clear bright skin"** Seymour, *Mary Shelley,* 85.

283. **"writings reinforced Owen's version"** Paul R. Bernard, "Irreconcilable Opinions: The Social and Educational Theories of Robert Owen and William Maclure," *Journal of the Early Republic,* vol. 8, no. 1 (spring 1988), 21–4.

283. **Both men "condemned political agitation"** Emma Barker, "Robert Owen and New Lanark," The Open University, www.open.edu/openlearn/mod/oucontent/view.php?id=1658.

286. **"seized by a military and civil force"** Byron, *The Parliamentary Speeches of Lord Byron* (London, 1824), 40.

287. **"His political disillusionment was total"** MacCarthy, *Byron,* 200.

287. **"I presume the illuminations"** Marchand, *Byron's Letters and Journals,* vol. 2.

288. **"soothe and moderate the public mind"** Document found in the Home Office Papers at HO 42/166.

288. **"a closer and more efficient combination"** Hammond and Hammond, *Skilled Labourer,* 230.

290. **"burglary under the color of Luddism"** *Leeds Mercury,* June 20, 1812.

290. **They joined dozens of others** *Leeds Mercury,* October 3 and 10, 1812.

291. **the ad he posted** It's worth looking at the expanded text of the advertisement for all the machinery in John Goodair's factory, which ran in the *Manchester Mercury* of September 1 and 15, 1812, for a sense of the full scope of the technologies and machinery in a cutting-edge factory at the time:

> An extensive assortment of very valuable MACHINERY, and EFFECTS, comprising picking machines, carding engines, drawing and fly frames, throstles, mules, warping mills and machines, dressing machines, cup presses, patent looms, straps, cans, skips, bobbins, iron safe, desks, counters, shelves, beams, scales and weights. A large quantity of joiner's, machine maker's and smith's tools, cast-iron, brass, new cards, and a great variety of useful and requisite implements and materials for spinning and manufacturing concerns.
>
> The principal part of the above property is nearly new, was made by mechanics of the greatest eminence, and has been lately in use [...]

The Trial

292. **Six days after the dawn** The trial of George Mellor, Will Thorpe, and Thomas Smith was widely covered by the press. A complete documentation of the courtroom proceedings was printed as *An Historical Account of the Luddites of 1811, 1812, and 1813, with Report of Their*

Trials at York Castle, from the 2nd to the 12th of January, 1813, before Sir Alexander Thompson and Sir Simon Le Blanc, Knights, Judges of the Special Commission (Huddersfield, UK: John Cowgill, 1862).

292. **"The prisoners were all young men"** *An Historical Account,* 21.

293. **"Mr. Webster!" [George] cried** The dialogue, as stated in the text, in this section, between George and the priest, is sourced from Sykes and Walker, *Ben o' Bill's.*

295. **"were made up of men generally unaffected"** William Leman Rede, *York Castle in the Nineteenth Century,* 550–1.

296. **It was Walker who'd drained** *Leeds Mercury,* August 10, 1812.

296. **"Mellor said, the method"** *Report of the Trials at York,* 30.

297. **despite some gaps and contradictions** Some historians do doubt George Mellor's guilt, and attribute his unwillingness to say anything to his steadfast dedication to the Luddite oath.

298. **The only hope** This section follows the trial records and contemporaneous newspaper accounts.

300. **"and the dear ones there"** This quote is from Sykes and Walker, *Ben o' Bill's.*

300. **"I forgive all the world"** George's final words were made famous by the account that ran in the *Leeds Mercury.* Some heard them differently, and a debate over what he actually said on the scaffold ensued in the local newspapers in the weeks following his death. At least one attendee claimed they heard Mellor refer to himself and his fellow Luddites as "us poor murderers," but this suggestion was roundly denied by most others. That the debate raged in these publications gives us some idea of the level of interest and scrutiny of the event by readers at the time — it was a sensational event, but readers also felt passionately about the cause and the characters. One wrote in multiple times defending Mellor from the assertion that he'd confessed, though they admitted to not having been there.

301. **"I have read with astonishment"** William Cartwright's letter to Blackburn is archived in the Home Office Papers at HO 42/132.

303. **After the executions, a Quaker** Thomas Shiltoe's written account is available at the Huddersfield Archives in Yorkshire. The text is also available in the *Luddite Bicentenary* at "A Quaker missionary, Thomas Shillitoe, visits the families of George Mellor, Jonathan Dean & John Walker."

303. **"Harriet Shelley, writing on her husband's behalf"** Russell Smith, "Frankenstein in the Automatic Factory," *Nineteenth-Century Contexts,* vol. 41, no. 3 (2019), 303–19.

304. **A statue of George III** *Leeds Mercury,* April 3, 1813.

304. **It read, "Blood for blood"** *Leeds Intelligencer,* January 25, 1813.

The Invention of the Luddites

305. **"If the Luddites had never existed"** Theodore Roszak, "In Defense of the Living Earth," foreword to Stephanie Mills, ed., *Turning Away from Technology: A New Vision for the Twenty-First Century* (San Francisco: Sierra Club Books, 1997).

305. **Google's online search** As of October 2022, Google defined Luddite as:"Lud·dite /ˈlədˌīt/, noun, DEROGATORY: a person opposed to new technology or ways of working; 'a small-minded Luddite resisting progress.'"

306. **the CEOs, business consultants, and corporate scribes** The articles cited in the footnote are available online at Inc.com, CNBC.com, Forbes.com (behind a paywall), and Reason.com.

307. **"those who acquire it"** The first chapter of Neil Postman's 1992 book *Technopoly: The Surrender of Culture to Technology* (New York: Alfred A. Knopf, 1992) is "The Judgment of Tha-

mus," and opens with a discussion of the debate between Thamus and Theuth, and the question of whether technology is blunting certain capacities for human potential.

308. **"a return of the Luddites"** William Safire, "Return of the Luddites," On Language, *New York Times,* December 6, 1998.

308. **In 2014, conservative commentators** Steven Greenhut, "Uber Faces Backlash from New Luddites," *Chicago Tribune,* March 31, 2014.

310. **"the wicked misrepresentations"** A copy of the proclamation is included in *An Historical Account of the Luddites* (Huddersfield, UK, 1862), 134.

311. **"We proceeded to the mournful house"** The details of this visit are drawn from the same set of writings Thomas Shiltoe made during his previous call.

312. **while George was in prison** A copy of the only confirmed letter written by George Mellor — and one of the only signed letters by any Luddite in existence — is kept in the Home Office Papers at HO 42/132, and another, which I viewed, is in the West Yorkshire Archives. A copy of the full text is transcribed at the *Luddite Bicentenary* (November 30, 1812). Addressed to "Mr. Thomas Hellice, Wool Stapler, Lockwood Nr Huddersfield," the letter reads as follows:

> I now take the Liberty of informing you that I am in good health as by the Blessing of God I hope they will find you all Pl. to give my respects to my cousin and tell him to stick fast by what he [swore] the first time before Rattiffe and I hope his wife will do the same, that I left there house before five oClock and I did not leave any thing at their House and if the Boy swore any thing tell my Cousin to contradict him & say he told him a different Storey that there had been a man and left them & he did not know him and as for the Girl she cannot sware any thing I know that will harm me and tell the Boys to stick by what they said the first time if not [they] are proved forsworn tell him and his wife I hope they will befrend me and never mind thier work for I if I come home I will do my best for them Remember a Soul is of more value than work or Gould — I have heard your are Potitioning for a Parlimenter Reform and I wish thees names to be given as follows — G.M. Mark Hill James Haigh Josh Thornton Wm Thorp Geo Rigg Saml Booth John Hogdin C. Cockcroft Jas Brook Jno Brook Geo Brook James Brook C. Thornton Jonathan Deane Jno Walker Joshua Schorsfiend Jno [Schersfield] Thomas Smith James Starkey — Anthony Walker Joseph Greewood Thomas Green Benjamin Sigg Geo Ludge Wm Hodgson Geo Brook William Barnard Geo Beamont David Moerhouse William Whitehead Joseph Fisher Jon Batley Jon Lum Jon Shore Benjamin Hinchcliff Geo Hanlin Jon Laild Jon Fawset James Whitehouse give my respects to all enquiring Friends and acccept These few Lines from your Friend —

313. **"Sit down"** This section continues to draw from *A Memoir of Robert Blincoe, an Orphan Boy,* recorded by John Brown. Blincoe's visit to Ol' Beckka is recounted on page 56.

315. **From Amazon's Mechanical Turk** Phil Jones discusses the phenomenon of tech companies enabling their "automated" products by employing low-wage workers around the world, often in the global South, and even sourcing labor from refugee camps, in *Work without the Worker: Labour in the Age of Platform Capitalism* (London: Verso, 2021).

315. **"ghost work"** This is the topic of a fine book by researchers Mary L. Gray and Siddharth Suri: *Ghost Work: How to Stop Silicon Valley from Building a New Global Underclass* (Boston: Houghton Mifflin Harcourt, 2019).

315. **"not only did that device"** Alexandra Mateescu and Madeleine Clare Elish studied the impacts of self-checkout and published the findings in a paper called "AI in Context: The

Labor of Integrating New Technologies," *Data & Society*, January 30, 2019. I wrote about their work and the history of automated self-checkout for *Gizmodo* in "Why Self-Checkout Is and Has Always Been the Worst."

What the Entrepreneurs Won

317. **"Mr. William Cartwright"** The festivities of the day, including the address read by Cartwright, were carried exhaustively by the *Leeds Mercury*, June 12, 1813.
317. **"warmest sentiments of gratitude"** This speech is, to me, a striking example of exactly how the class divide was widening, and how the entrepreneurs and owners started thinking of themselves as opposed to their neighbors, the workmen who needed to labor at their machines; the owners celebrated, essentially, victory in battle over the workmen's attempt to maintain autonomy over their lives.
318. **"A full year has now passed over us"** *Leeds Intelligencer*, April 19, 1813.
319. **By March, though, a quiet** Home Office Papers, HO 40/2/3. "Mr Cartwright feels so much confidence in the alter'd disposition of the People, as to have desired me if I thought proper to reduce his Mill Guard," Captain Raynes wrote to the Home Office.
320. **"Central to Ure's book"** Steve Edwards, "Factory and Fantasy in Andrew Ure," *Journal of Design History*, vol. 14, no. 1 (2001), 17–33.
320. **"I think it's worth it"** Harry Davies, Simon Goodley, Felicity Lawrence, Paul Lewis and Lisa O'Carroll, "The Uber Files: Uber Broke Laws, Duped Police, and Secretly Lobbied Governments, Leak Reveals," *Guardian* (London), July 11, 2022.
322. **nothing if not persistent** Gravener Henson's post-Luddite life is described in William Felkin's autobiography, in his own letters kept in the *Record of the Borough of Nottingham*, and in Thompson, *Making*. Thompson dedicates significant space to highlighting Henson's persistent efforts to enact reform, and gives him the honor of being one of three major unsung figures in ushering in the era of class-consciousness in England.
324. **Henson was also rumored to have written** The full title of Henson's book is *The Civil, Political, and Mechanical History of the Framework-Knitters in Europe and America, Including a Review of the Political State and Condition of the People of England, and an Account of the Rise, Progress, and Present State of the Machinery for Superseding Human Labour, in the Various Manufactures of Europe, since the Fifteenth Century, Exhibiting the True and Real Cause of the Present Convulsed State of the Western World.* It was published in Nottingham in 1831.
325. **At the end of 1813, Charles enlisted in the army** Charles may have watched those rockets' red glare illuminating the battlefield, though they failed to seriously puncture the forts' defenses. The missiles were designed by the British military scientist William Congreve, who had studied and developed rocketry for England throughout the first decade of the 1800s, after seeing them effectively used against the East India Company as it undertook colonial expansion on the subcontinent. Congreve's efforts were so celebrated that not only was he anointed to the Royal Society, but he became close friends of the Prince Regent, who directly supported his work. The rockets proved "very effective in burning down much of the town of Boulogne" in an 1806 attack, and in setting fire to large swaths of Copenhagen in 1807. They were a terror tactic, and a successful one. In 1811, as the Luddite uprisings were gaining momentum in England, Congreve was appointed a personal equerry by the Prince Regent.
328. **"Brethren, the time has come"** Quoted in Matt Sandler, *The Black Romantic Revolution: Abolitionist Poets at the End of Slavery* (London: Verso, 2020), 25.
328. **"Hereditary bondsmen!"** The famous line in *Childe Harold* appears in stanza LXXVI.

328. **"This little bit of British Romantic poetry"** Sandler, *Black Romantic Revolution,* 27.

329. **"Byron is in many ways"** Clara Tuite, "Lord Byron's Preposterous Liberalism: Perversity, or the Fear That Pleases," *Occasion,* vol. 11 (2019).

329. **"our one great interpreter"** John Buchan, ed., *A History of English Literature* (New York: Thomas Nelson, 1948), 433. Some further thoughts here worth considering, courtesy of Romantic scholar Sarah Wootton of Durham University: "Byron has a significant contribution to the ideology of liberalism: Byron was associated with liberty, rebellion, and revolution throughout the nineteenth century, and writers such as Thomas Carlyle and John Stuart Mill considered him seriously as a political poet in the early Victorian period. Mill's progressive liberalism did not stretch to Byron's 'transgressive eloquence,' however, a poetics that spoke to, instead of on behalf of, the masses." Sarah Wootton, *Byronic Heroes in Nineteenth-Century Women's Writing and Screen Adaptation* (London: Palgrave Macmillan, 2016).

Part V: The Modern Prometheus

333. **To pass the time** The anthology was called *Fantasmagoriana,* a French translation of German ghost stories, published in 1812.

333. **The tales, about cursed spirits** Mary recalled being struck by a tale called "The History of the Inconstant Lover," "who, when he thought to clasp the bride to whom he had pledged his vows, found himself in the arms of the pale ghost of her whom he had deserted," and another one about "the sinful founder of his race, whose miserable doom it was to bestow the kiss of death on all the younger sons of his fated house, just when they reached the age of promise."

333. **"His proposition"** Mary Shelley, preface to *Frankenstein* (1831 ed.).

334. **"As the daughter of two persons of distinguished literary celebrity"** Mary Shelley, preface to *Frankenstein* (1831 ed.).

334. **stories of an obsessed young alchemist** Seymour, *Mary Shelley,* 111:

> Any local hoping to earn a few coins with a good story would have told the young travelers about Konrad Dippel [a pastor's son born at Castle Frankenstein]. After studying alchemy at university, he became a fashionable physician whose dream was to buy and live in his birthplace. (He liked to sign himself as Dippel Frankensteina, Dippel of Frankenstein.) Chased out of Strasbourg after allegations that he had been robbing graveyards for his anatomical experiments, Dippel was convinced that he could bring a body back to life by injecting it with a concoction of blood and bone, often made from both mammal and human corpses.

335. **Byron later published the story** Byron's poem, "A Fragment," was published on June 17, 1816.

336. **("not devoid of imagination")** Betsy Morais explores the legacy and technology philosophy of Augusta Ada Byron in her *New Yorker* piece, "Ada Lovelace, the First Tech Visionary," *The New Yorker,* October 15, 2013, as does Claire L. Evans in her book *Broad Band: The Untold Story of the Women Who Made the Internet* (New York: Portfolio, 2018). "I do not believe that my father was (or ever could have been) such a Poet as I shall be and Analyst; (& Metaphysician)," Lovelace wrote in one of her letters to Babbage, per Evans, 15.

336. **" My sister is now with me"** Marchand, *Byron's Letters and Journals,* vol. 3.

336. **a band of Luddites launched their last major attack** *Nottingham Review,* June 29, 1816.

336. **"Are you not near the Luddites?"** Marchand, *Byron's Letters and Journals,* vol. 3, 349.

337. **the "vulgar masters"** Shelley, *Queen Mab* (1813), 5.64–78.

338. **the prospect of a bloody revolution made her "shudder"** Betty T. Bennett, ed., *The Letters of Mary Wollstonecraft* (Baltimore: Johns Hopkins University Press, 3 vols., 1980–1988), vol. 1, 49. Mary's sympathy for the Luddites, and her passion for social reform, were tempered by a fear of

revolutionary violence, as Paul O'Flinn observed: "Her politics,…in short, are those of a radical liberal agonising in the face of the apparent alternatives of 'anarchy and oppression'" (Paul O'Flinn, "Production and Reproduction: The Case of *Frankenstein,*" in *Popular Fictions: Essays in Literature and History* [London: Methuen, 1986], 194–213). The phrase "anarchy and oppression" is in Percy Shelley, "An Address to the People on the Death of the Princess Charlotte" (1817).

339. **"Frankenstein's monster shares many characteristics"** Russell Smith, "Frankenstein in the Automatic Factory," *Nineteenth-Century Contexts,* vol. 41, no. 3 (2019), 303–319.

340. **"These workers have the potential"** Anne K. Mellor, *Mary Shelley, Her Life, Her Fiction, Her Monsters* (New York and London: Routledge, 1989), 112.

340. **"Frankenstein, as creator," writes Edith Gardner,** Edith Gardner, "Revolutionary Readings: Mary Shelley's *Frankenstein* and the Luddite Uprisings," *Iowa Journal of Cultural Studies,* vol. 13 (1994), 70–91.

341. **Mary Shelley said that the last burst** The full quote describing that burst of inspiration comes from the preface to the 1831 edition:

> Many and long were the conversations between Lord Byron and Shelley, to which I was a devout but nearly silent listener. During one of these, various philosophical doctrines were discussed, and among others the nature of the principle of life, and whether there was any probability of its ever being discovered and communicated. They talked of the experiments of Dr. Darwin, (I speak not of what the Doctor really did, or said that he did, but, as more to my purpose, of what was then spoken of as having been done by him,) who preserved a piece of vermicelli in a glass case, till by some extraordinary means it began to move with voluntary motion. Not thus, after all, would life be given. Perhaps a corpse would be reanimated; galvanism had given token of such things: perhaps the component parts of a creature might be manufactured, brought together, and endued with vital warmth.

341. **In his seminal history** Brian Wilson Aldiss, *Billion Year Spree: The History of Science Fiction* (London: Weidenfeld and Nicolson; New York: Doubleday, 1973), 29.

341. **Today, Victor Frankenstein** Notable recent *Dr. Frankenstein*s include Oscar Isaac's AI startup founder in *Ex Machina*, Eric Andre's tech bro parody in *The Mitchells vs. the Machines*, James Spader's Ultron in *The Avengers 2*, and so many more.

341. **In the *Terminator* franchise** Like Cyberdyne Systems's HQ in the *Terminator* films, Yahoo! and LinkedIn are both based in Sunnyvale, California, as is a division of Lockheed Martin, one of the nation's biggest defense contractors.

342. **This is a fear, even today** Even the most influential AI in cinema history, HAL-9000 from *2001: A Space Odyssey,* cannot escape ominous corporate ties; fans have pointed out that HAL is just one letter in each case removed from IBM, and HAL states at one point that he was activated in Urbana, Illinois, where a real-life HAL Communications corporation is centered. Even if Arthur C. Clarke and Stanley Kubrick did not intend for it to be the case, HAL is inseparable from the monolithic IBM of the '60s (IBM consulted on the film, and its logo is even visible on some of the technology products featured) and works as a critique of the amount of control we have ceded to computers.

342. **In the *Communist Manifesto*** Karl Marx and Frederick Engels, *Manifesto of the Communist Party,* trans. Samuel Moore, in *Selected Works* (3 vols., Moscow: Progress Publishers, 1973), vol. 1, via *Marxist Internet Archive,* www.marxists.org/archive/marx/works/1848/communist-manifesto/index.htm.

343. **The final portion of the novel** It might be noted as well that Mary Shelley grounded her novel in a realistic-feeling premise, presenting it as steely-eyed truth: "The event on which

the interest of the story depends is exempt from the disadvantages of a mere tale of spectres or enchantment," as the book's meta-narrator, Marlow, explains in the opening pages. (Marlow is the captain who, at the story's conclusion, finds Dr. Frankenstein alone and freezing in the Arctic.) This choice, too, is interesting: Shelley comments on Frankenstein through the perspective of another male explorer, who rescues him, and is impressed and awed by him, even after hearing about all the hell and suffering he has caused. This is how tech CEOs are commonly treated, even after they've proven capable of unleashing disaster — sympathetically and patiently. Pundits and policy makers cannot help but applaud them, if they are bold enough in their ambition, even as they cause suffering. Perhaps Mary Shelley saw the danger in this reverence, how it feeds a culture that begets similarly reckless innovators. Ultimately, *Frankenstein* is about what happens when we fail to tend to those who fall under the weight of such ambitions, such recklessness: disaster. Furthermore, Shelley's explorer is lonely and isolated, and his hero-worship of the great and troubled doctor, in whom he presumably sees himself, presages the cult of Silicon Valley introverts recognized in pop culture as troubled but brilliant.

343. **"Learn from me"** At the end of the first chapter, Frankenstein's family is described as egalitarian, balanced — "neither of us possessed the slightest pre-eminence over the other; the voice of command was never heard amongst us; but mutual affection engaged us all to comply with and obey the slightest desire of each other."

345. **"broadened our view"** Peter Linebaugh, *Ned Ludd and Queen Mab: Machine-Breaking, Romanticism, and the Several Commons of 1811–12* (Oakland, CA: PM Press, 2012), 8.

What the Luddites Won

346. **"lie down and die"** Taken from Frank Peel's sarcastic remarks, first quoted on page 44.

346. **"You don't see a flood"** Quotations here and below are from an interview with Adrian Randall conducted by the author in 2020.

347. **they sent threatening letters** Here's one from 1811, near the height of the Luddite uprisings, that was sent to the *London Gazette* and published on September 7, 1811. "Blood and Vengance against Your Life and Your Property for taking away our Labour with Your Threshing Machine. Seven of us near your Dwelling House have agreed that if you do not refrain from Your Threshing Machine we will Thresh Your Rick with Fire & Bathe Your Body in Blood."

347. **"of all the machine-breaking movements"** Eric Hobsbawm and George Rudé, *Captain Swing* (London: Lawrence and Wishart, 1969), 460.

348. **in "the present tense"** David Noble, *Progress without People: New Technology, Unemployment, and the Message of Resistance* (Toronto: Between the Lines, 1995).

348. **"discredited once and for all"** Geoffrey Bernstein. Unpublished paper cited by David Noble, "General Ludd and Captain Swing: Machine Breaking as Tactic and Strategy" (1981).

348. **There was no natural, united drive toward progress** In order to answer for this, the study of political economy sprang forth; as Maxine Berg explains in *Machinery Question*, the business class "had to find an explanation for the economic and social impact of the machine. Expressions of wonder at the technical perfection of the machinery were not adequate. It was thus that the middle class took to itself a 'scientific' theory, political economy." These new political economists "issued long and turgid justifications of the introduction of machinery," and were "above all others…either optimistic or blind, and possibly both, to the conditions of the working class." In other words, they forged the quasi-scientific rationale that continues to justify and elevate technological progress above many other political priorities, even today. The political economists answered the machinery question by

emphasizing the unassailable benefits of technology and automation first, and considering the plight of the worker second — just as is done today.

349. **"Violence and threats worked"** Binfield, *Writings*, 50.

349. **"Since machine-breaking"** Raymond Boudon, *The Analysis of Ideology*, trans. Malcolm Slater (Chicago: University of Chicago Press, 1989), 96.

349. **"England's loss"** John Baker, quoted in David Noble, "Present Tense Technology: Technology's Politics" (part 3, in *Democracy*, vol. 3, no. 3 [fall 1983], 71–93), 79. "Where the worker responses were active, positive, and assertive on their own immediate interests," Baker noted, "these attitudes flowed through the rest of society with rather positive consequences for most institutions of society."

349. **Australia led the world** Discussed in Noble, *Progress without People*.

350. **"planted time bombs that damaged the Vitaphone"** Gavin Mueller, *Breaking Things at Work: The Luddites Were Right about Why You Hate Your Job* (London and New York: Verso, 2021), 53.

350. **"a string of wildcat strikes"** Noble, *Progress without People*, 27.

350. **"There is a time when the operation"** Marco Savio. Footage can be found on YouTube, https://www.youtube.com/watch?v=lsO_SlA7E8k, as archived by the Human Rights Foundation.

350. **In 1982, a tech worker** "*Sabotage*: The Ultimate Video Game" appeared in issue no. 5 of *Processed World*, a radical magazine based in San Francsico that took aim at the tech sector and office culture. It's collected in Chris Carlsson, ed., *Bad Attitude: The "Processed World" Anthology* (London: Verso, 1990).

351. **The media-shy novelist** Thomas Pynchon, "Is It OK to Be a Luddite?," *New York Times*, October 28, 1984.

Fear Factories

354. **"The term Factory"** Ure, *Philosophy of Manufactures*, 13.

354. **"Even before the use"** Thompson, *Making*, 281.

355. **"To enter the mill"** Thompson, *Making*, 306.

355. **Remote workers must log on to increasingly** Kate Dwyer, "Don't Worry, We're Not Actually Monitoring Your Productivity," *New York Times*, August 19, 2022.

356. **"Jobs are not just the source of money"** Anne Case and Angus Deaton, *Deaths of Despair and the Future of Capitalism* (Princeton, NJ: Princeton University Press, 2020), 8.

356. **"The separation of workplace and home"** Andrea Komlosy, *Work: The Last 1,000 Years*, trans. Jacob K. Watson with Loren Balhorn (London and New York: Verso, 2018), 1.

356. **"The greater evil"** Letter from Joseph Coope, transcribed in Randall, *Before the Luddites*, 218.

356. **"the family was together"** Thompson, *Making*, 307.

357. **"Increasingly it became the practice"** Randall, *Before the Luddites*, 202.

357. **"Many who were masters"** Pat Hudson, *The Genesis of Industrial Capital: A Study of West Riding Wool Textile Industry, c. 1750–1850* (Cambridge: Cambridge University Press, 1986), 35.

357. **"The factory was recognized"** Randall, *Before the Luddites*, 219.

357. **"In the early 1900s"** Saima Akhtar, "Employers' New Tools to Surveil and Monitor Workers Are Historically Rooted," *Washington Post*, May 6, 2021.

358. **"Amazon uses such tools"** Nandita Bose, "Amazon's Surveillance Can Boost Output and Possibly Limit Unions — Study," Reuters, September 15, 2020.

358. **These Amazon workers are paid so little** Lauren Kaori Gurley, "A Homeless Amazon Warehouse Worker in New York City Tells Her Story," *Vice,* June 18, 2021.

358. **Tesla CEO Elon Musk, meanwhile** Paris Marx, *Road to Nowhere: What Silicon Valley Gets Wrong about the Future of Transportation* (London and New York: Verso, 2022).

358. **"Yes, excessive automation at Tesla"** Elon Musk, Twitter post, April 13, 2018, 12:54 p.m., http://twitter.com/elonmusk: "Yes, excessive automation at Tesla was a mistake. To be precise, my mistake. Humans are underrated."

358. **"fauxtomation"** Astra Taylor, "The Automation Charade," *Logic,* no. 5 (August 1, 2018), https://logicmag.io/failure/the-automation-charade/.

359. **"The gig economy promises flexibility"** Alexandrea J. Ravenelle, *Hustle and Gig: Struggling and Surviving in the Sharing Economy* (Berkeley: University of California Press, 2019), 15.

360. **The economic historian Louis Hyman** The story of the rapid corporate assault on workers' benefits and the transition to part-time work is the focus of Louis Hyman's book *Temp: How American Work, American Business, and the American Dream Became Temporary* (New York: Viking, 2018).

360. **It's difficult to definitively gauge** Yuki Noguchi, "Freelanced: The Rise of the Contract Workforce," *All Things Considered,* NPR, January 22, 2018.

361. **Contract work is on the rise** Daisuke Wakabayashi, "Google's Shadow Work Force: Temps Who Outnumber Full-Time Employees," *New York Times,* May 28, 2019.

361. **Journalism, law, and marketing** Noguchi, "Freelanced."

361. **In 2021, *Vice* reported** Edward Ongweso Jr., "The Military Is Creating a 'Gig Eagle' App to Uber-ize Its Workforce," *Motherboard/Vice,* May 20, 2021.

361. **"subscription law enforcement service"** Joseph Cox, "'Find This Fuck': Inside Citizen's Dangerous Effort to Cash In on Vigilantism," *Motherboard/Vice,* May 27, 2021.

361. **"These are very, very precarious workforces"** I interviewed Veena Dubal multiple times for this book and for related stories. These quotes are from our interviews.

Part VI: The Owners of the New Machine Age

367. **"Companies do not care"** Douglas Schifter, Facebook post, February 5, 2018.

368. **He was proud** Ginia Bellafante, "A Driver's Suicide Reveals the Dark Side of the Gig Economy," *New York Times,* February 6, 2018.

368. **With his large frame** Jessica Bruder, "Driven to Despair," *New York,* May 2018.

The Great Comet Returns

369. **Closed for Business signs** "Retail Trade Employment: Before, During, and After the Pandemic," Bureau of Labor Statistics, *Beyond the Numbers,* vol. 11, no. 4 (April 2022).

370. **By the end of the '00s** According to *Forbes* in 2007, "The high-tech hub of San Jose leads the list of *Forbes* 400 members per capita." San Francisco was second.

372. **France's interior minister** "French Government Vows to Shut Down Uber," *Deutsche Welle,* June 26, 2015.

372. **"We are facing extinction"** Douglas Schifter, *Black Car News,* quoted in Jessica Bruder, "Driven to Despair."

The New Tech Titans

373. **"We're going to see our already record-high inequality"** This and many subsequent quotes are taken from an interview I conducted with the then presidential candidate in 2018.

374. **Mass production of self-driving** "Ease the Transition to Self-Driving Vehicles," Yang2020 .com.

374. **"There's going to be a lot of passion"** *The Joe Rogan Experience*, episode 1245, February 12, 2019. YouTube video, 1:52:02, youtu.be/cTsEzmFamZ8?t=1390.

375. **Yang even proposed policy solutions** Recall, some weavers were hoping to institute a tax on cloth produced by machinery to help pay unemployment funds for the workers they displaced. (See page 000.)

375. **He would use the proceeds** "How a New Hampshire Family Spent Andrew Yang's 'Freedom Dividend,'" Reuters, February 10, 2020.

375. **With money pouring into AI** There's no better example of this kind of prediction than the book *The Second Machine Age: Work, Progress, and Prosperity in a Time of Brilliant Technologies* (New York: Norton, 2014), by MIT economists Erik Brynjolfsson and Andrew McAfee.

375. **"fourth industrial revolution"** Njuguna Ndung'u and Landry Signé, "The Fourth Industrial Revolution and Digitization Will Transform Africa into a Global Powerhouse," Brookings Institution, January 8, 2020.

375. **"I'm hanging out with the tech wizards"** *Joe Rogan Experience*, YouTube video, 46:37.

376. **A 2020 report from the Economic Policy Institute** Lawrence Mishel and Jori Kandra, "CEO Compensation Surged 14 Percent in 2019 to $21.3 Million," *Economic Policy Institute*, August 18, 2020. "CEOs now earn 320 times as much as a typical worker."

376. **"the richest 0.1 percent"** David Leonhardt and Yaryna Serkez, "America Will Struggle after Coronavirus. These Charts Show Why," *New York Times*, April 10, 2020.

376. **"Why does everybody suddenly hate billionaires?"** Roxanne Roberts, "Why Does Everybody Suddenly Hate Billionaires? Because They've Made It Easy," *Washington Post*, March 13, 2019.

376. **Polls conducted in 2021** Theodore Schleifer, "What Americans Really Think about Billionaires during the Pandemic," *Vox*, Mar 30, 2021.

Fear Factories Redux

377. **At the beginning of March** Christian Smalls's story has been reported on at length in the media, with key stories coming from the *New York Times*, *Motherboard/Vice*, and *Bloomberg*, among many others.

378. **Then one of his colleagues tested positive** Ginia Bellafante, "'We Didn't Sign Up for This': Amazon Workers on the Front Lines," *New York Times*, April 3, 2020.

378. **"He's not smart"** Paul Blest, "Leaked Amazon Memo Details Plan to Smear Fired Warehouse Organizer: 'He's Not Smart or Articulate,'" *Vice News*, April 2, 2020.

378. **Talk of unions was rippling** Lauren Kaori Gurley and Joseph Cox, "Inside Amazon's Secret Program to Spy On Workers' Private Facebook Groups," *Motherboard/Vice*, September 1, 2020; and Lauren Kaori Gurley and Janus Rose, "Amazon Employee Warns Internal Groups They're Being Monitored for Labor Organizing," *Motherboard/Vice*, September 24, 2020.

379. **"Amazon, along with Walmart"** Luis Feliz Leon, "Workers of the World Unite against Amazon," *In These Times*, April 2021.

379. **Until then, Amazon will continue** Jodi Kantor, Karen Weise, and Grace Ashford, "The Amazon That Customers Don't See," *New York Times*, June 15, 2021.

380. **Analysts at *MarketWatch* estimate** Rex Nutting, "Amazon Is Going to Kill More Jobs than China Did," *MarketWatch*, March 15, 2017.

380. **one June 2021 report** "Amazon Warehouse Injuries '80% higher' than Competitors, Report Claims," *BBC News,* June 2, 2021. The report was commissioned by the Strategic Organizing Center (SOC).

380. **When he was declared dead** Michael Sainato, " 'Go Back to Work': Outcry over Deaths on Amazon's Warehouse Floor," *Guardian,* October 18, 2019.

Gig Workers Rising

382. **"I hope Tammy didn't fall asleep"** I have attended a number of protests and strike actions with gig workers in Los Angeles and San Francisco and spoken with numerous workers and organizers about the growing movement. I joined the Gig Workers Rising on a Zoom call on Election Night 2020 as the results for Prop 22 came in.

382. **But the reason pay was so high** In one particularly striking example, Uber raised $3.5 billion from Saudi Arabia's Sovereign Wealth Fund in 2016.

383. **Yet the drivers who make it all possible** Because Uber and Lyft do not make their earnings or payment data publicly available, it is hard to confirm figures, but a number of studies analyzing driver earnings in the wild have been carried out. Here's a reputable one: James A. Parrott and Michael Reich, "A Minimum Compensation Standard for Seattle TNC Drivers," *Report for the City of Seattle,* July 2020, carried out in part by UC Berkeley's Center on Wage and Employment Dynamics.

The New Luddites

388. **"We need a Luddite revolution"** Ben Tarnoff, "To Decarbonize We Must Decomputerize: Why We Need a Luddite Revolution," *Guardian,* September 18, 2019.

388. **The science-fiction author Cory Doctorow** Cory Doctorow, "Science Fiction Is a Luddite Literature," *Locus,* January 3, 2022; also mentioned in a post on *Medium,* November 15, 2021, https://doctorow.medium.com/science-fiction-is-a-luddite-literature-e454bf5a5076.

388. **The tech worker and author** Wendy Liu, *Abolish Silicon Valley: How to Liberate Technology from Capitalism* (London: Repeater, 2020).

389. *Techlash* **was shortlisted** "Word of the Year 2018: Shortlist," *Oxford Languages,* https://languages.oup.com/word-of-the-year/2018-shortlist/. As of 2023, the *OED Online* defines techlash as: "Originally: opposition to digital or computer technology. In later use: *spec.* a strong and widespread negative reaction to the far-reaching power and influence of large technology companies, esp. in relation to their control of personal data, social media, regulation of online access and content, etc."

391. **"My hope is that recognizing Luddism"** Mueller, *Breaking Things at Work,* 135.

393. **"It's not about money"** The Century Foundation, panel discussion, "What Can Labor Do to Build on This Unusually Promising Moment?," July 28, 2022, Steven Greenhouse, Christian Smalls.

394. **At Amazon, even before the ALU's historic victory** I attended a handful of events Smalls headlined to talk about the labor movement, and spoke with him about organizing at Amazon in Los Angeles. The quotes about Amazon Go stores are taken from an event at Stories bookstore in LA.

398. **He published the post to his Facebook page** Jessica Bruder, "Driven to Despair," *New York,* May 2018.

399. **But like the Luddites** By the end of the next year, the city had enacted a cap on the number of Uber and Lyft drivers that could operate in the çity, extending a lifeline to Schifter's taxi-driving peers.

NOTES

Afterword

400. **On the one hand, the answer is no** Aaron Benanav, *Automation and the Future of Work* (London and New York: Verso, 2020).

401. **Since the Luddites' day** Parts of this epilogue are adapted from an essay that originally appeared in *Gizmodo* in 2019.

403. **"The automation discourse"** Quotes are from an interview I conducted with Aaron Benanav in 2020.

404. **"Young Rossum is the modern scientist"** Karel Čapek sat for an interview with the London *Saturday Review* in 1921.

404. **"The automation theory"** Here's Aaron Benanav on why we keep gravitating to automation theory, and why it's so inconvenient for economists and policymakers even if, in fact, it's not true that robots are coming to take our jobs — and we'd then have to fundamentally reorganize our economy on a scale far greater than anything most economists want to consider:

> If on the other hand, if what I am saying is true, and the economy is running more slowly, then the story becomes a lot more complicated. We can't just change distribution. We also have to change production. We have to figure out how to redistribute not just incomes but also work. We have to figure out how to make the economy less dependent on investment or we create a different system for managing investment. Anything of that sort would require going against the interests of very powerful people, with a vested interest in things not changing. Many people benefit from stagnation, because it means that workers are insecure and hence more docile, and politics are beholden to businesses as "job creators," and so on.

406. **A Brookings Institution study** Mark Muro, Jacob Whiton, and Robert Maxim, "Automation Perpetuates the Red-Blue Divide," *Brookings,* March 19, 2019.

407. **"The American economy runs on poverty"** Ezra Klein, "What the Rich Don't Want to Admit about the Poor," *New York Times,* June 13, 2021.

408. **digital piecework** Veena Dubal, "Digital Piecework," *Dissent,* fall 2020.

SELECTED BIBLIOGRAPHY

Aldiss, Brian Wilson. *Billion Year Spree: The True History of Science Fiction.* Garden City, NY: Doubleday; London: Weidenfeld and Nicolson, 1973.

Armstrong, Isobel, and Virginia Blain, eds. *Women's Poetry in the Enlightenment: The Making of a Canon, 1730–1820.* London: Palgrave Macmillan, 1999.

Axon, William Edward Armytage. *Lancashire Gleanings.* London, 1883.

Babbage, Charles. *On the Economy of Machinery and Manufactures.* London: Charles Knight, 1832.

Bailey, Brian. *The Luddite Rebellion.* New York: New York University Press, 1998.

Ball, Charles. *Slavery in the United States: A Narrative of the Life and Adventures of Charles Ball, a Black Man, Who Lived Forty Years in Maryland, South Carolina and Georgia, as a Slave, Under Various Masters, and Was One Year in the Navy with Commodore Barney, During the Late War.* New York, 1837. Reprinted as *Fifty Years in Chains; or, The Life of an American Slave.* New York, 1859.

Bamford, Samuel. *Passages in the Life of a Radical.* London, 1841.

Benanav, Aaron. *Automation and the Future of Work.* London and New York: Verso, 2020.

Benjamin, Ruha. *Race after Technology: Abolitionist Tools for the New Jim Code.* Cambridge, UK, and Medford, MA: Polity, 2019.

Bennett, Betty T., ed. *The Letters of Mary Wollstonecraft Shelley.* 3 vols. Baltimore: Johns Hopkins University Press, 1980–1988.

Berg, Maxine. *The Machinery Question and the Making of Political Economy, 1815–1848.* Cambridge: Cambridge University Press, 1980.

Binfield, Kevin. "Industrial Gender: Manly Men and Cross-Dressers in the Luddite Movement." In Jay Losey and William Dean Brewer, eds., *Mapping Male Sexuality: Nineteenth-Century England.* Teaneck, NJ: Fairleigh Dickinson University Press, 2000.

Binfield, Kevin, ed. *Writings of the Luddites.* Baltimore: Johns Hopkins University Press, 2004.

Blackner, John. *The History of Nottingham, Embracing Its Antiquities, Trade, and Manufactures, from the Earliest Authentic Records, to the Present Period.* Nottingham, UK: Sutton and Son, 1815.

Boldrin, Michele, and David K. Levine. *Against Intellectual Monopoly.* Cambridge: Cambridge University Press, 2008.

Boudon, Raymond. *The Analysis of Ideology.* Translated by Malcolm Slater. Chicago: University of Chicago Press, 1989.

Brontë, Charlotte. *Shirley, A Tale.* London, 1849.

Brooke, Alan, and Leslie Kipling. *Liberty or Death: Radicals, Republicans and Luddites 1728–1828.* Huddersfield, UK: Huddersfield Local History Society, 2012.

Brown, John. *A Memoir of Robert Blincoe, an Orphan Boy; Sent from the Workhouse of St. Pancras, London, at Severn Years of Age, to Endure the Horrors of a Cotton-Mill, through his Infancy and Youth, with a Minute Detail of His Sufferings, Being the First Memoir of the Kind Published.* Manchester, UK: J. Doherty, 1832.

SELECTED BIBLIOGRAPHY

Brynjolfsson, Erik, and Andrew McAfee. *The Second Machine Age: Work, Progress, and Prosperity in a Time of Brilliant Technologies*. New York: W. W. Norton, 2014.

Carlsson, Chris, ed. *Bad Attitude: The "Processed World" Anthology*. London: Verso, 1990.

Carlyle, Thomas *Chartism*. London: James Fraser; Boston: Little, Brown, 1840.

Case, Anne, and Angus Deaton. *Deaths of Despair and the Future of Capitalism*. Princeton, NJ: Princeton University Press, 2020.

Crouzet, François. *Formation of Capital in the Industrial Revolution*. London: Methuen, 1972.

Crump, William Bunting, and Gertrude Ghorbal, *History of the Huddersfield Woollen Industry*. Huddersfield, UK, 1935.

Darvall, F[rank]. O[gley]. *Popular Disturbances and Public Order in Regency England: Being an Account of the Luddite and Other Disorders in England During the Years 1811–1817 and of the Attitude and Activity of the Authorities*. Oxford: Oxford University Press, 1934.

David, Saul. *Prince of Pleasure: The Prince of Wales and the Making of the Regency*. New York: Grove, 1999.

Defoe, Daniel. *A Tour thro' the Whole Island of Great Britain*. London, 1724.

Dick, Philip K. *Do Androids Dream of Electric Sheep?* Garden City, NY: Doubleday, 1968.

Erickson, Carolly. *Our Tempestuous Day: A History of Regency England*. New York: William Morrow, 1986.

Evans, Claire L. *Broad Band: The Untold Story of the Women Who Made the Internet*. New York: Portfolio, 2018.

Felkin, William. *A History of the Machine-Wrought Hosiery and Lace Manufactures*. Cambridge, UK, 1867.

Fitton, R. S., *The Arkwrights: Spinners of Fortune*. Manchester, UK: Manchester University Press. 1989.

Fitton, R. S., and Alfred P. Wadsworth. *The Strutts and the Arkwrights, 1758–1830: A Study of the Early Factory System*. Manchester, UK: Manchester University Press, 1958.

Franklin, John Hope, and Loren Schweninger. *Runaway Slaves: Rebels on the Plantation*. New York: Oxford University Press, 1999.

Furniss, Tom. "Mary Wollstonecraft's French Revolution." In *The Cambridge Companion to Mary Wollstonecraft*, ed. Claudia Johnson. Cambridge: Cambridge University Press, 2002.

Gardiner, William. *Music and Friends; or, Pleasant Recollections of a Dilettante*. 2 vols. London, 1838.

Gardner, Edith. "Revolutionary Readings: Mary Shelley's *Frankenstein* and the Luddite Uprisings." *Iowa Journal of Cultural Studies*, vol. 13 (1994), 70–91.

Gaskill, Elizabeth Cleghorn. *The Life of Charlotte Brontë*. Cambridge: Cambridge University Press, 2010.

Godwin, William. *An Enquiry Concerning Political Justice*. London, 1793.

Gray, Mary L., and Siddharth Suri, *Ghost Work: How to Stop Silicon Valley from Building a New Global Underclass*. Boston: Houghton Mifflin Harcourt, 2019.

Hammond, J. L., and Barbara Hammond. *The Skilled Labourer: 1760–1832*. London: Longmans, Green, 1919.

Henson, Gravener. *The Civil, Political, and Mechanical History of the Framework-Knitters in Europe and America, Including a Review of the Political State and Condition of the People of England, and an Account of the Rise, Progress, and Present State of the Machinery for Superseding Human Labour, in the Various Manufactures of Europe, since the Fifteenth Century, Exhibiting the True and Real Cause of The Present Convulsed State of the Western World*. Nottingham, UK, 1831.

An Historical Account of the Luddites of 1811, 1812, and 1813, with Report of Their Trials at York Castle, from 2nd to the 12th January, 1813, before Sir Alexander Thompson and Sir Simon Le Blanc, Knights, Judges of the Special Commission. Huddersfield, UK: John Cowgill, 1862.

SELECTED BIBLIOGRAPHY

Hobsbawm, Eric. *The Age of Revolution: 1789–1848.* New York: Vintage, 1996.

Hobsbawm, Eric. *Industry and Empire: The Making of Modern English Society, from 1750 to the Present Day.* New York: Pantheon, 1968.

Hobsbawm, Eric. *Labouring Men: Studies in the History of Labour.* London: Weidenfeld and Nicolson, 1964; Charlottesville: University of Virginia Press, 1965. Reprint, New York: Basic Books, 2009.

Hobsbawm, Eric. "The Machine Breakers." *Past and Present,* no. 1 (February 1952), 57–70. Reprinted in Hobsbawm, *Labouring Men,* 5–22.

Hobsbawm, Eric, and George Rudé. *Captain Swing.* London: Lawrence and Wishart, 1969. Reprint, London and New York: Verso, 2014.

Honeyman, Katrina. *Child Workers in England, 1780–1820: Parish Apprentices and the Making of the Early Industrial Labour Force.* London: Routledge, 2007.

Hudson, Pat. *The Genesis of Industrial Capital: A Study of West Riding Wool Textile Industry, c. 1750–1850.* Cambridge: Cambridge University Press, 1986.

Hyman, Louis. *Temp: How American Work, American Business, and the American Dream Became Temporary.* New York: Viking, 2018.

Isaac, Mike. *Super Pumped: The Battle for Uber.* New York: W. W. Norton, 2019.

Jackson, George. *The Bath Archives: A Further Selection from the Diaries and Letters of Sir George Jackson, from 1809 to 1816.* 2 vols. London, 1873.

Jones, Frederick L., ed. *The Letters of Percy Bysshe Shelley.* 2 vols. Oxford: Clarendon Press, 1964.

Jones, Phil. *Work without the Worker.* London: Verso, 2021.

Kessler, Sarah. *Gigged: The End of the Job and the Future of Work.* New York: St. Martin's Press, 2018.

Komlosy, Andrea. *Work: The Last 1,000 Years.* Trans. Jacob K. Watson with Loren Balhorn. London and New York: Verso, 2018.

Linebaugh, Peter. *Ned Ludd and Queen Mab: Machine-Breaking, Romanticism, and the Several Commons of 1811–12.* Oakland, CA: PM Press, 2012.

Liu, Wendy. *Abolish Silicon Valley: How to Liberate Technology from Capitalism.* London: Repeater, 2020.

MacCarthy, Fiona. *Byron: Life and Legend.* New York: Farrar, Straus & Giroux; London: John Murray, 2002.

Malm, Andreas. *How to Blow Up a Pipeline.* London and New York: Verso, 2021.

Marchand, Leslie A., ed. *Byron's Letters and Journals.* 12 vols. London: John Murray; Cambridge, MA: Belknap Press of Harvard University Press, 1973–82.

Marshall, Florence Ashton. *The Life and Letters of Mary Wollstonecraft Shelley.* 2 vols., London, 1889.

Marx, Karl, and Friedrich Engels. *The Communist Manifesto* (originally *Manifesto of the Communist Party*). London, 1848.

Marx, Paris. *Road to Nowhere: What Silicon Valley Gets Wrong about the Future of Transportation.* London and New York: Verso, 2022.

Mathias, Peter. *The First Industrial Nation: An Economic History of Britain, 1700–1914.* Second ed. London: Methuen, 1983.

Mayor, Adrienne. *Gods and Robots: Myths, Machines, and Ancient Dreams of Technology.* Princeton, NJ: Princeton University Press, 2018.

McAfee, Andrew, and Erik Brynjolfsson. *The Second Machine Age: Work, Progress, and Prosperity in a Time of Brilliant Technologies.* New York: W. W. Norton, 2016.

Mellor, Anne K. *Mary Shelley, Her Life, Her Fiction, Her Monsters.* New York and London: Routledge, 1989.

Mills, Stephanie, ed. *Turning Away from Technology: A New Vision for the Twenty-First Century.* San Francisco: Sierra Club Books, 1997. With a foreword ("In Defense of the Living Earth") by Theodore Roszak.

Moore, Thomas. *The Life of Lord Byron, with his Letters and Journals.* 2 vols. Philadelphia: Thomas Wardle, 1840.

Mueller, Gavin. *Breaking Things at Work: The Luddites Were Right about Why You Hate Your Job.* London and New York: Verso, 2021.

Munger, Mike. "Division of Labor," *EconLib,* n.d., www.econlib.org/library/Enc/DivisionofLabor.html.

Noble, David. *Progress without People: New Technology, Unemployment, and the Message of Resistance.* Toronto: Between the Lines, 1995.

Noble, David. "Present Tense Technology: Technology's Politics." *Democracy: A Journal of Political Renewal and Radical Change,* vol. 4, no. 2 (1983), 8–24.

Oakes, James. *Slavery and Freedom: An Interpretation of the Old South.* New York: Alfred A. Knopf, 1990.

O'Flinn, Paul. "Production and Reproduction: The Case of Frankenstein." In *Popular Fictions: Essays in Literature and History* (London: Methuen, 1986), 194–213.

O'Mara, Margaret. *The Code: Silicon Valley and the Remaking of America.* New York: Penguin Press, 2019.

Peel, Frank. *The Risings of the Luddites, Chartists, and Plugdrawers.* Heckmondwike, UK, 1880. Reprint, E.P. Thompson, ed. *The Risings of the Luddites: Chartists, and Plug-Drawers.* London: Routledge, 1968; fourth ed., 2019.

Pellew, George. *The Life and Correspondence of Henry Addington, First Viscount Sidmouth.* 3 vols. London: John Murray, 1847.

[Polidori, John William]. *The Vampyre; a Tale.* London, 1819.

Postman, Neil. *Technopoly: The Surrender of Culture to Technology.* New York: Alfred A. Knopf, 1992.

Randall, Adrian. *Before the Luddites: Custom, Community, and Machinery in the English Woollen Industry, 1776–1809.* Cambridge: Cambridge University Press, 1991.

Randall, Adrian. *Riotous Assemblies: Popular Protest in Hanoverian England.* Oxford: Oxford University Press, 2006.

Ravenelle, Alexandrea J. *Hustle and Gig: Struggling and Surviving in the Sharing Economy.* Berkeley: University of California Press, 2019.

Raynes, Francis. *An Appeal to the Public: Containing an Account of Services Rendered during the Disturbances in the North of England in the Year 1812* (1817).

Rede, William Leman. *York Castle in the Nineteenth Century: Being an Account of All the Principal Offences Committed in Yorkshire from the Year 1800 to the Present Period, with the Lives of the Capital Offenders.* London, 1831.

Reid, Robert. *Land of Lost Content: The Luddite Revolt.* London: Heinemann, 1986.

Romilly, Samuel. *The Life of Sir Samuel Romilly Written by Himself, with a Selection from His Correspondence.* 2 vols. London: John Murray, 1842.

Sale, Kirkpatrick. *Rebels against the Future: The Luddites and Their War on the Industrial Revolution; Lessons for the Computer Age.* Reading, MA: Addison-Wesley, 1995.

Sandler, Matt. *The Black Romantic Revolution: Abolitionist Poets at the End of Slavery.* London: Verso, 2020.

Say, Jean-Baptiste. *A Treatise on Political Economy.* Trans. Charles Robert Trinsop. London, 1821. First published as *Traité d'économie politique, ou simple exposition de la manière dont se forment les richesses* (1803).

Seymour, Miranda. *Mary Shelley.* New York: Grove Press; London: John Murray, 2000.

Shelley, Mary Wollstonecraft. *Frankenstein; or, The Modern Prometheus.* London, 1818; rev. ed., London, 1831.

Shelley, Mary Wollstonecraft. *The Last Man.* London, 1826.

Smith, Russell. "Frankenstein in the Automatic Factory." *Nineteenth-Century Contexts,* vol. 41, no. 3 (2019), 303–319.

Smith, Sydney, ed. *Peter Plymley's Letters, and Selected Essays.* London, 1886.

Strickland, Mary, ed. *A Memoir of the Life, Writings, and Mechanical Inventions of Edmund Cartwright.* London, 1843.

Sykes, D. F. E., and George Henry Walker. *Ben o' Bill's, The Luddite: A Yorkshire Tale.* London: Simpkin, Marshall, Hamilton, Kent, 1898.

Thomis, Malcolm I. *The Luddites: Machine-Breaking in Regency England.* Newton Abbot, UK: David and Charles, 1970.

Thomis, Malcolm I., and Jennifer Grimmett. *Women in Protest, 1800–1850.* London: Croom Helm, 1982.

Thompson, E. P. *The Making of the English Working Class.* London: Gallancz, 1965. Reprint, New York: Vintage, 1966.

Uglow, Jenny. *In These Times: Living in Britain through Napoleon's Wars, 1793–1815.* New York: Farrar, Straus & Giroux, 2015.

Ure, Andrew. *The Philosophy of Manufactures: Or, An Exposition of the Scientific, Moral, and Commercial Economy of the Factory System of Great Britain.* London, 1835.

Williams, Eric. *Capitalism and Slavery.* Chapel Hill: University of North Carolina Press, 1994. First published in 1944.

Willis, Kirk. "The Role in Parliament of the Economic Ideas of Adam Smith, 1776–1800." *History of Political Economy,* vol. 11 (Winter 1979), 505–44.

Wilson, Ellen Gibson. *Mass Politics in England Before the Age of Reform: The Great Yorkshire Election of 1807.* Lancaster, UK: Carnegie Publishing, 2015.

Wollstonecraft, Mary. *A Vindication of the Rights of Men.* London, 1790.

Wollstonecraft, Mary. *A Vindication of the Rights of Woman.* Boston, 1792.

Wootton, Sarah. *Byronic Heroes in Nineteenth-Century Women's Writing and Screen Adaptation.* London: Palgrave Macmillan, 2016.

Yang, Andrew. *The War on Normal People: The Truth about America's Disappearing Jobs and Why Universal Basic Income Is Our Future.* New York: Hachette, 2018.

INDEX

INDEX

INDEX

ABOUT THE AUTHOR

Brian Merchant is the technology columnist for *The Los Angeles Times,* and the author of the national bestseller *The One Device: The Secret History of the iPhone.* He's the co-founder of Terraform, Vice's science fiction outlet, and the founder of Gizmodo's Automaton project examining AI and the future of work. His writing has appeared in *The New York Times, Wired, the Atlantic, Harper's, Fast Company,* and beyond. He lives in L.A.